PLANNING FOR CLIMATE CHANGE

PLANNING FOR CLIMATE CHANGE

STRATEGIES FOR MITIGATION AND ADAPTATION FOR SPATIAL PLANNERS

Edited by Simin Davoudi, Jenny Crawford and Abid Mehmood

earthscan

publishing for a sustainable future

London • Washington, DC

First published by Earthscan in the UK and USA in 2009
Reprinted 2010

ISBN: 978-1-84407-662-8

Typeset by MapSet Ltd, Gateshead, UK
Cover design by Rob Watts

For a full list of publications please contact:
Earthscan
Dunstan House
14a St Cross St
London, EC1N 8XA, UK
Tel: +44 (0)20 7841 1930
Fax: +44 (0)20 7242 1474
Email: earthinfo@earthscan.co.uk
Web: **www.earthscan.co.uk**

Earthscan publishes in association with the International Institute for Environment and Development

A catalogue record for this book is available from the British Library

Library of Congress Cataloging-in-Publication Data

Planning for climate change : strategies for mitigation and adaptation for spatial planners / edited by Simin
Davoudi, Jenny Crawford and Abid Mehmood.
    p. cm.
  Includes bibliographical referecnes and index.
  ISBN 978-1-84407-662-8 (hardback)
  1. Regional planning. 2. Spatial behavior. 3. Climatic changes. I. Davoudi, Siin. II. Crawford, Jenny. III.
Mehmood, Abid.
    HT391.P5443 2009
    307.1'2–dc22

                                                                              2009009652

At Earthscan we strive to minimize our environmental impacts and carbon footprint through reducing waste,
recycling and offsetting our CO_2 emissions, including those created through publication of this book. For more
details of our environmental policy, see www.earthscan.co.uk.

This book was printed in the UK by TJ International, an ISO 14001 accredited company.
The paper used is FSC certified and the inks are vegetable based.

The paper used for this book is FSC-certified and
totally chlorine-free. FSC (the Forest Stewardship Council)
is an international network to promote responsible
management of the world's forests.

Contents

Part 1 The Challenge of Climate Change: Adaptation, Mitigation and Vulnerability

Part 2 Strategic Planning Responses

Part 3 Implementation, Governance and Engagement

List of Figures, Tables and Boxes

Figures

Tables

Boxes

List of Contributors

Jillian Anable is a senior lecturer at the Centre for Transport Research, University of Aberdeen. Her work focuses on transport and climate change with particular emphasis on the application of behavioural and psychological theory to the understanding of travel behaviour change. She is co-topic leader for transport in the UK Energy Research Centre, has advised the Commissions for Integrated Transport and the Scottish Government on transport and climate change and recently led an Evidence Base Review on Public Attitudes to Climate Change and Transport for the Department for Transport.

David Banister is Professor of Transport Studies at Oxford University and Director of the Transport Studies Unit. During 2009, he is also acting Director of the Environmental Change Institute at Oxford University. Until 2006, he was Professor of Transport Planning at University College London. He has also been Research Fellow at the Warren Centre in the University of Sydney (2001–2002) on the Sustainable Transport for a Sustainable City project and was Visiting VSB Professor at the Tinbergen Institute in Amsterdam (1994–1997). He was a visiting professor at the University of Bodenkultur in Vienna in 2007. He is a trustee of the Civic Trust and Chair of their Policy Committee (2005–2009). He has published 18 books and over 200 papers in refereed journals and as contributions to books.

Harriet Bulkeley is a Reader at the Department of Geography, Durham University. Her research interests are in the nature and politics of environmental governance, and focus on: environmental policy processes; climate change; and urban sustainability. She is co-author (with Michele Betsill) of *Cities and Climate Change* (Routledge 2003), and has published widely including articles in *Political Geography, Environment and Planning A, International Studies Quarterly, Global Environmental Politics* and *Environmental Politics*. She is an editor of *Environment and Planning C* and editor of 'Policy and Governance' for WIREs Climate Change. She currently holds an ESRC Climate Change Leadership Fellowship, coordinates the Leverhulme International Network *Transnational Climate Change Governance*, and in 2007 was awarded a Philip Leverhulme Prize.

Jason Byrne is a lecturer in Urban and Environmental Planning at Griffith University. A member of Griffith's Urban Research Programme, his research interests focus on urban political ecologies of greenspace, environmental justice and geographies of the nature–society interface. He has collaborated with researchers in the USA looking at equitable access to urban parks and multiple use trails, and is now collaborating with researchers from Zhejiang University in Hangzhou, China on how urban greenspace can help adapt cities to climate change. Jason is an associate fellow with the University of Southern California's Centre for Sustainable Cities and a fellow of the Johns Hopkins University's International Fellows in Urban Studies Association. Jason has previously worked as an urban planner in Western Australia.

Andrew Coleman works in the Environment Agency's Head Office 'Planning and Environmental Assessment' team, concentrating on ensuring that the Agency's concerns are reflected in national, regional and local planning, Strategic Environmental Assessment and Environmental Impact Assessment in England and Wales. In recent years he has particularly been involved in the development of national policy guidelines. He is the Royal Town Planning Institute's (RTPI) Climate Change and Water Task Group leader. He has been a chartered town planner for 20 years and an environmental planner for

13 years. Before joining the Environment Agency, Andrew worked as a planner and environmental manager in the public and private sectors in the UK and the Caribbean. He is also an affiliate of the Institute of Environmental Management and Assessment.

Jenny Crawford is Head of Research for the RTPI. Trained as an urban and regional planner, with an initial training in ecological sciences, she has 20 years of experience in government, community and academic planning in the United Kingdom. She has worked for local authorities in Scotland and the UK government's Sustainable Communities Pilot Programme, gaining extensive experience in rural development and coastal management initiatives, European projects and funding, and working with community-based organizations. Since joining RTPI in 2002, she has worked with policy makers and researchers in the UK and Ireland to develop joint research projects and develop the links between research and planning practice. Her work has included reviews of planning for waste management, and urban–rural settlement, water management and planning for climate change. She manages the UK Contact Point for the European Spatial Planning Observation Network (ESPON).

Simin Davoudi is Professor of Environment Policy and Planning at the School of Architecture, Planning and Landscape and Co-Director of the Institute for Research on Environment and Sustainability (IRES) at Newcastle University. She has held: Wibaut visiting professorship at University of Amsterdam, presidency of the Association of the European Schools of Planning (AESOP) and coordination of the Planning Research Network and membership of the Expert Panel on the UK government's Housing Markets and Planning Analysis, the Research and Knowledge Committee of the RTPI, the Expert Group for the EU DG Environment and DG Regional Policy, the Advisory Board of the Irish Social Sciences Platform, Swedish School of Planning at Blekinge Institute of Technology, North East Region Academic Panel and expert groups for Irish and Austrian EU presidency seminars. Her latest book is *Conceptions of Space and Place in Strategic Planning* (2009, Routledge).

Roland Ennos is a biomechanics researcher in the Faculty of Life Sciences, University of Manchester. He has strong interests in the physical processes involved in urban ecology. Following extensive work on the mechanics of root anchorage and the effect of wind and other environmental factors on plant growth, in recent years he has turned to the effect of plants on their environment. In particular he has collaborated with the Centre for Urban and Regional Ecology (CURE) to investigate the effects of greenspace on urban climate and ecological performance.

Thomas B. Fischer (PhD Manchester, Dipl.-Geogr. Berlin) is a Reader and Director of Research/Director of the MA in Environmental Management and Planning in the Department of Civic Design at Liverpool University, UK. He is a visiting professor in the Research Centre for South East Asia (SEA) at Nankai University, Tianjin, China, chair of the SEA section of the International Association for Impact Assessment (IAIA) and an associate research member of the Viessmann Research Centre on Modern Europe at Wilfrid Laurier University, Ontario, Canada. Thomas has published widely on environmental assessment and planning and is author of the Earthscan books *SEA in Transport and Land Use Planning* (2002) and *Theory and Practice of Strategic Environmental Assessment* (2007). Furthermore, he is co-editor of the *Earthscan Handbook of SEA* (2008).

Susannah Gill is a Green Infrastructure Planning Officer for The Mersey Forest, one of England's Community Forests. She is an honorary research fellow in the School of Environment and Development, University of Manchester, where she completed her PhD on 'Climate Change and Urban Greenspace' in 2006. This formed part of the 'Adaptation Strategies for Climate Change in the Urban Environment' project as part of the Engineering and Physical Sciences Research Council and UK Climate Impacts Programme consortium 'Building Knowledge for a Changing Climate'. A paper from her PhD won the Planning Research Network prize in 2006. She is currently working on a strand of the North West Climate Change Action Plan on green infrastructure for climate change mitigation and adaptation.

Brendan Gleeson is Professor of Urban Policy at Griffith University and a leading commentator on urban Australia. He has authored, co-authored or co-edited seven books and has written numerous opinion pieces for the *Sydney Morning Herald*, the *Courier Mail* and the *Canberra Times*. He is co-author (with Nicholas Low) of *Justice, Society and Nature: An Exploration of Political Ecology* (1998), which received the prestigious Harold and Margaret Sprout award in 1999 from the International Studies Association. In 2006 Gleeson's *Australian Heartlands: Making Space for Hope in the Suburbs* won the inaugural John Iremonger Award for Writing on Public Issues. He has most recently been appointed as a fellow of the Academy of Social Sciences. He currently lives in the Brisbane suburbs with his partner and their two children.

Nick Green is a research fellow in the Centre for Urban and Regional Ecology, University of Manchester. His primary interests are regional planning and analysis; network theory; cartography and visualization of regional dynamics; and complexity theory. He holds a PhD in planning studies from University College London, where he studied with Professor Sir Peter Hall and Professor Mike Batty, and is a member of the Royal Town Planning Institute and the Royal Institution of Chartered Surveyors. Dr Green is a member of the Town and Country Planning Association (TCPA) Policy Council, convenor of the TCPA's Regional Task Team, and is a member of the RTPI's UK Spatial Development Framework Steering Group.

Claire Haggett is Lecturer in Human Geography in the Centre for Environmental Change and Sustainability at the University of Edinburgh. Her work centres on understanding the relationship between people and the environment; and in particular on understanding opposition to renewable energy, and the wider implications of its implementation on people, communities and landscapes. She has held a number of ESRC research grants focusing on the development of renewable energy, and addressing the way forward for planners, stakeholders and policy-makers.

Jim Hall holds the Chair in Earth Systems Engineering in the School of Civil Engineering and Geosciences at Newcastle University. Professor Hall's research focuses upon risk-based adaptation decision making. A civil engineer with a background in flood and coastal engineering, Professor Hall links these subjects with methods for decision making under uncertainty and analysis of long term climate and socio-economic changes. In 2001 he received the Institution of Civil Engineer's George Stephenson Medal and the Frederick Palmer Prize for his work on risk-based benefit assessment of coastal cliff recession. In 2004 he was awarded the Institution of Civil Engineers' Robert Alfred Carr Prize for his work on broad-scale assessment of flood risk. Professor Hall is Deputy Director of the Tyndall Centre for Climate Change Research and is a member of the Adaptation Sub-Committee to the UK Climate Change Committee.

Kirsten Halsnæs is the head of the Climate Centre at the Technical University of Denmark, with a PhD in economics. She worked with the UNEP Risø Centre from 1992 until 2008. Kirsten specializes in development and environmental economics, with a particular focus on climate change issues in developing countries, and has played a leading role in several international projects on climate change adaptation and mitigation in developing countries. She has been a coordinating lead author for the IPCC Third and Fourth Assessment Reports and has published a number of international journal papers and books.

John Handley is Director of CURE in the School of Environment and Development at the University of Manchester. CURE brings together researchers from a variety of disciplines to explore ways of improving the sustainability and liveability of city regions. John is an environmental scientist who has worked in universities, local government and the NGO sector. He has special expertise in landscape planning, restoration ecology and the dynamics of urban systems. His current work is focused on the development of adaptive strategies for climate change.

Jeff Howard is an assistant professor in the School of Urban & Public Affairs, University of Texas at Arlington. His research focuses on the role of scientific, technical and professional expertise and expert knowledge in democratic decision making to achieve environmental sustainability, with emphasis on situations of high uncertainty and high decision stakes. He is developing case studies on urban planning, industrial chemistry, industrial ecology and higher education curricula. His doctoral training is in the interdisciplinary field of science and technology studies, and he was the recipient of a US Environmental Protection Agency Science to Achieve Results graduate fellowship for research on environmental policy.

Michael Howes is a senior lecturer in Griffith University's School of Environment. He also convenes the Group on Ecological Modernization and Sustainability within Griffith's Urban Research Programme. Michael's research centres on the issue of how governments try to make business more sustainable. He has undertaken specific projects on climate change policies, sustainable development policies, environment protection institutions, public environmental reporting, eco-efficiency programmes and environmental regulation. Before becoming an academic Michael worked for several years as an industrial chemist and technical manager in the manufacturing sector. More recently he sat on the board of Queensland Conservation Council and chaired the Technical Advisory Panel for the government's National Pollutant Inventory.

Stephen Jay (PhD Manchester) is a senior lecturer in environmental planning at Sheffield Hallam University in the UK. He has researched and written on planning and environmental assessment aspects of the development of energy infrastructure. Most recently he has studied the present growth of offshore wind farms around the UK coastline and in other European waters. He is currently carrying out an investigation into initiatives to prepare marine plans through the emerging concept of marine spatial planning.

Allan Jones (MBE) is an energy and climate change consultant with experience in local and regional government in energy, water, waste, transport and climate change sectors. Allan is currently working for the City of Sydney, Australia on a number of energy and climate change projects and the delivery of the climate change targets in Sustainable Sydney 2030. Allan is also a director/trustee of the Sustainable Environment Foundation that promotes the conservation, protection and improvement of the physical and natural environment. From 2004 to 2008 he was Chief Executive Officer of London Climate Change Agency Ltd. Before that Allan was Woking Borough Council's Director of Thameswey Ltd. He remains Chairman of the Surrey Energy and Environment Group and Chairman of the South East Home Energy Conservation Act Forum.

Thanasis Kizos is an agronomist with a PhD in environmental science and is Lecturer in Rural Geography in the Department of Geography of the University of the Aegean. His research and publications cover: rural landscape change and rural development, with a focus on islands; sustainable development with a focus on rural areas and specific concern on issues of rural livelihoods; farm household dynamics; and micro (farm) scale landscape change.

Richard Langlais is a senior research fellow at Nordregio – Nordic Centre for Spatial Development. He is responsible for the theme of 'social dimensions of climate change' and leads several projects that explore municipal response in the Nordic countries and in the rest of Europe. His previous positions include directorships and professorships at the Laboratory of Environmental Protection, TKK (Helsinki University of Technology), and at the Arctic Centre, in Rovaniemi, Finland.

Nethe Veje Laursen is an economist at DTU Climate Centre at Risø National Laboratory for Sustainable Energy. Her recent work includes projects on how development is vulnerable to climate change, and assessments of the costs of climate-proofing development planning. Currently, she is working on assessing the costs and potentials of climate change mitigation in fast growing developing countries. Ms Laursen received her Masters in Environment and Development Economics from Copenhagen University in 2006.

Abid Mehmood is a postdoctoral researcher at IRES and School of Architecture, Planning and Landscape, Newcastle University. He has a PhD in planning. His particular works and interests include the evolutionary perspectives to spatial planning and socio-economic development especially in small islands, peripheral regions and developing countries. He has worked on a number of UK and European research projects on urban and regional development planning and social innovation, as well as environment and sustainability.

Peter Newman is the Professor of Sustainability at Curtin University in Perth, Australia. In 2006–2007 he was a Fulbright senior scholar at the University of Virginia, Charlottesville where he completed his new book *Resilient Cities: Responding to Peak Oil and Climate Change* with Tim Beatley and Heather Boyer. In early 2008 he published *Cities as Sustainable Ecosystems*. Peter's book with Jeff Kenworthy *Sustainability and Cities: Overcoming Automobile Dependence* was launched in the White House in 1999. Peter is very involved in the practice of public policy. In Perth, he is best known for his work in saving, reviving and extending the city's rail system. Peter is on the Board of Infrastructure Australia which provides $20 billion of infrastructure to Australian cities. In 2001–2003 Peter directed the production of Western Australia's Sustainability Strategy in the Department of the Premier and Cabinet. In 2004–2005 he was a sustainability commissioner in Sydney advising the government on planning issues. He was a councillor from 1976 to 1980 in the City of Fremantle where he still lives.

Paul Nolan is Director of the Mersey Forest. He has been involved in both the commercial and community forest sectors for over 18 years, initially working in large scale commercial forest management in south-west Scotland and now in community forestry. He has degrees in forestry and forest products technology. Paul is a member of the North West Regional Advisory Committee for Forestry, the National Forestry Commission Grants Advisory Committee and the steering group for the Regional Forestry Framework, and is currently Chair of the National Community Forest Programme and has helped to establish both the North West Green Infrastructure Unit and the Green Infrastructure Think Tank.

Rafael E. Pizarro is Lecturer in Sustainable Urban Planning at the University of Sydney, Australia and Visiting Professor at the Berlin University of Technology (TU Berlin), Faculty of Planning, Building and Environment. He teaches postgraduate courses in sustainable urban environments, urban design and development control, and advanced urban design-planning studios. His current research and teaching focus on designing cities for peak oil and climate change. He is co-editor of *Dialogues in Urban Planning: Towards Sustainable Regions* (Sydney University Press) and of *Southern California and the World* (Praeger). He has previously taught at various universities in his native Colombia, at the University of Southern

California in Los Angeles and at the Universidad de Sevilla in Spain. He has also been an architect in Bogotá and a city planner for the City of Phoenix Planning Department in Arizona (USA).

Pamela Robinson is an assistant professor in the School of Urban and Regional Planning. She holds a PhD in geography from the University of Toronto and a Masters of Urban Planning from Queen's University (Kingston). She is an editor of the forthcoming book *Urban Sustainability – Imperatives, Reconciliation and Planning*. Her research on Canadian municipal response to climate change has been published in Canadian and international academic journals. In 2004 she received the Canada Mortgage and Housing Corporation Award for Teaching Excellence for her teaching in urban sustainability. She is the Director of the Ryerson Centre for the Advance of Scholarship of Teaching and Learning. Her current research explores new forms of civic engagement in urban planning practice in North America.

Yvonne Rydin is Professor of Planning, Environment and Public Policy at University College London's Bartlett School of Planning, where she specializes in environmental policy, and governance and sustainability issues at the urban level. Her recent research projects cover work on sustainable construction and planning in London and the implementation of local sustainability indicators. Other work within London has involved study of institutional change and interest representation concerning planning for sustainability in the Greater London Authority. Other interests include community engagement on sustainable development and the discursive aspects of environmental planning.

Ioannis Spilanis is an economist graduated at the University of Athens (1979). After his post-graduate studies in 'European and International Studies' at the University of Grenoble (France) he got a PhD on the topic of 'Tourism and regional development: The Greek case' (1985). He has worked on regional planning in the Ministry of the Aegean (1987–1990). He is currently Assistant Professor of the Department of Environmental Studies in the University of the Aegean, where he is teaching the relationship between development and environment. Dr Spilanis' research is on regional and local development and is focused on planning sustainable development and tourism, in particular for islands within the Laboratory of Local and Islands' Development.

Wendy Steele is a lecturer in Urban and Environmental Planning in the School of Environment, Griffith University (Australia). Her research interests are in the areas of planning policy, practice and education for sustainability, with a focus on: the politics and practice of place shaping/making within contemporary governance contexts; networked infrastructure; and critical planning pedagogy. A member of Griffith's Urban Research Programme, Wendy has worked on a number of Australian research projects on performance-based planning, land-use reform initiatives, planning education and child-friendly cities and communities.

Olivier Sykes (PhD Liverpool, MCD Liverpool) is a lecturer in Spatial Planning in the Department of Civic Design at the University of Liverpool, UK. He is currently a visiting lecturer in Spatial Planning at the Université de Bretagne Occidentale, Brest, France and a member of the European Spatial Planning Observation Network (ESPON) UK Network Advisory Group. His main areas of research interest are European influences on the practice of spatial planning, concepts employed within European spatial planning, the institutional and policy frameworks for planning in the UK, and regionalism and regional planning in European states. Olivier has published on these topics in a variety of journals and contributes a bi-monthly article on European affairs and spatial planning to the journal of the Town and Country Planning Association.

Jochem de Vries is a lecturer at the department of Human Geography, Planning and International Development Studies of the University of Amsterdam. He holds a PhD in planning from the same institution. He previously held academic positions at Delft University of Technology and the University of Groningen and was a guest-researcher at the Netherlands Institute for Spatial Research in The Hague. He specializes in the relationship between water management and planning, cross-border planning and regional planning for metropolitan regions.

Stephen M. Wheeler is associate professor in the Landscape Architecture Program at the University of California at Davis. He is author of *Planning for Sustainability: Towards Livable, Equitable, and Ecological Communities* and co-editor (with Timothy Beatley) of *The Sustainable Urban Development Reader*, both published by Routledge. In addition to UC Davis, he has taught at the University of New Mexico and the University of California at Berkeley, from which he received his Masters and PhD degrees in city and regional planning. His areas of interest include planning for climate change, sustainable development and urban design.

Elizabeth Wilson is Reader in Environmental Planning in the School of the Built Environment at Oxford Brookes University, UK. Her research and teaching interests lie in sustainability, the response of spatial planning to climate change, issues of policy discourses and implementation, and futures thinking. She has recently collaborated on European and UK research projects on biodiversity, spatial planning and climate change. She was a member of the research consortium for the ASCCUE (Adaptation strategies for climate change in the urban environment) project, and has published a number of articles on planning and climate change, as well as guidance on climate change adaptation in sustainable communities, and the UK government's *The Planning Response to Climate Change*. With Jake Piper, she is currently writing a book for Routledge on *Spatial Planning and Climate Change*.

Maarten Wolsink graduated in Political Science and Social Science Methods and holds a PhD in psychology (University of Amsterdam) on a thesis about the social acceptance and implementation of wind power. From 1993 he worked as associate professor in the Department of Environmental Science on energy, environmental issues and environmental policies. Since 1999, he has been associate professor at the Department of Geography and Planning, where he is reading environmental geography and planning. His main research topics are renewable energy innovation, environmental conflict, energy and waste policies, and water management in relation to spatial planning, and climate change mitigation and adaptation.

Preface and Acknowledgements

Simin Davoudi, Jenny Crawford and Abid Mehmood

There is a compelling scientific consensus that since the dawn of the industrial era human activity has become increasingly responsible for the changing climate. While research and publications on the science of climate change have grown substantially in recent decades, less attention has been paid to the role of spatial planning in developing both mitigation measures to reduce emissions, and adaptation measures to ameliorate the effects of climate change. The literature in the field of planning is limited and disparate. This is despite the fact that climate change is changing the context of planning and shaping its priorities. It has strengthened the environmental dimension of spatial planning and has become a new rationale for coordinating actions and integrating different policy priorities.

This book, therefore, aims to map out the main challenges for spatial planning that have been created or amplified by climate change, in order to encourage further debate and deeper engagement by planning academics and practitioners. The book has a wide ranging scope and an international coverage. It draws on the expertise and experience of both academics and practitioners from five continents. The 23 contributions to the book are structured under 3 main parts. Part 1 'The Challenge of Climate Change' aims to unpick the complexity and contested nature of climate change with particular emphasis on urban forms and the role of spatial planning. It explores three main themes: mitigation–adaptation interface, urban form and development patterns, and vulnerability to climate change. Part 2 'Strategic Spatial Responses' describes how strategic frameworks and planning processes have been responding to the climate change challenge in terms of both mitigation and adaptation. It includes examples from different parts of the world that illustrate the development of new paradigms for spatial planning and the challenges of policy integration and diversity. Part 3 'Implementation, Governance and Engagement' focuses on emerging tools and methods for spatial planning and the significance of governance. The themes explored in this part of the book include the role of scenarios and modelling, policy implementation, and governance and public engagement.

Working on this book has been a remarkable experience for us in terms of the learning involved not only in the intellectual engagement, but also in crossing the academy–practice boundaries that are increasingly perpetuated by powerful institutional, professional and social practices in the field of spatial planning, and indeed elsewhere. We hope that this book opens up new channels of shared learning among researchers, practitioners, educators and decision makers about what is one of the most challenging and cross-cutting global issues of the 21st century.

We would like to thank Tamsine O'Riordan, Earthscan Commissioning Editor until 2008, for encouraging us to embark on this stimulating, though demanding, task before handing over to Claire Lamont and Alison Kuznets, Editorial Assistants at Earthscan, who since then have maintained the publication momentum and provided us with tireless advice and support. We would also like to thank all the authors for their valuable contributions and for their positive responses to our frequent editorial requests.

List of Acronyms and Abbreviations

ADB	Asia Development Bank
AESOP	Association of the European Schools of Planning
AfDB	African Development Bank
ARB	Air Resources Board
ARK	Spatial Planning and Adaptation Strategy
BERR	Department of Business, Enterprise and Regulatory Reform
BMZ	Federal Ministry for Economic Cooperation and Development
BOM	Bureau of Meteorology
BREEAM	Building Research Establishment Environmental Assessment Method
Cal EPA	California Environmental Protection Agency
CAT	Climate Action Team
CCHP	combined cooling, heat and power
CCSP	Climate Changes Spatial Planning
CCP	Cities for Climate Protection
CEQA	California Environmental Quality Act
CFMP	catchment flood management plan
CHP	combined heat and power
CLG	Department of Communities and Local Government (same as DCLG)
CNG	compressed natural gas
COAG	Council of Australian Governments
CPACC	Caribbean Planning for Adaptation to Climate Change
CPMR	Conference of the Peripheral and Maritime Regions (of Europe)
CSIRO	Commonwealth Scientific and Industrial Research Organisation
CURE	Centre for Urban and Regional Ecology
DCLG	Department for Communities and Local Government (UK)
DEA	Danish Energy Authority
Defra	Department for the Environment, Food and Rural Affairs (UK)
DEM	digital elevation model
DETR	Department of the Environment, Transport and the Regions
DFID	Department for International Development
DGIS	Directorate General for International Cooperation (the Netherlands)
DoE	Department of the Environment (UK)
DTI	Department of Trade and Industry (UK)
EC DG	European Commission Directorate General
EEA	European Environment Agency
EEZ	exclusive economic zone
EIR	Environmental Impact Report
EM	ecological modernization

EPA	US Environmental Protection Agency
ESDP	European Spatial Development Perspective
ESPON	European Spatial Planning Observation Network
ETAAC	Environmental Technical Assistance Advisory Committee
EURISLES	European Islands System of Links and Exchanges
EU ETS	EU Emissions Trading Scheme
FCM	Federation of Canadian Municipalities
FUA	Functional Urban Area
GDP	gross domestic product
GHG	greenhouse gas
GIS	geographical information system
GOD	Green–Oriented Development
GTA	Greater Toronto Area
GVA	gross value added
GWP	global warming potential
HECA	Home Energy Conservation Act
HIP	Housing Information Pack
IAIA	International Association for Impact Assessment
IEA	International Energy Agency
ICLEI	International Council on Local Environmental Initiatives
IPC	Infrastructure Planning Commission
IPCC	Intergovernmental Panel on Climate Change
IPPR	Institute of Public Policy Research
IRES	Institute for Research on Environment and Sustainability
KK	Klimat Kommuner (The Swedish Network of Municipalities on Climate Change)
KLIMP	Local Climate Investment Programme
LCA	Life Cost Analysis
LCCA	London Climate Change Agency
LCCP	London Climate Change Partnership
LDA	London Development Agency
LDD	local development document
LEED	Leadership in Energy and Environmental Design
LIMDEF	limiting deficit
LIP	Local Investment Programme
LNG	liquefied natural gas
MAP	Mediterranean Action Plan
MDG	Millennium Development Goals
MDSF	Modelling and Decision Support Framework
MEGA	Metropolitan European Growth Area
mt	million tonnes
NFCDD	National Flood and Coastal Defence Database
OECD	Organization for Economic Cooperation and Development
PBL	Netherlands Environmental Assessment Agency
PCP	Partners for Climate Protection
PHEV	plug–in electric hybrid vehicle
POD	Pedestrian–Oriented Development
PPG	Planning Policy Guidance
ppm	parts per million

PPP	public private partnership
PPS	Planning Policy Statement
PUR	Polycentric Urban Region
RCM	Regional Climate Model
RFRA	regional flood risk appraisal
RSS	regional spatial strategy
RTPI	Royal Town Planning Institute
SAP	Standard Assessment Procedure
SEKOM	The Swedish Eco Municipalities
SEPA	Scottish Environmental Protection Agency
SFRA	strategic flood risk assessment
SIDS	Small Island Developing States
SPG	Supplementary Planning Guidance
SPSLCM	South Pacific Sea Level and Climate Monitoring project
SRES	Special Report on Emissions Scenarios
SuDS	sustainable drainage systems
SWD	soil water deficit
TCPA	Town and Country Planning Association
TEN-Ts	trans-European Transport Networks
TIA	Territorial Impact Assessment
TOD	Transit-Oriented Development
TSPEU	Territorial States and Perspectives of the European Union
UHI	urban heat island
UKCIP	UK Climate Impacts Programme
UMT	Urban Morphology Type (mapping)
UN	United Nations
UNDP	United Nations Development Programme
UNEP	United Nations Environment Programme
UNFCCC	United Nations Framework Convention on Climate Change
VKT	vehicle kilometres of travel
VMT	vehicle miles travelled
WAG	Welsh Assembly Government
WHO	World Health Organization

Part 1

The Challenge of Climate Change: Mitigation, Adaptation and Vulnerability

Introduction to Part 1

Simin Davoudi

When in his speech at the Royal Meteorological Association in 1938 a British engineer, called Guy Callendar, claimed to have proven that the world was warming he was considered to be an eccentric. With the onset of the 'cooling world' in the 1950s to 1970s, the idea of global warming was further pushed towards intellectual oblivion (*The Economist*, 2006). It was not until the 1980s that it was retrieved and turned into one of the most significant arguments of our time. The establishment of the Intergovernmental Panel on Climate Change in 1988 and the increasingly undisputed findings of its reports have left little doubt that the climate is changing and, more importantly, human activity is responsible for it. It is now widely acknowledged that mitigating for greenhouse gases (GHGs) that are causing global warming (known as mitigation measures) and adapting our built and productive environments to withstand its extreme consequences (known as adaptation measures) have become the most formidable challenges faced by society. Nevertheless, there remain major uncertainties as well as disagreements, often coloured by political motivations, over the right course of action. At the heart of such debates is an emerging tension between mitigation and adaptation measures which raises difficult conundrums for spatial planners and decision makers. The contributions to this part of the book aim to unpick the complexity and contested nature of tackling climate change with particular reference to urban forms and the role of spatial planning. Three themes in particular underpin the chapters in this part:

- mitigation–adaption interface;
- urban form and development patterns;
- vulnerability to climate change.

Mitigation–adaption interface

The first three chapters in this part share a common concern over the divergent implications of mitigation and adaptation approaches. Davoudi et al, in Chapter 1, outline the origin of the divide and suggest ways of developing synergies between them. They also contextualize the theme of the book by providing an overview of the science of climate change, the global policy responses to its challenges and the ways in which spatial planning has addressed these in the wider context of the sustainability agenda. They raise concerns that while the discursive shift from sustainability towards climate change may refocus the spatial planning agenda on ecological priorities, the current economic recession may once again bring planning under pressure to set aside its sustainability goals in the interest of economic growth.

The tensions and the synergies between mitigation and adaptation measures are analysed in detail in Chapter 2 by Howard. The chapter contends that the interrelationships between the two have been largely neglected in planning literature and raises concerns that the 'adaptation turn' in planning may be at the cost of paying less attention to measures that reduce GHG emissions. Howard echoes Davoudi et al in advocating an integrated approach to adaptation and mitigation while pointing out the problem of mismatch in terms of spatial and temporal scales. He urges planners to adhere to three key principles: mitigation has priority; mitigation is a primary form of adaptation; and effective local adaptation requires long term global perspective. In his integrated model of climate change adaptation, Howard emphasizes that, 'the most desirable form of adaptation is adaptation that is not necessary'.

Chapter 3, by Pizarro, follows this theme with particular emphasis on exploring the inherent tensions between adaptation and mitigation with specific reference to urban forms. While the author confirms that the sprawling, car-dependent suburb is not an appropriate urban form to reduce GHG emissions, he challenges the view that such an urban form is equally inappropriate for adapting to climate change problems in all geographical locations. It is therefore argued that in a hot–humid region, for example, spreading out the buildings on the landscape and hence facilitating circulation of breezes can make the city more resilient to high temperatures and humidity but will do little for reducing travel and hence the mitigation objectives. The chapter demonstrates the complexity of adapting urban forms to the impacts of global warming. This clearly rules out the possibility of one-size-fits-all solutions. Instead, Pizarro advocates a minimum set of variables that needs to be taken into account in the design of urban form in different climatic zones. In conclusion, he argues that, 'in certain locations the need to adapt to extreme climatic events exceeds the need to mitigate'. This raises the question, posed by Howard, that if a city 'cannot adapt to climate change without undermining its ability to aggressively participate in climate mitigation, should it not be written off as intrinsically unsustainable?'

Urban form and development patterns

The question of how best to accommodate new development formed the main thrust of the much-cited Michael Breheny's work in the early 1990s. Drawing on this work particularly, but not exclusively, Chapter 4 by Green and Handley compares three main types of settlement patterns (urban infill, urban extension and entirely new settlement), which can be adopted to accommodate new development. These alternative forms are then examined in relation to their social, economic and environmental advantages and disadvantages. While the authors make little direct reference to climate change, their review of the environmental performance of these alternative settlement patterns contributes to our understanding of their ability to mitigate GHG emissions. The authors rightly remind the readers that it is not just the pattern of new settlement, but also how individual settlements interact with one another that matters. Furthermore, mitigating climate change depends not just on sustainable design and production but also on sustainable consumption of urban space. In other words, sustainable behaviour and lifestyle are as important as settlement patterns.

Chapters 5 and 6 deal with one of the most critical relationships in the debate about development pattern: the relationship between travel, land use and urban form. Chapter 5, by Banister and Anable, provides a review of trends in the transport sector in the UK, demonstrating that little has been done to reduce the rate of growth of carbon emissions from this sector. The chapter looks at the options available in the four markets identified – city travel, long distance travel, freight and aviation – summarizing the progress that has been made at the UK and EU levels. These analyses include comments on land use and planning among other policy instruments. The authors emphasize the need for behavioural change and argue that the planning system can have an important role in that. They mention a number of areas where the impacts of land use factors on travel distance are significant: location of new development, particularly housing, at the regional level; density; mixed use development; settlement size; accessibility to public transport hubs; and (un)availability of parking space.

Chapter 6, by Newman, continues this theme but grounds the debate in the context of oil vulnerability and peaking of supply. The author argues that subsequent waves of industrialization and de-industrialization have led to oil-dependent economies and the emergence of car-dependent cities. Newman argues that, 'the crash of September 2008 signals the end to the urban economy around oil' and opens up opportunities for cities to move towards a greater degree of resilience. Five transport-related areas are identified as key in facilitating such a move. These are elaborated largely within the context of the United States and Australia. One area with particular relevance to spatial planning is transit-, pedestrian- and green-oriented development.

Vulnerability to climate change

The impact of climate change differs in different parts of the world and depends not only on the level of exposure but also the adaptive capacity of people and places, as discussed in Chapter 1. Hence, tackling climate change requires a sound understanding of the vulnerability of places and social groups. The final two chapters in this part provide an account of adaptation challenges in some of the most vulnerable areas. However, as Halsnaes and Laursen argue, in Chapter 7, assessing and managing vulnerability is complex and impinges upon a number of disciplinary areas such as development and poverty, public health, climate, geography, political ecology and risk management. Indeed, focusing on reducing vulnerability brings to the fore issues of poverty eradication and other internationally agreed Millennium Development Goals. The authors draw on the findings from the climate screening

of the Danish Climate and Development Action Programme (Danida) in Ghana, Uganda and Bangladesh to assess the extent to which the activities included in the Programme were vulnerable to climate change. They demonstrate the disproportionate impact of climate change on poorer countries and poorer populations within them, suggesting that climate vulnerability reflects both natural and social factors. They raise particular concern about a lack of institutional and governance capacity in developing countries for effective integration of climate adjusted risks in urban development decisions.

Similar concerns are raised in Chapter 8 by Kizos et al but in the context of the Aegean Islands in Greece. They highlight the importance of equity issues, arguing that small islands are not only especially vulnerable to impacts such as sea level rise and loss of biodiversity, but also have less scope for and leverage on mitigation measures because of their size, accessibility and isolation. Reflecting on the shortcomings of the planning system in dealing with climate change vulnerability in these islands, they advocate two principles for a more responsive planning system: one is a qualitative, rather than quantitative, approach to new development; and the other is a proactive, rather than reactive, approach to policy making. The emphasis on issues of inequity in terms of the balance between mitigation and adaptation in the final two chapters raise significant questions about the extent to which approaches to climate change response can follow a single model.

Reference

The Economist (2006) 'The heat is on', 9 September, survey, p3

1

Climate Change and Spatial Planning Responses

Simin Davoudi, Jenny Crawford and Abid Mehmood

> *The climate change issue is part of the larger challenge of sustainable development. As a result, climate change policies can be more effective when consistently embedded within broader strategies designed to make national and regional development paths more sustainable.*
>
> (IPCC, 2001, p4)

Introduction

Understanding the impacts of cyclical cooling and warming of the Earth's climate has made an important contribution to our knowledge of the evolution and distribution of populations and ecosystems. Incorporating this understanding into contemporary human development processes is, however, a major challenge. We realize that human use of the atmosphere as a carbon sink has systemic impacts that translate into significant social, economic and environmental costs. Given that these costs are hugely unpredictable in terms of location, nature and scale, we face risks that we had not factored into our decision making processes. Our physical connections and interdependencies with nature

are being demonstrated in inescapably practical terms. These realizations are changing both the context and the nature of spatial planning at all levels.

The relationship between energy use, development and climate has renewed the focus of planning analysis and policy on the complexity and uncertainty of environmental, social and economic systems. This is forcing a reassessment of how planners envisage development and the scope and appraisal of planning interventions. Climate change therefore raises profound professional, technical, theoretical and ethical issues for planners. Climate change awareness is now shaping the sustainable development debate, further strengthening the critiques of dominant development pathways and raising interest in alternative development policy responses at different scales and in different places. It advocates searching for new opportunities, new tools and new rationales. Planners are being asked to reconcile, trade and, indeed, overturn short-term and long-term expectations for development. They need to address questions such as: what will low carbon, 'climate-proof' settlement look like in terms of urban form and infrastructure; what are the barriers to effective planning for such development; what are the implications for governance, from transnational to local levels, and the relationship between these levels; who will bear the risks and what are the implications for equity and social development? Current evidence and research raise yet more questions and many of the related projections are bleak. However, planners like to cast themselves as being 'in the business of hope': believing that knowledge and debate are powerful levers to finding policy and implementation solutions that meet complex social, economic and environmental needs. This has been an important motivation for this book.

This chapter aims to set the context for subsequent chapters by providing an overview of how the science of climate change is informing policy and the frameworks that are emerging in response at both transnational and national levels. It asks: how do we know that the world is warming and that human activities are responsible for it; what will be the main impacts of climate change; who are the main emitters of greenhouse gases (GHGs) and who are going to suffer most from the effects of a changing climate? The chapter then outlines the global policy framework before focusing on the nature of spatial planning and its contribution to climate change responses.

The science of climate change

The United Nations Framework Convention on Climate Change (UNFCCC) uses the term 'climate change' to refer specifically to 'a change of climate which is attributed directly or indirectly to human activity that alters the composition of the global atmosphere and which is in addition to natural climate variability observed over comparable time periods'. The Intergovernmental Panel on Climate Change (IPCC) uses the term with respect to 'any change in climate over time, whether due to natural variability or as a result of human activity' (IPCC, 2001, p21). Importantly, the changes we face are a result of both processes as a range of natural and human factors drives changes in atmospheric concentration of GHGs[1] and aerosols, solar radiation, and land surface proper-

ties. These in turn alter the energy balance of the climate system – exerting warming or cooling influences on global climate. These changes are expressed in terms of *radiative forcing*.[2] Increases in GHG, including carbon dioxide (CO_2), methane (CH_4) and nitrous oxide (N_2O), tend to warm the Earth's surface. IPCC's *Fourth Assessment Report* is unequivocal that the Earth's climate has warmed by 0.74 degrees Celsius (°C) since 1900, through increases in GHG emission (IPCC, 2007a). Between 1970 and 2004, global *human-induced* GHG emissions have grown by 70 per cent.

Complex systems such as climate have an inherent tendency to maintain states of equilibrium. As a result, some impacts of anthropogenic (man-made) climate change may be slow to become apparent. At the same time, effects are likely to last. Thus, even after GHG concentrations are stabilized, anthropogenic warming and sea level rise will continue for centuries due to the timescales associated with climate processes and feedbacks. For instance, if concentrations of GHG and aerosols could be held at year 2000 levels, the IPCC (2007a) estimates that a 0.2°C warming would still be expected over the next 20 years. Beyond certain thresholds, some impacts could be irreversible. For example, 'major melting of the ice sheets and fundamental changes in the ocean pattern could not be reversed over a period of many human generations' (IPCC, 2001, pp16–17).

The IPCC's forecasts for future climate change are based on the use of a range of alternative emissions scenarios. For the next two decades its best estimates are for an overall warming of about 0.4°C. Depending on the level at which global carbon emissions peak and begin to fall, increases of between 1.4 and 5.8°C are projected for the period 1990 to 2100. This is two to ten times larger than the observed warming during the 20th century (IPCC, 2007a). Indeed the IPCC warns that the projected rate of increase in the 21st century 'is very likely to be without precedent during at least the last 10,000 years' (IPCC, 2001, p8), and that 'GHG forcing in the 21st century could set in motion large-scale, high-impact, non-linear, and potentially abrupt changes in physical and biological [as well as social and economic] systems over the coming decades to millennia', some of which 'could be irreversible' (IPCC, 2001, p14).

Anthropogenic emissions

In 2008, the level of GHG in the atmosphere was about 430 parts per million (ppm) compared with 280ppm before the Industrial Revolution. This is estimated to reach 550ppm by 2050 at the current rate of increase, but given that the levels are rising faster than expected, the 550ppm could be reached as early as 2035 (Stern, 2007). The emission of CO_2, which is the most important anthropogenic GHG, increased by 80 per cent in that time (IPCC, 2007a, p5). Global increases in CO_2 concentrations are due mainly to fossil fuel use and, to a lesser extent, land-use change. Increases in CH_4 concentrations are predominantly due to agriculture and fossil fuel use. The sectors that were most responsible for growth in GHG emissions between 1970 and 2004 include the energy supply sector (contributing to an increase of 145 per cent),

transport (120 per cent), industry (65 per cent) and land use, land use change and forestry (40 per cent).[3] Between 1970 and 1990, direct emissions from agriculture grew by 27 per cent and from buildings by 26 per cent. The latter has remained at roughly the 1990 levels thereafter. However, when taking into account the energy use of the buildings, the total direct and indirect emissions amount to 75 per cent (IPCC, 2007b). Figure 1.1 shows the distribution of GHG emissions in 2000 by sector.

Two important drivers of the rise in energy-related emissions are global population growth (up by 69 per cent) between 1970 and 2004 and the increase in per capita income (up by 77 per cent). These figures refer to global averages – the contribution of individual countries to global warming varies substantially between the rich and the poor. In 2004, for instance, high-income nations accounted for 20 per cent of world population, produced 57 per cent of gross domestic product (GDP) and generated 46 per cent of global GHG emissions (IPCC, 2007b). Per capita emissions from developing countries in 2004 were one quarter of per capita emissions from developed countries. While a progressive

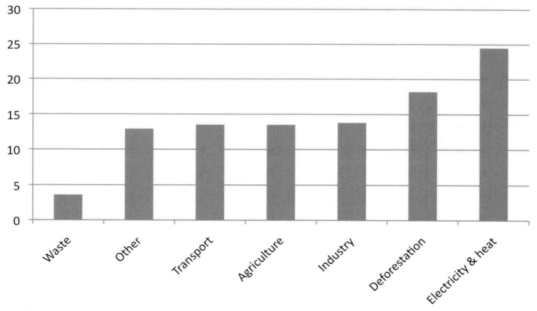

Source: Adapted from *The Economist* (2007, p4, survey)

Figure 1.1 World GHG emissions by sector in 2000, percentage

decoupling of income growth from GHG emissions has taken place through measures such as reducing the energy intensity (33 per cent decrease in energy used per unit of GDP), the level of improvement has not been sufficient to counteract the global rise in emissions.

The scale and location of emissions are also highly differentiated below the national level with an ongoing debate about the role of cities gaining increasing currency. Satterthwaite (2008, pp539–540), for example, has challenged the assertion that, 'cities are responsible for about 75 per cent of the heat-trapping greenhouse gases that are released into our atmosphere', and instead estimates that the figure is nearer 30–40 per cent. Any such estimates, of course, mask the effects of huge variations in relative wealth. For example, total GHG emissions ranged from 44.3 million tonnes (mt) in London in 2006 to 64.8mt in Mexico City in 2000 and a mere 1.8mt in Dhaka in 1999. The per capita emissions were respectively: 6.18, 3.6 and 1.7 tonnes; i.e. much higher in the wealthy city of London than in Mexico City or Dhaka (Romero-Lankao, 2007; Dodman, 2009).

In determining spatial differences in emissions, however, a crucial point is the issue of attribution – that is, how are the geographical boundaries of settlements defined for the purposes of carbon emissions? Do they, for instance, correspond with the administrative (municipality), the metropolitan (contiguous built up area) or the functional (city–region) boundaries? (See Davoudi, 2008, for detailed discussions.) Boundary definition has major implications for attributing GHG emissions to cities and 'non-cities'. Often major emitters such as power stations, landfill sites, or even large factories are located in 'rural' areas. Furthermore, activities such as aviation, shipping and other major transportation do not respect physical boundaries and while they cannot be directly attributed to 'cities', they are likely to be driven by city-based consumption. Overall, it is misleading to focus on a particular settlement type (such as cities) in attributing GHG (or CO_2) emissions, because as Satterthwaite (2008, p547) stresses, 'the driver of most anthropogenic carbon emissions is the consumption patterns of middle- and upper-income groups, regardless of where they live, and the production systems that profit from their consumption'. However, this is not to suggest that the spatial dimensions of settlements are not key drivers of emissions, as explored in detail in subsequent chapters in this volume.

Impacts of climate change

As global temperature increases, the models reviewed by the IPCC show an increasing risk of extreme weather events, including destructive storms, floods and droughts. They predict the melting of both sea ice and glaciers and changes in season that are being corroborated by measurements on the ground. Different global regions are expected to experience different changes as a result of global warming. For example, while Europe is expected to experience an increase in inland flash floods, Africa will see a rise in arid and semi-arid land (IPCC, 2007a). Projected patterns of warming will have increasingly significant impacts on various terrestrial, marine and coastal ecosystems, as well as on water resources, particularly in dry regions and agriculture in low latitudes and low-lying coasts. Some of these impacts are irreversible. For example, 20–30 per cent of species assessed so far are likely to be at increased risk of extinction if the rise in global average warming exceeds 1.5–2.5°C relative to 1980–1999 (IPCC, 2007a). Projected changes would transform the physical geography of the world with millions of people facing starvation, water shortages or homelessness. Even a one metre rise in sea level[4] would flood 17 per cent of Bangladesh's land mass and threaten coastal cities such as London and New York (*The Economist*, 2006a, p8, survey).

The nature and intensity of impact will vary depending on the vulnerability of different places. Vulnerability is a function of both *exposure* and *sensitivity*. The former refers to the character, magnitude and rate of climate change and variability to which places are exposed. The latter refers to places' adaptive capacity. Hence, *vulnerability* is the extent to which people, places, economic sectors and infrastructures are prone to the adverse affects of climate change. As will be discussed later, adaptive capacity is as important as the level of exposure in determining the extent to which places can attenuate climate stresses.

The level of vulnerability differs not only between places, but also between population groups. Differences in demographic and socioeconomic profiles affect the level of vulnerability considerably. Hence, children and the elderly are often the most vulnerable groups, as are those who already suffer from poor health or are unable to cope with injuries and illnesses caused by the impact of climate change. Similarly, those who lack the capacity to reduce the direct and indirect impacts of climate change on their well-being are also among vulnerable groups. These include lower-income groups with little resources at their disposal to, for example, move to safer areas, insure their assets or gain access to adequate water, electricity, sanitation, sewage and other basic utilities (Satterthwaite et al, 2007; Chapter 7). Previous incidents have shown the disproportionate impacts of climate extremes on vulnerable groups. For example, most of the 20,000 lives claimed by the European heat wave of 2003 were among the poor and isolated elderly; as were the majority of the 1101 people who died in Louisiana following Hurricane Katrina in August 2005 (Wilbanks et al, 2007).

The global policy context

The global policy context for climate change, and other global environmental issues, has been predominantly shaped by the United Nations (UN). Its 1972 Conference on the Human Environment in Stockholm prompted the creation of the UN World Commission on Environment and Development in 1983, which produced their famous Brundtland Report, *Our Common Future*, in 1987. One year later, the UN Environment Programme along with the World Meteorological Organization established the IPCC to assess published scientific evidence about human impacts on climate and the options for mitigation and adaptation. Since then, the IPCC's periodic reports (the fourth of which

was published in 2007) have become an authoritative reference for tracking climate change and its impacts. Another significant UN conference was the 1992 Earth Summit in Rio de Janeiro which led to the establishment of the UNFCCC. The convention became the driving force behind the Kyoto Protocol which was adopted in 1997 and came into force in 2005. Together the UNFCCC and Kyoto Protocol have established a global policy framework for climate change which underlies an array of national policies. They have also created an international carbon market and set up new institutional mechanisms to provide the foundation for future climate policies.

As of 2008, 180 nations had ratified the Kyoto Protocol, which sets binding targets to reduce GHG emissions to an average of 5 per cent against 1990 levels over the period 2008 and 2012, when the first Kyoto Protocol ends. The exact target for each member state varies depending on their historic emission levels and capacity to change. The UK, for example, is committed to achieving a 12.5 per cent reduction. More importantly, the largest per capita polluter in the world – the United States – failed to sign up to any mandatory targets. This remained the case even after the UN Climate Change Conference, in December 2005 in Montreal, where negotiations over post-2012 emission reductions were taking place. By contrast, the EU has fully supported the Protocol. In 2005, its European Climate Change Programme set up the EU Emissions-Trading Scheme (EU ETS)[5] aimed at cutting emissions from the EU's major polluting industries and meeting Kyoto targets. However, progress towards meeting the Kyoto targets has varied across the EU and over time. For example, the UK put forward its own ambitious target of cutting CO_2 emissions from their 1990 level by 20 per cent by 2010, but failed to meet it. By 2006 it became clear that CO_2 emissions had been rising every year since 2002 (*The Economist*, 2006b, p25). However, more recently, the UK Climate Change Act, 2008, introduced legally binding GHG emission reduction targets, through action in the UK and abroad, of at least 80 per cent by 2050, and reductions in CO_2

emissions of at least 26 per cent by 2020, against a 1990 baseline. Several other international organizations (such as the World Bank) have also responded to the call to tackle climate change by putting forward policy measures, financial assistance and awareness raising activities. While climate change is a global problem requiring coordinated global action, climate change responses are enacted and governed at multiple scales. The role of sub-national government is particularly critical in formulating and implementing spatial planning policies. At all levels, attention has been focused on the two key areas of adaptation and mitigation, as elaborated on below.

Climate change mitigation and adaptation

The IPCC defines *mitigation* as 'anthropogenic [human] intervention to reduce the sources or enhance the sinks of greenhouse gases'; and *adaptation* as 'adjustment in natural or human systems in response to actual or expected climatic stimuli or their effects, which moderates harm or exploits beneficial opportunities' (IPCC, 2007c, p869). While mitigation measures aim to avoid the adverse impacts of climate change in the long term, adaptation measures are designed to reduce unavoidable impacts of climate change in the short and medium terms. This is because even if concentrations of GHGs could be fixed at 2005 levels, the world could be committed to a long-term eventual warming of 2.4°C. Therefore, strategies need to be in place for adaptation to temperature increases of at least 2°C (Committee on Climate Change, 2008).

As an integral part of sustainable development, mitigation of, and adaption to, climate change are closely linked and both have the same purpose: reducing undesirable consequences of climate change. However, for historical reasons, they have been split in both scientific and policy discourses. This is clearly reflected in the IPCC's definition of the terms mentioned above and is also reflected in the structure of its Working Groups. During the initial climate change negotiations, adaptation was not only treated separately from mitigation, but also was given little attention. This, according to Swart and Raes (2007, p289), was because a focus on adaptation was considered, particularly in Europe, as distracting attention away from mitigation. Mitigation was given priority partly because climate change itself was conceived as an environmental problem similar to, for example, ozone depletion or acid rain which could be handled by setting targets and timetables (Munasinghe and Swart, 2004). Larger uncertainties about adaptation measures also played a part in initially paying limited attention to adaptation. Furthermore, mitigation was seen as the problem of developed countries (as the main emitters), while adaptation was considered as the problem of developing countries (as the main victims). Such artificial dualism began to lose its credibility as the global impact of climate change was increasingly demonstrated. It also became clear that climate change can be more usefully 'framed as a *developmental* rather than an *environmental* problem' (Swart and Raes, 2007, p289, our emphasis) given its fundamental roots in current production and consumption patterns. Hence, it is now widely acknowledged that climate change is unavoidable and both natural ecosystems and human societies will be affected by its unmitigated impacts. As Swart and Raes (2007, p301) put it: 'the question is not whether the climate has to be protected from humans or humans from climate, but how both mitigation and adaptation can be pursued in tandem'. They propose five ways to develop links between adaptation and mitigation measures, as follows:

1   Avoid trade-offs between the two and in designing adaptation measures take into account the consequences for mitigation strategies (see Chapter 2).
2   Identify synergies between the two in response to options within specific policy sectors, notably through spatial planning and design.
3   Enhance both adaptive and mitigative response capacity simultaneously and put such capacity into action particularly in developed countries (see Chapter 18).

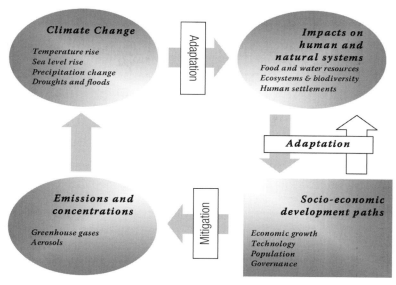

Source: Adapted from IPCC (2001, p3)

Figure 1.2 Climate change – an integrated framework

4   Build institutional links between the two and bridge the communication gap between policy makers.
5   Mainstream climate policies into the overall sustainable development policies at all levels of governance (see Chapter 11).

An integrated view of climate change, as adopted by the IPCC, considers the dynamics of non-linear cause and effect relationships across all sectors, as depicted in Figure 1.2. The solid arrows show the cycle of cause and effect among the four quadrants and the blank arrow indicates societal responses to the impacts of climate change.

The existence of inertia and uncertainty in climate, ecological and socio-economic systems requires precautionary principles and safety margins to be taken into account when setting strategies, policies, targets and timetables. The combined effect of inertia and irreversibility in the interacting climate, ecological and socio-economic systems mean that anticipatory mitigation and adaptation measures are essential to minimize the time lag between policy and action and between technological development and its uptake. As Stern emphasizes, 'there is a high price to delay. Weak action in the next 10–20 years would put stabilisation (of GHG levels) at even 550ppm beyond reach – and this level is already associated with risks' (quoted in *The Times*, 2006, p7).

The roles of spatial planning

Despite major uncertainties, the above summary has shown that the knowledge about causes and impacts of climate change has advanced substantially. There is also widespread recognition that the spatial configuration of cities and towns and the ways in which land is used and developed have significant implications for both adaptation to the adverse impacts of climate change and reduction of the emissions that are causing the change. Settlement forms and their impacts on the use of natural resources and levels of emissions are influenced by many complex factors, including available building technologies, land and property markets, the investment strategies of public and private institutions, public policies (related to, for example, planning, housing, transport, environment and taxation), institutional traditions, social norms and cultures, and individual lifestyle choices and behaviour. Spatial planning interventions are therefore one factor among many in shaping settlement forms.

We use the term spatial planning in its broader sense to refer to actions and interventions that are based on 'critical thinking about space and place' (Royal Town Planning Institute (RTPI), 2003). It involves not only legislative and regulatory frameworks for the development and use of land, but also the institutional and social resources through which such frameworks are implemented, challenged and transformed. In this context, spatial planning is understood as place-based problem-solving aimed at sustainable development. It involves the processes through which options for the development of places are envisioned, assessed, negotiated, agreed and expressed in policy, regulatory and investment terms.

National (and sometimes regional) planning systems vary greatly in terms of their priority, the scope and extent of their powers, their regulatory tools and the resources with which they work. Hence, their capacity to perform and deliver varies from place to place and from time to time. Despite this diversity, mitigation of carbon emissions and adaptation to climate change impacts are increasingly recognized as major priorities for the development and delivery of spatial planning policy in many jurisdictions. Indeed, recognition of the complexity, uncertainty and irreversibility demonstrated by climate science is changing the nature and framing of spatial planning, with an increasing expectation for it to play a part in mitigation and adaptation efforts.

Spatial planning policies

Responding to climate change involves an iterative risk management process that includes both adaptation and mitigation and takes into account climate change damages, co-benefits, sustainability, equity and attitudes to risk (IPCC, 2007b). While there are strong interactions between mitigation and adaptation objectives, they each call for different or complementary planning tools. Indeed, integration of, and

conflicts between, mitigation and adaptation priorities have become the focus of growing debate (see Chapters 2 and 3). At the same time, the importance of spatial and temporal scales in analysis has become critical (see Chapter 17). Mitigation policies, that deliver major cuts in the carbon emissions of built form and human activity, are necessarily led or coordinated at international or national levels, but sub-national innovation and leadership are essential to their delivery. Aspirations to achieve low or zero-carbon development can drive innovation, new partnerships and competitive advantages for areas. Major, though inconsistent, advances have been made in agreeing emissions targets. However, such targets raise critical locational issues in terms of the capacity of jurisdictions, at all levels, to comply (see Chapters 7 and 8). At the very least, it must be expected that area-based development pathways that can meet these targets should be identified through spatial planning processes. This requires assessments of the potential for renewable energy production and increases in the efficiency with which energy is both distributed and used. It also requires understanding of the potential for carbon sequestration, the most commonly recognized forms of which, so far, are forestry and habitat restoration and conservation (e.g. wetlands). Identifying such development paths also requires understanding of the networks of actors whose engagement and behaviours (whether organizational or individual) underpin delivery. It must also be based on a sound understanding of the markets, networks and technologies involved.

As mentioned above, an important area of mitigation for which spatial policy can provide a powerful lever is the shaping of settlement forms and patterns which play a major, complex role in energy use and efficiency (see Chapters 3, 4 and 5). At the same time, mitigation strategies require setting new standards for the materials, construction and management used for buildings and infrastructure, as well as new approaches to waste and water management and infrastructure in order to harness low-energy and closed-loop processes which cut the materials and energy intensity of development.

Policies for climate change adaptation require the development of techniques to explore and achieve consensus around the risks associated with possible change. The understanding of impacts in terms of probability requires investment in modelling, not only on the basis of physical measurements but also in terms of stakeholder engagement. Such models help reduce uncertainty and prioritize issues. Risk assessments (as discussed in Chapters 15 and 16) support decision-making on the allocation of land and design for resilience. In this context, processes of scenario building frame the scoping and weighting of risks, the involvement of stakeholders and the identification of options (Chapters 17 and 18). The focus of adaptation policy is not only on the direct allocation of land use but also on the details of locationally specific design, management and control. It also highlights the importance of ecological functions of land in, for example, flood regulation and temperature control (Chapters 15, 16 and 19).

Spatial planning processes

As demonstrated throughout this book, politics, values, governance, legislation and institutional capacity are integral to spatial planning. Indeed, spatial planning is a fundamental component of governance and a key determinant of governance capacity to respond effectively to climate change and other sustainable development challenges. Seen in this light, spatial planning processes provide key arenas in which integrated approaches to adaptation and mitigation can be designed, trade offs between these and other social and economic goals can be negotiated, conflicts of interests can be mediated and intra- and inter-generational equity concerns can be considered. Furthermore, climate change is part of the larger challenge of sustainable development, and climate policies will be more effective if they are embedded in broader strategies designed to make development paths more sustainable. This further reinforces the role of spatial planning in general, and spatial strategies and plans in particular, in integrating and coordi-

nating related policy, investment and regulation. However, this role is often undermined or indeed resisted by vested interests. In the UK, for instance, while it is argued that, 'the *concept* of sustainable development has been adopted more extensively and more firmly on a statutory basis in the planning system than in any other field' (Owens, 1994, p87), this has not always been matched by its outcomes in terms of dominant development processes. Planning's capacity to deliver cuts in carbon emissions has remained constrained by not only its own limitations, but also other policy, fiscal and investment responses. Examples include the taboo on raising fuel taxes, relatively low levels of investment in public transport and renewable energy, the 'predict and provide' response to air travel and poor integration of transport planning within spatial development frameworks. It is argued that this reflects a weak ecological modernization approach in the UK planning system which asserts that a balance between economic, environmental and social objectives can be found, without clarifying limits, priorities and imperatives (see also Chapters 11, 14 and 21). Such an approach has allowed government, at various levels, to avoid politically difficult choices (Davoudi, 2000; Davoudi and Layard, 2001). Some argue, for instance, that this balancing principle, which underpins most planning decisions, dooms the environment to incremental erosion (Levett, 1999).

Conclusion

The extent to which the climate change agenda has, in fact, been able to introduce a systematic shift in spatial planning towards ecological priorities remains to be seen. On the one hand, the discursive shift from sustainability to climate change, which has become increasingly apparent, can be seen as a catalyst for a refocusing of the spatial planning agenda on ecological issues. It has encouraged planners to rethink their processes, methods, skills and even perception of what constitute 'good places'. Progress has been made in embedding some hard-won requirements for environmental and social sustainability into planning frameworks through mechanisms such as sustainability appraisals of plans and policies. Furthermore, there has been a proliferation of governmental reports, national planning policy statements and emerging legislation at both national and international levels demonstrating a widespread recognition of the pivotal role of spatial planning in delivering climate change mitigation and adaptation policies (see contributions from different countries in this volume).

On the other hand, however, most of the progress has been made in a long period of unprecedented economic growth fuelled by an incredibly buoyant property, and particularly housing, market. This period has now come to a halt. The developed world is in an economic recession, the like of which has not been experienced since the great depression of the 1930s. Thus, the critical question is, how the downturn is going to affect the balance of priorities in spatial planning decisions. If history is anything to go by the answer is not promising. In 1979, faced with the 1980s economic downturn, Michael Heseltine (the then UK Environment Secretary) declared that: 'thousands of jobs every night are locked away in the filing trays of planning departments' (Heseltine, 1979, p27) portraying planning as an obstructive and technocratic bureaucracy which would stifle wealth-creating private enterprise by unnecessary regulatory curbs (including environmental regulation) on development applications (Ward, 1994). As a result, planning policies which aimed at protecting the high street, green spaces and communities were discarded in favour of creating more jobs. Far too often social distribution and environmental interests were sidelined in favour of economic imperatives in the plan making processes (Davoudi et al, 1996). Today, planning is likely to come increasingly under similar pressures to set aside its sustainability goals, which may be perceived as 'luxurious embellishments to developments rather than forming an integral and vital part of their success' (Hartley, 2009, p16).

However, as Stern has argued, 'with strong, deliberate policy choices it is possible to "decarbonise" both developed and developing economies on the scale required for climate stabilisation, while maintaining economic growth in both' (quoted in *The Times*, 2006, p7). Indeed, there are synergies to be made between economic and ecological concerns if a long term perspective is developed. It is in this context that spatial planning can play a pivotal role not just as a technical means by which climate change policies can be delivered, but also as a democratic arena through which negotiations over seemingly conflicting goals can take place, diverse voices can be heard, and place-based synergies can be aimed for. This is a kind of planning that 'is less and less about technical matters' and more and more about the 'critical appreciation and appropriation of ideas' (Friedmann, 1998, p250). As the contributions to this book demonstrate, however, this also requires spatial planners to contribute high levels of knowledge, expertise and skill in building capacity for addressing climate change issues in uncertain times.

Notes

1   Greenhouse gases (GHGs) are the natural and anthropogenic gaseous components of the atmosphere which absorb and emit radiation at specific wavelengths within the spectrum of infrared radiation emitted by the Earth's surface, atmosphere and clouds. This causes the greenhouse effect and gradual warming of the Earth. The primary GHGs are: carbon dioxide (CO_2), nitrous oxide (N_2O), methane (CH_4) and ozone (O_3). In addition, the Kyoto Protocol considers sulphur hexafluoride (SF6), hydrofluorocarbons (HFCs) and perfluorocarbons (PFCs) as GHGs.

2   Radiative forcing is a measure of the influence that a human or natural factor has on the global climate. Positive forcing tends to warm the surface while negative forcing tends to cool it. Other complex aspects of radiative forcing include cloud formation and the role of nitrogen oxides. Increase in aerosols in the atmosphere tends to have cooling effects but these are poorly understood (Committee on Climate Change, 2008).

3   The term 'land use, land use change and forestry' refers to the aggregated emissions from deforestation, biomass and burning, decay of biomass from logging and deforestation, decay of peat and peat fires, and excludes carbon uptake/removal.

4 Sea levels are rising because firstly water expands as it warms and secondly glacier ice is melting.
5 ETS works like any other commodity except that the trade is not in carbon but instead in certificates establishing the level of carbon which has not been emitted by the seller and hence can be bought by potential buyers. The carbon price was established by the Commission but remained volatile in the first phase of the scheme (2005–2008) because the allowance given to the industry was set at a high level. This was reduced in the second phase and hence pushed up the price of carbon, which stood at €20 per tonne in 2007 (*The Economist*, 2007, p10, survey)

References

Committee on Climate Change (2008) *Building a low-carbon economy – the UK's contribution to tackling climate change*, The Stationery Office, London

Davoudi, S. (2000) 'Sustainability: A new "vision" for the British planning system', *Planning Perspectives*, vol 15, no 2, pp123–137

Davoudi, S. (2008) 'Conceptions of the city region: A critical review', *Journal of Urban Design and Planning*, vol 161(DP2), pp51–60

Davoudi, S. and Layard, A. (2001) 'Sustainable development and planning: An introduction to concepts and contradictions', in A. Layard, S. Davoudi and S. Batty (eds) *Planning for a Sustainable Future*, Spon, London, pp7–19

Davoudi, S., Hull, A. and Healey, P. (1996) 'Environmental concerns and economic imperatives in strategic plan-making', *Town Planning Review*, vol 64, no 4, pp421–436

Dodman, D. (2009) 'Blaming cities for climate change? An analysis of urban greenhouse gas emissions inventories', *Environment and Urbanization*, vol 21, no 1, April

Friedman, J. (1998) 'Planning theory revisited', *European Planning Studies*, vol 6, no 3, pp245–250

Hartley, L. (2009) 'Rocks and hard places', *Planning*, no 1800, 9 January, pp16–17

Heseltine, M. (1979) 'Secretary of State's address', *Report of Proceedings of the Town and Country Planning Summer School 1979*, Royal Town Planning Institute, London, pp25–29

IPCC (2001) 'Summary report', *Climate Change 2001: Synthesis Report*, Cambridge University Press, Cambridge UK

IPCC (2007a) 'Summary for policymakers', *Climate Change 2007: Synthesis Report*, Cambridge University Press, Cambridge UK

IPCC (2007b) 'Summary for policymakers', in *Climate Change 2007: Mitigation, Contribution of Working Group III to the Fourth Assessment Report of the Intergovernmental Panel on Climate Change*, Cambridge University Press, Cambridge UK

IPCC (2007c) *Climate Change 2007: Impacts, Adaptation and Vulnerability, Contribution of Working Group II to the Fourth Assessment Report of the Intergovernmental Panel on Climate Change*, Cambridge University Press, Cambridge UK

Levett, R. (1999) 'Planning for a change', *Town and Country Planning*, September

Muir-Wood, R. and Zapata-Marti, R. (2007) 'Industry, settlement and society', in M. L. Parry, O. F. Canziani, J. P. Palutikof, P. J. van der Linden and C. E. Hanson (eds) *Climate Change 2007: Impacts, Adaptation and Vulnerability. Contribution of Working Group II to the Fourth Assessment Report of the Intergovernmental Panel on Climate Change*, Cambridge University Press, Cambridge, UK

Munasinghe, M. and Swart, R. (2004) *Primer on Climate Change and Sustainable Development: Facts, Policy Analysis and Applications*, Cambridge University Press, Cambridge, UK

Owens, S. (1994) 'Land, limits and sustainability: A conceptual framework and some dilemmas for the planning system', *Transactions of the Institute of British Geographers*, vol 19, pp430–456

Romero-Lankao, P. (2007) Are we missing the point? Particularities of urbanization, sustainability and carbon emission in Latin American cities, *Environment and Urbanization*, vol 19, no 1, pp159–175

RTPI (2003) Final Report of RTPI Education Commission, www.rtpi.org.uk/download/236/Education-Commission-Final-Report.pdf, accessed 26 February 2009

Satterthwaite, D. (2008) 'Cities' contribution to global warming: Notes on the allocation of greenhouse gas emissions', *Environment and Urbanization*, vol 20, no 2, pp539–549

Satterthwaite, D., Huq, S., Pelling, M., Reid, A. and Romero-Lankao, P. (2007) 'Building climate change resilience in urban areas and among urban populations in low- and middle-income nations', International Institute for Environment and Development (IIED) Research Report, commissioned by Rockefeller Foundation

Stern, N. (2007) *The Economics of Climate Change: The Stern Review*, Cambridge University Press, Cambridge

Swart, R. and Raes, F. (2007) 'Making integration of adaptation and mitigation work: Mainstreaming into sustainable development policies?', *Climate Policy*, vol 7, pp288–303

The Economist (2006a) 'Those in peril by the sea', 9 September, pp5–8, survey

The Economist (2006b) 'Hot under the collar', 1 April, p25

The Economist (2007) 'Trading thin air', 2 June, pp10–12, survey

The Times (2006) 'Stern Report: If we act now, we can avoid the very worst', 31 December, pp6–7

Ward, S. (1994) *Planning and Urban Change*, Sage, London

Wilbanks, T. J., Romero-Lankao, P., Bao, M., Berkhout, F., Cairncross, S., Ceron, J.-P., Kapshe,

2

Climate Change Mitigation and Adaptation in Developed Nations:

A Critical Perspective on the Adaptation Turn in Urban Climate Planning

Jeff Howard

Now we must learn to live in the real world.
George Monbiot (2008)

The planning community's recent attention to climate change adaptation is well deserved, for reasons laid out clearly by this volume's contributors and the Intergovernmental Panel on Climate Change (IPCC, 2007a). Even in purely economic terms, failing to attend to the local and regional planning implications of climate change that is now unavoidable could be devastating (Ackerman et al, 2008). Crucially, however, the mitigation challenge is not diminishing but growing. Even as planners and other local decision makers are becoming more cognizant of the extraordinary dangers that climate change poses – prompting recognition that adaptation measures are urgent – most communities are failing to take sufficient steps to prevent climate destabilization from presenting even larger dangers in the future. As planners make the 'adaptation turn', then, it is fundamentally important that they also push harder than

ever for changes in land use, transportation systems, energy systems, water systems, and built environment that will dramatically reduce emissions of greenhouse gases (GHGs). Planning finds itself at a juncture where both climate change adaptation and climate change mitigation are more important than ever – and where the relationship between them deserves careful consideration.

I am concerned that the planning community's initial approach to the adaptation challenge sometimes appears to betray muddy thinking about the relationship between the need to adapt to climate change that is *already* unavoidable and the need to minimize the magnitude of change that *becomes* unavoidable. In this chapter I argue that adaptation and mitigation are organically related in ways that have important ramifications for effective 'climate planning' and especially for adaptation planning. Planners are on the front lines of local and regional decision making about climate. Failing to intelligently and responsibly approach the interface between mitigation and adaptation could tragically extend planners' historical culpability in the emergence of climate change via its roots in urban consumption and sprawl economies. At the same time, careful attention to the dynamics of this interface presents an opportunity for planning to begin confronting and making amends for this culpability and for the professional, practical, theoretical and ethical failings that underlie it – while helping to pull our communities, and humanity itself, away from the brink of climate catastrophe.

Central to this argument is the idea that, for theoretical, moral and thoroughly practical reasons, planners should regard climate change mitigation as the most fundamental and urgent form of climate change adaptation.

Mitigation/adaptation in the planning literature

Like Hamin and Gurran (2009), I am struck by the lack of systematic attention in the planning literature to tensions between and potential complementarity of mitigation and adaptation. While much of planning's effort on climate change over the past decades has focused (often ineffectually) on mitigation, the recent adaptation turn is being rapidly extended without sufficient effort to closely examine the conceptual and practical relationship between the two. The rather thin fashion in which the relationship has been treated in recent planning literature suggests a need to begin exploring this territory in greater depth.

In a wide range of literatures, mitigation and adaptation have largely been treated as separate projects (IPCC, 2007a, ch18; Davoudi et al, Chapter 1). To date this appears to be the case in planning literature and conference presentations as well, which, in considering adaptation, ordinarily have focused little on its intersection with mitigation. In a session on adaptation at the 2008 Joint Congress of the American Collegiate Schools of Planning and the Association of European Schools of Planning, in Chicago, for example, a group of Canadian scholars identified tree planting and high-albedo hard surfaces as measures to adapt to urban heat waves without finding it useful to point out that these are simultaneously basic means to mitigate climate change (Chan et al, 2008). The report on which the presentation was based warns that increased use of air conditioning as an adaptation measure would exacerbate climate change (Chan et al, 2007, p46); but in both the presentation and the report the authors considered it reasonable to discuss adaptation at length without directly examining its relationship to mitigation.

Presumably it was clear to everyone in the conference room that tree planting and high-albedo rooftops serve the purposes of both adaptation and mitigation, while air conditioning serves one and undermines the other. And clearly there is value in a segment of the planning literature focusing principally or exclusively on adaptation. But at a time when scholars properly criticize planning's fixation on mitigation to the near exclusion of adaptation, it is striking that planners, to a considerable extent, now seem to be trying to make up for years of neglecting adaptation by merely turning more attention to adaptation and proceeding as if tensions and complementarities between the two activities have little bearing on how to design effective strategies for either.

Adapting a Venn diagram that Hamin and Gurran (2008) offered at the Chicago Congress, we can depict the relationship between mitigation and adaptation as shown in Figure 2.1. In A, the 'sweet spot', tactics such as tree planting accomplish both mitigation and adaptation; in *B* and *C*, tactics such as using renewable energy (*B*) and rainwater harvesting (*C*) serve one purpose but neither support nor hinder the other; in *D*, tactics such as using biodiesel to reduce reliance on fossil fuel might undermine adaptation (in this case, via a minor increase in urban air pollution); and in *E*, tactics such as air conditioning serve adaptation while thwarting mitigation. This is perhaps a reasonable approximation of how planners typically conceptualize the relationship between mitigation and adaptation.

Note, however, that the set of overlapping circles alone makes no distinction between the short and long term or between the local and

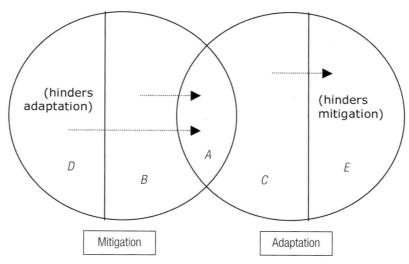

Figure 2.1 Simple conceptualization of the relationship between mitigation and adaptation

global scale. As IPCC (2007a, p750) and others have noted, efforts to approach mitigation and adaptation in an integrated fashion are impeded by 'an obvious mismatch in terms of scale, both spatially and temporally' (McEvoy et al, 2006, p187). For reasons that I explore more fully later in this chapter, three implications, corresponding to the arrows in the diagram, emerge when the local and global and the short and long term are considered simultaneously:

1   Any local mitigation measure that does not directly and immediately hinder local adaptation should be recognized to facilitate or reduce the need for adaptation on a global scale in the long run. Hence, some local measures that are located in *B* if only local, short-term outcomes are considered would instead be in *A* if considered in light of their long-term consequences at the global scale.
2   The same goes for even local mitigation measures that impede adaptation in the short term. They, too, would serve to reduce the need for adaptation globally in the long run and hence would be located not in *D* but in *A*.
3   Similarly, any local adaptation measure that does not directly and immediately facilitate mitigation arguably does not constitute effective adaptation in the long run.

Overall, when one accounts for the second-order, long-term, non-local effects of local actions, measures in *B* and *D* tend to migrate toward *A*, and measures in *C* tend to migrate toward *E*. That is, *B*, *C* and *D* shrink, while *A* and *E* grow. In this view, climate planning measures tend to fall into two categories: On one hand are those that serve to both mitigate climate change and make it easier to adapt to it. On the other hand are attempts to adapt in ways that undermine mitigation – and, in the process, perversely, undermine the long-term prospects for successful local adaptation.

While Pizarro's formulation (see Pizarro, Chapter 3) makes an important contribution to the literature on the mitigation/adaptation interface in planning (see also Hamin and Gurran, 2009), it appears to make this latter, perverse move. He contends that in hot, humid climates, locally effective adaptation will necessitate non-compact development and preclude the high-density, transit-friendly development widely regarded as the backbone of good mitigation planning design. In such climes, sprawling suburban development might well be effective at the local scale in the short term; but at a large scale in the longer term, even a selective embrace of local sprawl inevitably would exert pressure against effective mitigation – and, hence, against the possibility of effective long-term adaptation.

This chapter offers a contribution toward a more comprehensive mapping of the mitigation/adaptation interface, attempting to conceptualize the relationship in a way that broadly accounts for the long-term, second-order and non-local effects of local planning decisions. In a sense, a central question I ask is: How can local planners pursue adaptation, commonly understood as preparing for climate change that is 'unavoidable' (e.g. Ludwig, 2007), without inadvertently making more-severe climate change unavoidable in the long run? The answer, I propose, requires recognizing mitigation as a fundamental form of adaptation. While this idea perhaps is implicit in the planning literature, it seems important at this juncture in the climate planning dialogue to bring it into the open and consider its ramifications. It might be obvious to most planners that, as Janetos (2007) suggests, 'We cannot and must not view adaptation and mitigation as competing with each other – this would be irresponsible public policy.' But I believe there is a clear danger that many communities will be inclined to do precisely this. As understanding of the implications of climate change continues to penetrate the popular consciousness, many localities' response – at least in countries where national-level mitigation mandates are weak or nonexistent – will likely be to fixate on adaptation and give mitigation short shrift. In important respects, this would be analogous to learning that the house is on fire but, instead of fighting the fire, trying to devise methods to live in the flaming structure. Wilbanks puts his finger on the matter. An 'integrated perspective' of mitigation and adaptation will be especially important, he argues, 'if … commitments to mitigation are threatened by the apparent attractiveness of adaptation' (Wilbanks, 2005, p541). I argue it is imperative that planners anticipate this self-destructive local response and steel themselves to take responsibility for pre-empting it. I also propose that planners take this as a potent opportunity to bootstrap the field and the profession into a political, intellectual and institutional mode capable – at long last – of helping to move contemporary society decisively toward ecologically sound development.

Basis for planning of mitigation/adaptation

Urban planning at the mitigation/adaptation interface can and should proceed from the following understandings.

Climate change is accelerating dangerously

Even the dramatic revelations contained in IPCC's *Fourth Assessment Report* (IPCC, 2007b) – revelations that prompted a sea-change in public and official cognizance of the threat – were unduly optimistic. As the blockbuster reports were being released, strong evidence of an accelerating threat was emerging (Rosenthal and Kanter, 2007). 'Sadly, even the most pessimistic of the climate prophets of the IPCC panel do not appear to have noticed how rapidly the climate is changing,' James Lovelock remarked (quoted, Rosenthal and Kanter, 2007). Polar ice caps are melting more rapidly than projected, in turn threatening to further escalate the pace of warming; emissions of carbon dioxide and other GHGs have risen faster than anticipated; and carbon uptake by biological and geophysical systems is not keeping up (Eilperin, 2008; Global Carbon Project, 2008; Hood, 2008; Weiss et al, 2008).

Mitigation planning at the local level has insufficient traction

Wilson (Chapter 17) is correct that 'adaptation to climate change has been a relatively slow area of policy to develop' relative to mitigation. One might wish that academic and practising planners' efforts on mitigation would by now have created intellectual and institutional inertia making it difficult for adaptation planning to be promoted and implemented in ways that would undermine mitigation. But by almost any standard mitigation to date is far from adequate globally, has failed to gain significant traction in most nations, provinces and localities, and often is insufficient or all but absent in high-carbon-emitting nations and provinces where it is needed most urgently. In light of the fact that, decades after predictions of climate change began to solidify (see Reiss, 2001) and more than a decade after the Kyoto Protocol was initiated, the scale of global emissions of GHGs continues to increase rapidly (Eilperin, 2008), mitigation planning's modest head start over adaptation planning provides scant basis for confidence that local decision makers, groggily awakening to the climate threat, will not begin to fixate on the palpable need to adapt without regard for their comparatively abstract and politically more problematic obligation to mitigate. The world economic crisis that emerged in the autumn of 2008, putting future GHG emission cuts in jeopardy (Hanley, 2008), almost certainly heightens this danger (as elaborated by Davoudi et al, Chapter 1).

Without effective global mitigation, local adaptation is impossible

There is good reason to believe that the magnitude of the adaptation challenge will be even larger than many in the planning community seem to expect – and without effective mitigation, impossible. First, it is likely to require pervasive change. Adaptation often appears to be conceived as effectively responding to a relatively well-defined set of climate-driven changes that sophisticated climatology models indicate can be expected in a given region: shoreline communities will have to be moved or defended from rising sea levels, more reservoirs will have to be built to prepare for drought, and so forth – projects that will be sometimes challenging but overall fairly well delimited. However, even under the most favourable modelling scenarios, the required changes are likely to pervade every corner of contemporary society. Rodney White, a perceptive observer of the dilemmas of urban sustainability, points out that 'most places on earth will undergo climatic changes that will affect every aspect of people's lives' (White, 2002, p105). Changes in air temperature, water temperature, sea level, precipitation patterns, agricultural

productivity, disease vectors, storm frequency and intensity – 'All of these predicted changes have immediate consequences for urban environmental management and the building of ecological cities. The changes will clearly make the task much more difficult' (White, 2002, p106).

Second, even under the most favourable warming scenarios and in rich countries best able to respond, adaptation efforts will sometimes (often?) encounter unfortunate limits in society's ability to respond effectively to ecological regime shifts and other climate impacts (IPCC, 2007a, p733). Even with significant alterations of urban development designed to perpetuate prevailing economic patterns, the status quo will be forced to shift as new barriers emerge; and meanwhile, biodiversity and other aspects of ecological integrity will be further impoverished.

Third, crucially, the most favourable warming scenarios are increasingly unlikely, because warming is increasingly unlikely to be slow and steady. If the acceleration due to escalating net emissions continues, climate change almost inevitably will push biological and geophysical systems past critical thresholds, or tipping points, in their complex and often poorly understood dynamics (Lenton and Schellnhuber, 2007; Hansen, 2008). In most cases, passing these thresholds is expected to provide positive feedback further escalating global warming, perhaps beyond all hope of effective mitigation and initiating a complex cascade of impacts at every scale. The deep climate record is replete with episodes in which gradual warming precipitated rapid warming, which in some cases precipitated rapid cooling, subjecting Earth's organisms to a brutal whipsaw effect (Hansen et al, 2007; Hansen, 2008). A US National Research Council committee warns that 'abrupt climate change' can occur 'so rapidly and unexpectedly that human or natural systems have difficulty adapting to it' (Committee on Abrupt Climate Change 2003, p1). Concerns about runaway effects are heightened by recent modelling indicating that climate change resulting from GHG emissions that have already occurred will persist for at least a millennium (Solomon et al, 2009).

With characteristic blandness, IPCC notes that 'the options for successful adaptation diminish and the associated costs increase with increasing climate change' (2007a, p19). Further, '[u]nmitigated climate change would, in the long term, be likely to exceed the capacity of natural, managed and human systems to adapt' (IPCC, 2007a, p20). Even in a scenario of continuing, gradual but relentless warming, the vision of 'effective climate-proofing of our towns and cities' (McEvoy et al, 2006, p185) begins to sound distinctly naive. And in the positive-feedback-driven-runaway-change scenario it is simply absurd. Effective global-scale mitigation is the only available means to improve the odds of avoiding widespread, calamitous local outcomes (Orr, forthcoming).

Global mitigation is impossible without local mitigation

Does anyone except those expecting revolutionary technological breakthroughs or advocating vast experimental carbon-sequestration projects (carrying unprecedented and largely unknowable risks) imagine that adequate mitigation can be achieved at the global scale if the vast majority of communities do not participate? Does anyone imagine that local-scale mitigation would need to occur merely in places such as London where local planning is wedded to national planning priorities and not also in places like libertarian San Diego? Mitigation at the local level constitutes a necessary (if not sufficient) foundation for global mitigation, which, in turn, provides the only reasonable prospect for a stabilized global climate and hence a stable basis for adaptation at the community level.

Local adaptation has the potential to undermine local mitigation

It is clear that some adaptation measures double as effective mitigation, as the example of urban tree planting cited earlier illustrates; but it is equally clear that some – such as significant expansion of the use of air conditioning powered by electricity generated with fossil fuels – do not. It is reasonable to suspect that the latter category includes adaptation measures likely to be preferred by communities already deeply shaped by or committed to sprawl and technology-intensive, high-consumption lifestyles. The potential for such retrograde measures to significantly increase local GHG emissions will be greater if, as White (2002) suggests, adaptation will require cities the world over to undertake not merely scattered alterations but 'major modifications to their infrastructure and to the flows of people, water, and materials' (p106). This potential also would increase with the magnitude of climate change impacts that communities experience. All in all, it seems quite plausible that a community's adaptation, if not done intelligently and responsibly, could seriously undercut the effectiveness of its (in most cases no more than nascent) mitigation efforts and conceivably overwhelm them.

The mitigation/adaptation interface is suffused with important asymmetries

The relationship between adaptation and mitigation is in important respects dominated by mitigation, as Figure 2.1 illustrates. In the 'sweet spot', responsible adaptation must 'seek out' the territory of mitigation, so to speak. Moreover, all classes of mitigation, including mitigation that appears to undermine adaptation, tend toward the sweet spot, while adaptation tends to be polarized: either sweet spot or mitigation-hindering, with scant middle ground.

Further, local and global responsibilities are asymmetrical. It can be argued that the global community has a responsibility to protect local communities from climate impacts via both mitigation and adaptation. This seems to be one of the key assumptions behind IPCC's (2007a) understanding of the importance of international efforts to initiate and facilitate mitigation and adaptation. But for rich nations, at least, obligation often is seen as flowing primarily in the other direction, as in ICLEI–Local Governments for Sustainability's Cities for Climate Protection programme (www.iclei.org/

index.php?id=800): local communities are understood to have obligations to the global. Arguably, a local community bears responsibility to the global in proportion to the scope of the environmental externalities it has exported. The more it has externalized and the less mitigation it has undertaken, the more it is shirking its responsibility, adding global insult to global injury.

Finally, responsibilities for present and future are asymmetrical. As the Brundtland formulation of sustainable development and the growing literature on generational environmental equity emphasize, present generations bear a substantial responsibility for maintaining environmental conditions that will support the well-being of future generations (World Commission on Environment and Development, 1987; Barry, 2003). The climate impacts of 20th- and early 21st-century GHG emissions will redound upon our progeny for centuries to come (Solomon et al, 2009), and we have a profound, transparent and thus largely ignored responsibility to act in ways that will protect the environmental interests of our great great great grandchildren (see Wilson, Chapter 17). If planners do not act accordingly, they will join the ranks of experts who 'will be seen as absolute pariahs' (Peter Newman, quoted in Salzman, 2007, p27).

Principles for planning of mitigation/adaptation

Operating on the basis of these understandings, planners in the developed nations would see the mitigation/adaptation interface to be structured by several principles or heuristics.

Principle 1: Mitigation has priority

The accelerating pace of climate change heightens, rather than reduces, the need to closely couple adaptation and mitigation. It clearly makes both of these tasks more urgent and underscores the need for planners to operate as much as possible in the 'sweet spot' and avoid adaptation that undermines mitigation.

Crucially, however, in the event of an unavoidable conflict between the two, mitigation should routinely be given priority. Adaptation planning should eschew use of fossil fuels, or at least uses that result in net increases in GHG emissions. ICLEI's Climate Resilient Cities campaign (ICLEI, n.d.) seems to strike the right balance: cities, it contends, 'must continue to develop and implement effective emissions reductions plans in order to slow and ultimately stop climate change, even as they consider plans to make the community more climate resilient'. Adaptation outside the sweet spot would represent an escalation of business as usual. For communities in rich nations, at least, adaptation in C of Figure 2.1 should be rare and adaptation in E should be regarded as all but forbidden.

Adaptation cannot reasonably be allowed to take precedence over a community's obligations to mitigate GHG emissions. Because global mitigation cannot be achieved without aggressive local mitigation, local adaptation must not be allowed to impede progress toward this goal. A community's adaptation measures must not be allowed to routinely externalize 'climate costs', an approach that would extend the pattern of naive and irresponsible development that has given rise to anthropogenic climate change in the first place. If effective adaptation in hot, humid Houston requires non-compact urban form (see Pizarro, Chapter 3), then Houston must: (a) work technological miracles to somehow make that seemingly unsustainable form carbon neutral; (b) purchase carbon credits or some other kind of absolution to compensate for this additional contribution to climate change; or (c) fail to effectively participate in global climate mitigation. One might reasonably ask whether such a city, if it cannot adapt to climate change without undermining its ability to aggressively participate in climate mitigation, should not be written off as intrinsically unsustainable.

To what extent will adaptation conflict with mitigation? IPCC notes that the question 'remain[s] largely unexplored' (IPCC, 2007a, p760). Although the panel places 'high priority' on adapting 'in ways that are synergistic with wider societal goals of sustainable development', however, its preliminary judgement in 2007 was that conflict between mitigation and adaptation will be fairly frequent but rarely significant (IPCC, 2007a, pp737, 760). McEvoy and colleagues are more pessimistic: 'It is entirely plausible that … despite their attractiveness, truly 'win–win' situations may be few and far between' (2006, p186; see also Hamin and Gurran, 2009).

The question is admittedly complex, but IPCC's nonchalance about the matter in 2007 must be interpreted in light of the panel's apparent underestimation of the pace of climate change. If the acceleration of climate change to a significant extent reflects the inadequacy of local mitigation measures, then the need to dramatically escalate those measures heightens the importance of avoiding adaptation whose adverse impacts on net GHG emissions might otherwise be judged insignificant. Moreover, adaptation to accelerated change will necessitate even more extensive and frequent modifications to infrastructure, settlement patterns, transportation systems and energy procurement, making the adverse impact of adaptation measures outside the sweet spot more frequent and more collectively significant. If this perspective is accurate, clashes at the mitigation/adaptation interface will be all too common and it is all the more important that planners be prepared to take a lead role in defusing them in a manner that honours the obligation of aggressive mitigation

Principle 2: Mitigation is the primary form of adaptation

The conception of the mitigation/adaptation interface presented so far holds a striking implication for our understanding of adaptation. It implies that mitigation – all mitigation – should be understood as a form of adaptation.

It is obvious how mitigation in the sweet spot of Figure 2.1 can be understood as adaptation, for here mitigation and adaptation work in harmony. But seemingly less adaptation-friendly forms of mitigation must be regarded in a similar light. Even mitigation that does not directly support adaptation in the short run serves the purposes of adaptation in the long run, if only indirectly. In fact, in an important sense even mitigation that in the short run *impedes* adaptation serves the purposes of adaptation in the long run. The reason is straightforward: the most desirable form of adaptation is adaptation that is made unnecessary. 'Since [the] speed of [climate] change is one of the reasons it presents a particular challenge,' McEvoy and colleagues point out, 'even a slowing of the rate of climate change could prove to be of substantial benefit to both human and wildlife communities. For humankind, it would influence the urgency and magnitude of adaptation necessary, and […] it would improve the prospects for animals and plants to adapt to new climate conditions' (McEvoy et al, 2006, p186; see National Leadership Summit on Energy and Climate Change, 2006, item on 'urgency'). As the old saying has it, an ounce of prevention is worth a pound of cure.

This understanding of adaptation is depicted in Figure 2.2, which borrows the labels and conceptions of Figure 2.1 but otherwise stands the earlier diagram on its ear. It incorporates the five zones of mitigation and adaptation into a unified depiction of adaptation. Here the original sweet spot is understood as the top position, A_1, in a larger sweet spot, A. The secondary position in the sweet spot, A_2, is mitigation formerly in B, adaptation-neutral in the short term but tending to obviate adaptation in the long term; the third position, A_3, formerly part of D, is mitigation that impedes adaptation in the short term but tends to obviate it in the long term. Zones C and E are unchanged, representing adaptation that is mitigation neutral and adaptation that hinders mitigation, respectively. By the logic outlined in the previous section, the zones and subzones would be ranked, in order of increasing priority, thus: $E \rightarrow C \rightarrow A_3 \rightarrow A_2 \rightarrow A_1$. Conceptualized in this way, the asymmetry of mitigation and adaptation is even more pronounced. Examples of planning activities in each of these zones are proposed in Table 2.1.

Some adaptation measures in A will be no more costly or difficult than measures in C or E. But planners will sometimes – perhaps often –

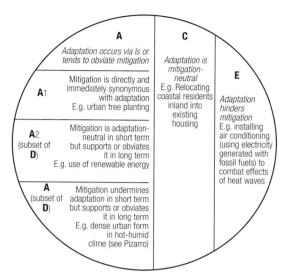

Figure 2.2 Integrated model of climate change adaptation

find themselves needing to convince local citizens and power brokers that expensive, inconvenient or culturally challenging forms of adaptation-via-mitigation in A_1, A_2, or A_3 are preferable to less expensive, more convenient or culturally innocuous adaptation in C or E. To compound the difficulty, planners will need to acknowledge that adaptation in A, although it might involve substantial local sacrifice, provides no guarantee that global mitigation will be sufficient to prevent a long string of climate impacts locally.

Adaptation in C and E provides no real guarantee of safety, either, of course. And unlike adaptation in C and E, adaptation in A, although not sufficient to guarantee global mitigation and long-term safety, is a prerequisite for them. For only by the vast majority of localities in the rich nations linking arms and adapting responsibly – via mitigation – will mitigation be sufficiently extensive to challenge the extraordinary inertia of a global climate system now being driven amok by anthropogenic emissions. The momentum will be insufficient if even a sizable fraction, such as those in Pizarro's hot, humid sprawling suburbs, selfishly or merely naively choose to opt out of the global mitigation project. A vast network of communities dedicated to aggressive local mitigation – including adaptation through mitigation – is the only reasonable hope for mitigation on a scale sufficient to produce global climate stability. And the prospect that the vast majority of potential nodes in this network will become actual nodes is necessary for the network to coalesce in the first place. Planners have done much to make this network necessary; now they are in a crucial position to make it possible.

Principle 3: Effective local adaptation requires a long-term, global perspective

Dawning awareness of the threat posed by global climate destabilization will do much to focus local decision makers on the need for decisive action to make their communities climate resilient. And adaptation is widely and justifiably understood to have a distinctly local focus (e.g. IPCC, 2007a). But no less than local mitigation or failure to locally mitigate, local adaptation will have global consequences. Adaptation undertaken in a blinkered fashion, with attention primarily to local well-being, would be a signal that our communities fail to grasp the global dynamics on which their own long-term welfare depends. It would sometimes, perhaps often, have negative impacts globally not only due to its effects on the community's contribution to

Table 2.1 Examples of five modes of adaptation to effects of climate change

	A1 – Mitigation is directly and immediately synonymous with adaptation Priority: 1	A2 – Mitigation is adaptation-neutral in short term but supports or obviates adaptation in long term Priority: 2	A3 – Mitigation undermines adaptation in short term but supports or obviates it in long term Priority: 3	C – Adaptation is mitigation-neutral[1] Priority: 4	E – Adaptation hinders mitigation[2] Priority: 5
Excessive heat	LEED-certified building standards	Reducing overall levels of consumption	Dense urban form in hot, humid clime (see Pizarro)		Increased use of air conditioning
Drought	Improved efficiency of water use	LEED-EB renovation of existing buildings Using construction materials with low embodied energy	Hydropower (potential conflict with agricultural irrigation)	Local rainwater harvesting and floodwater storage	New conventional reservoirs; importation of water from distant regions
Rising sea levels	Relocating coastal residents inland into LEED platinum urban housing	Reducing personal-vehicle VMT Reducing reliance on air travel Increasing reliance on mass transit and long-distance rail		Relocating coastal residents inland into existing conventional urban housing	Relocating coastal residents inland into new conventional suburban housing
Flooding	Restoration of wetlands				Building concrete dikes and levees
Degradation of urban air quality	Urban tree planting; open space and habitat protection		Use of biodiesel (minor increase in NO$_x$ emissions)	Passive filtering within buildings (e.g. via plants)	Active filtering within buildings via mechanical systems
Increase in incidence or severity of violent storms	Decentralized renewable energy generation; restoration of coastal wetlands			Improved storm warning systems & evacuation planning	Making structures storm-resistant through use of concrete and steel walls

Notes: 1 Most adaptations in column C assume use of renewable energy.

2 Most adaptations in column E assume use of fossil fuels. Some of these would be in column C if carried out with renewable energy.

global GHG emissions but due to its effects on the community's relationship to other communities and generations.

Although adaptation in *C* and *E* should be systematically avoided, it must be recognized to be less unacceptable in poor than in rich nations. In the logic of Rio and Kyoto, communities that have had a disproportionately large effect on the global climate have a comparably disproportionate responsibility to adapt via mitigation. Moreover, since any zone *C* and *E* adaptation necessary in poor communities would need to be compensated by even greater mitigation efforts elsewhere, the prohibition on such adaptation in rich nations must be regarded as all but absolute. If any community can claim a legitimate need to adapt by sprawling, as Pizarro proposes, it is not Jacksonville or Brisbane but Kampala or Djakarta.

Of course the imperative to adapt in a globally responsible manner is not merely a moral onus but a thoroughly practical one. As emphasized earlier, adaptation not deeply informed by the globally dynamic character of climate change and not seen as an integral part of responding globally – via local mitigation – would serve to feed rather than diminish the long-term threat the community faces. Moreover, it would undermine the prospect for globally responsible action by other communities. It would fail to model appropriate behaviour, either for neighbouring communities (whose contributions are also urgently needed) or for communities in poor nations. Just as poor nations have justifiably (and for the most part in vain) looked to the rich nations for leadership in mitigation, they will now be looking to these nations to model and promote responsible modes of adaptation. Failure to demonstrate global leadership on adaptation at this crucial hour would compound the rich nations' failure to adequately mitigate the climate-destabilizing effects of their own historical development, sending a signal that as far as their own communities are concerned the rule in the age of climate change will be: 'Each community is on its own, and devil take the hindmost'. It would signal a community's refusal to engage in the 'cosmopolitics' (Beck, 2007) that is our best hope for decisive, democratic global action.

It also must be emphasized that responsible, effective adaptation requires communities to greatly increase the temporal scale of their planning. Development in the 20th century was driven by technologies whose emissions are climatologically potent on the scale of centuries, perhaps millennia (Solomon et al, 2009). Instead of the 25- or 30-year time frame for 'long-term', strategic local planning that prevails in the United States, the challenge of climate adaptation illustrates the urgent need for – and provides a key opportunity for developing – a commitment to planning in a time frame of centuries (Tonn, 2004).

Failing to take a global, long-term view got us into this dilemma. Decisions about adaptation must be exploited as crucial opportunities to develop a habit of planning less on the basis of our own near-term needs than on the basis of the needs of our progeny and our contemporaries around the planet.

Climate planning for the real world

Global climate change must be understood as long-range, large-scale system feedback on the validity of a Western conceit deeply embedded in 20th century urban planning: that our technological systems, our communities and our species need be only superficially integrated in the planet's biological and geophysical systems. If planning is to systematically alleviate, rather than systematically exacerbate, the climate crisis, then the field and the profession must read this feedback as a devastating critique and as evidence of the need for a new theoretical and practical understanding of the world and planning's role within it (see White, 2002). The adaptation turn thus represents a crucial juncture. It is a major opportunity for planners to demonstrate that they comprehend the feedback and at last are prepared to begin vigorously exercising a kind of expertise that is urgently needed in an intensively technological, intensively urban civilization: precaution-oriented urban sustainability planning.

Planners operating in this more environmentally realistic mode will harness emerging public awareness of the local threat posed by global climate change to press for serious adaptation and, at the same time, far deeper public commitment to mitigation. Wilbanks points to the likelihood that 'in the coming years policymakers and stakeholders will increasingly be considering relative payoffs, tradeoffs, and complementarities of mitigation and adaptation strategies and actions' (Wilbanks, 2005, p541). In these deliberations it will be crucial for planners – practitioner and academic alike – to consistently and emphatically insist upon the practical urgency and moral imperative of adaptation-via-mitigation. Without this resolve, planning's adaptation turn threatens to degenerate into an open-ended commitment to adapt to escalating climate change while failing to do enough to arrest it.

This is not a task for which planning is constitutionally well equipped. The political and economic forces that powerfully shaped the field and profession during what Stephen Wheeler has called 'a century of disastrous planning' (personal communication, July 2008) are still present and still exerting their influence. This is especially true in the United States but to a considerable extent holds for planning worldwide. What will it take to bring about the needed shift in planning's operating assumptions and modus operandi? Pointing to the grave danger of a polar 'albedo flip', in which ongoing melting of the ice caps darkens the polar surface, allowing it to absorb more solar radiation, hastening the melting and in short order dramatically increasing the pace of global warming, Monbiot (2007) argues that the societal response necessary to counter climate change is a 'political albedo flip': rapid reversal of the political dynamics that are propelling humanity to the brink. The planning community is in a crucial location to help engineer this flip. And to do so it must – simultaneously – engineer a flip within planning itself.

Successfully promoting adaptation-via-mitigation will bring planning theory and practice into direct contact with three underlying needs.

First, we must institutionalize a recognition that even short-term local planning is simultaneously long-term global planning. Too many practising planners have failed to see or acknowledge this and to shoulder the responsibility to make good global climate policy rather than bad; meanwhile, too many planning academics have failed to inculcate such an awareness and build it deep into their curricula and the minds of their students. These failures have been instrumental in planning's capitulation to suburban sprawl, building codes oblivious to energy efficiency, transportation visions obsessed with the automobile, reliance on fossil fuelled power grids, and chronic neglect of most urban environmental externalities. This capitulation is a classic case of 'organized irresponsibility' (Beck, 1995), and planners must not compound it by capitulating to climate change adaptation that does not simultaneously serve the long delayed purpose of mitigation. The challenge will be all the more difficult if it means that planners must confront powerful economic players' penchant for exploiting disasters as opportunities to consolidate their wealth and political influence (see Klein, 2008).

Second, planners must resist the siren call of sophisticated, detailed, scientific analysis. The plodding, technocratic framework that took decades to produce consensus on anthropogenic climate change now threatens to morph into a plodding, technocratic framework that will take decades to produce consensus on how to marry mitigation and adaptation (see e.g. IPCC, 2007a, pp760–763 and pp770–771; Hall, Chapter 18). It is crucial to avoid the temptation to see highly sophisticated and fine-grained studies on the interrelationships between mitigation and adaptation *as a prerequisite for deciding how to proceed with systematic mitigation and adaptation*. Some of the details of Table 2.1 no doubt are flawed or simplistic; but sorting the good from the bad must not be a process that planners defer for an entire generation or more. Fine-grained analyses should be done only by individuals whose expertise and authority are not essential for moving our societies into a mode of vigorous mitigation and adaptation. The goal should be 'serviceable knowledge'. The watchword should be *precaution* (Tickner, 2003).

Third, we must insinuate sustainability planning throughout planning education and practice. Instead of a specialized form of planning, the purview of a relative handful of individuals, it should be allowed to deeply inhabit – and disruptively reconfigure – all forms of urban planning. Instead of being on the periphery of the curriculum, it should suffuse the core. Instead of remaining a series of exceptions to the rule, it should be the rule. Instead of being a kind of planning expertise, it should be a pivotal point of all planning expertise. It must transform planning in the 21st century as pervasively as market fetishism did in the 20th. The interface of climate change mitigation and climate change adaptation is a good place to start.

Acknowledgement

The author thanks Rafael Pizarro for a helpful suggestion and Kent Hurst for helpful conversations, research assistance and comments on drafts of the text.

References

Ackerman, F., Stanton, E. A., Hope, C., Alberth, S., Fisher, J. and Biewald, B. (2008) *The Cost of Climate Change: What We'll Pay if Global Warming Continues Unchecked*, Natural Resources Defense Council, New York, www.nrdc.org/globalwarming/cost/cost.pdf, accessed 29 October 2008

Barry, B. (2003) 'Sustainability and intergenerational equity', in A. Light and H. Rolston (eds) *Environmental Ethics: An Anthology*, Blackwell, Malden, MA

Beck, U. (1995) *Ecological Politics in an Age of Risk*, translated by A. Weisz, Polity, Cambridge, UK

Beck, U. (2007) 'In the new, anxious world, leaders must learn to think beyond borders', *The Guardian*, 13 July, p31

Chan, C. F., Lebedeva, J., Otero, J. and Richardson, G. (2007) 'Urban heat islands: A climate change adaptation strategy for Montreal (Final Report)', Climate Change Action Partnership, McGill University School of Urban Planning, Montreal

Chan, C. F., Lebedeva, J., Otero, J. and Richardson, G. (2008) 'Urban heat islands: A climate change adaptation strategy for Montreal', paper presented at Joint American Collegiate Schools of Planning and Association of European Schools of Planning Congress, Chicago, 6–11 July

Committee on Abrupt Climate Change (2003) 'Abrupt climate change: Inevitable surprises', National Research Council, Washington DC

Eilperin, J. (2008) 'Carbon is building up in atmosphere faster than predicted', *Washington Post*, 26 September, pA2

Global Carbon Project (2008) 'Carbon budget and trends 2007', www.globalcarbonproject.org/carbontrends/index.htm, accessed 4 October 2008

Hamin, E. M. and Gurran, N. (2008) 'Addressing climate change: Australian and U.S. planning responses', paper presented at Joint American Collegiate Schools of Planning and Association of European Schools of Planning Congress, Chicago, 6–11 July

Hamin, E. M. and Gurran, N. (2009) 'Urban form and climate change: Balancing adaptation and mitigation on two continents', *Habitat International,* vol 33, no 3, pp238–245

Hanley, C. J. (2008) 'UN: Financial chills are ill wind for climate', *Associated Press*, 9 October

Hansen, J. (2008) 'Global warming twenty years later: Tipping points near', briefing to the House Select Committee on Energy Independence & Global Warming, 23 June, www.columbia.edu/%7Ejeh1/2008/TwentyYearsLater_20080623.pdf, accessed 29 September 2008

Hansen, J., Sato, M., Kharecha, P., Russell, G., Lea, D. W. and Siddall, M. (2007) 'Climate change and trace gases', *Philosophical Transactions of the Royal Society A*, vol 365, pp1925–1954

Hood, M. (2008) 'Climate change gathers steam, say scientists', *Agence France-Presse*, 29 November

ICLEI Local Governments for Sustainability (n.d.) 'Climate resilient communities', www.iclei.org/index.php?id=6687, accessed 4 October 2008

Intergovernmental Panel on Climate Change (IPCC) (2007a) *Climate Change 2007: Impacts, Adaptation and Vulnerability*, Cambridge University Press, Cambridge

IPCC (2007b) 'Fourth Assessment Report: Climate Change 2007: Synthesis Report: Summary for Policymakers', United Nations Environment Programme

Janetos, T. (2007) 'Reporting and comments', presentation at Coping With Climate Change, National Summit, Ann Arbor, Michigan, 8–10 May, www.snre.umich.edu/files/janetoscoping presentation_pdf, accessed 26 September 2008

Klein, N. (2008) *The Shock Doctrine: The Rise of Disaster Capitalism*, Metropolitan/Holt, New York

Lenton, T. M. and Schellnhuber, H. J. (2007) 'Tipping the scales', *Nature Reports: Climate Change*, vol 1, no 7, pp97–98

Ludwig, F. (2007) 'Too much water or too little? Coping with the inevitable', *Planning*, vol 73, no 8, pp28–33

McEvoy, D., Lindley, S. and Handley, J. (2006) 'Adaptation and mitigation in urban areas: Synergies and conflicts', *Municipal Engineer*, vol 159, no 4, pp185–191

Monbiot, G. (2007) 'Stop doing the CBI's bidding, and we could be fossil fuel free in 20 years', *The Guardian*, 3 July, p29

Monbiot, G. (2008) 'This stock collapse is petty when compared to the nature crunch', *The Guardian*, 14 October, p31

National Leadership Summit on Energy and Climate Change (NLSECC) (2006) 'Wingspread principles on the U.S. response to global warming', www.summits.ncat.org/energy_climate/statement.php, accessed 20 November 2008

Orr, D. W. (forthcoming) 'Baggage: The case for climate mitigation', *Conservation Biology*

Reiss, B. (2001) *The Coming Storm*, Hyperion, New York

Rosenthal, E. and Kanter, J. (2007) 'Alarming UN report on climate change too rosy, many say', *International Herald-Tribune*, 18 November, www.iht.com/articles/2007/11/18/europe/climate.php, accessed 10 February 2008

Salzman, R. (2007) 'Greenhouse gurus: A conversation with two experts on the topic of the day', *Planning*, vol 73, no 8, pp24–27

Solomon, S., Plattner, G.-K., Knutti, R. and Friedlingstein, P. (2009) 'Irreversible climate change due to carbon dioxide emissions', *Proceedings of the National Academy of Science*, vol 106, no 6, pp1704–1709

Tickner, J. A. (ed) (2003) *Precaution, Environmental Science, and Preventive Public Policy*, Island Press, Washington DC

Tonn, B. E. (2004) 'Integrated 1000-year planning', *Futures*, vol 36, pp91–108

Weiss, R. F., Mühle, J., Salameh, P. K. and Harth, C. M. (2008) 'Nitrogen trifluoride in the global atmosphere', *Geophysical Research Letters*, vol 35, no 20, L20821

White, R. R. (2002) *Building the Ecological City*, CRC, Boca Raton, FL

Wilbanks, T. J. (2005) 'Issues in developing a capacity for integrated analysis of mitigation and adaptation', *Environmental Science & Policy*, vol 8, no 6, pp541–547

World Commission on Environment and Development (1987) *Our Common Future*, Oxford University, New York

3

Urban Form and Climate Change:
Towards Appropriate Development Patterns to Mitigate and Adapt to Global Warming

Rafael Pizarro

A lack of both conceptual and empirical information that explicitly considers both adaptation and mitigation makes it difficult to assess the need for and potential of synergies in climate policy.

(IPCC, 2007, p747)

Introduction: Mitigation, adaptation and urban form

The Intergovernmental Panel on Climate Change (IPCC) *Fourth Report* (2007) stresses the need for a fundamental transition in the structure and functioning of built environments to simultaneously mitigate climate change and adapt to the effects of global warming. Yet, *mitigation* and *adaptation* as discussed in Chapter 2, may actually be oppositional. For example, a development pattern that helps mitigate climate change may not be the best to adapt that settlement to the negative effects of global warming in its geographical location. The new chapter in the IPCC *Fourth Report*, 'Inter-Relationships Between Adaptation and Mitigation' (ch 18), readily acknowledges this challenge and identi-

fies the urgent need to create synergies between adaptation and mitigation while acknowledging the difficulty of such interrelationship. Furthermore, mitigation and adaptation appear as two sides of the same coin in the literature on sustainable development and climate change, making an implicit association between *sustainable urban form* and *disaster-resilient cities* (Berke, 1995; Bury, 1998; El-Masri and Tapple, 2002; Bulkeley and Betsill, 2005). The IPCC (2007) claims that, 'making development more sustainable can enhance both mitigative and adaptive capacity, and reduce emissions and vulnerability to climate change' (p22). It also argues that 'enhancing society's response capacity through the pursuit of sustainable development is [...] one way of promoting both adaptation and mitigation' (IPCC, 2007, p747). On the other hand, the literature on sustainable development has identified sustainable urban form with *compactness* as it is in the compact city where it is possible to achieve higher densities, mixed uses, pedestrian environments and mass public transit systems (Van der Ryb and Calthorpe, 1986; Elkin et al, 1991; Frey, 1999; Rudlin and Falk, 1999; Beatley, 2000; Jenks et al, 2002; Beatley and Wheeler, 2004; Girling and Kellet, 2005; Jenks and Dempsey, 2005; Jabareen, 2006).[1] The corollary from the connection between sustainable urban form, compactness and the IPCC postulates above is that an urban form that is compact, dense, high-rise and transit-oriented is best positioned to mitigate climate change and, at the same time, to respond to the impacts of global warming. In this chapter, however, I argue that while sustainable development is indeed a desirable planning goal, as it produces more energy-efficient cities that help mitigate global warming, the association between *compact urban form* and *disaster-resilient cities* (i.e. cities that have the capacity to adapt to global warming) can be misleading; one thing does not necessarily imply the other. Furthermore, the implicit association between those two concepts may actually hinder our understanding of the connections between urban form and climate change.

The chapter illustrates this perplexing planning paradox: an urban form that exacerbates climate change can, at the same time, adapt well to it. And, the other way around, an urban form that mitigates climate change may not adapt well to global warming. To this end, the chapter shows how the low-density, low-rise, spread-out, automobile-oriented suburb – a development pattern certainly most denounced in the literature as exacerbating climate change – may actually adapt better, in certain geographical locations, to the negative effects of global warming. But, also, the chapter illustrates how the dense, compact, mixed-use, transit-oriented development pattern – indeed a good urban form to mitigate global warming – may be the worst solution to adapt to certain climatic conditions triggered by climate change in some specific geographical locations.

The chapter also shows how the complexity of adapting urban form to the environmental stresses resulting from global warming rule out the possibility of a one-size-fits-all solution to the problem of adapting to climate change. It makes evident the potential problem of adopting a low-carbon development pattern (compact, dense, transit-oriented) for all geographical regions. The chapter shows how adapting to problems like extreme temperatures, flooding and strong winds (the most likely effects of global warming in many parts of the planet) needs to be addressed by a mixture of urban design strategies that sometimes are contradictory with each other and, worst of all, may not be the most appropriate to mitigate climate change. This assessment is achieved by reviewing the decades-old literature on climate and cities (Olgyay, 1963; Konya, 1980; Givoni, 1998) and applying them to performance of the spread-out suburb and the compact city under climatic duress. At the end of the chapter, I offer a matrix showing a minimum set of variables related to urban form to be taken into consideration when designing for each climatic zone of the planet.

Mitigation and adaptation through urban form: Compact versus sprawled development

There is no way to escape the fact that urban form in the 21st century needs to address simultaneously the problems of mitigation and adaptation. As the IPCC *Fourth Report* states, 'it is … no longer a question of whether to mitigate climate change or adapt to it. Both adaptation and mitigation are now essential in reducing the expected impacts of climate change on humans and their environment' (IPCC, 2007, p748). Further, a report from the Tyndall Centre for Climate Change Research points out that

> *When talking about how to respond to climate change we talk about both adaptation and mitigation to climate change. By adaptation we refer to activities that allow us to better cope with the impacts of climate change, and by mitigation we refer to reducing the emissions of the greenhouse gases that cause the problem. In the past many have argued that giving attention to the need for adaptations to climate change is defeatist or at best draws energy and resources away from addressing the root cause of the problem through mitigation. However there is now a widespread agreement among scientists that the problem is so serious that however fast we*

start to reduce emissions we will still face a significant amount of climate change during this century, so both adaptation and mitigation responses are required.

(Tyndall Centre for Climate Change Research, 2004, p14)

Mitigating climate change through urban form is, fortunately, more easily achieved than adapting to it. A settlement pattern that would not exacerbate the releases of greenhouse gases (GHGs) is one that substantially reduces the energy consumption of that settlement; especially when that energy comes from the burning of fossil fuels. In the case of transportation, for example, an urban form that mitigates climate change would be one that facilitates the movement of people between places of work, leisure, commerce or worship without having to drive private vehicles propelled by petrol. This mandate would translate into an urban form that is mixed-use, dense and compact enough (i.e. where buildings and activities are so close to each other and where the number of people residing and working in a given area is high enough) so that movement of people can be achieved by transporting the same number that would move in private vehicles but on a mass-transit system (e.g. bus, tram or train), even if the system is also fuelled by the burning of a fossil fuel. The energy needed per capita to move the same number of people will be lower in the mass-transit system.

Now, in certain geographical locations, the urban form resulting from the mass-transit system will serve well the purposes of mitigation and adaptation. In a hot–dry climate, for example, the resilience of an urban form to withstand high temperatures in the summer season without having to resort to air conditioners (with their concomitant high energy inputs) depends on how close together the buildings are so that they can cast shadows on each other to help lower the temperature. So, this urban form will not only help to mitigate global warming by encouraging mass public transit and low-energy inputs for electrical cooling devices but will also help to adapt it to the spectre of soaring high summer temperatures.

Unfortunately, this coincidence is not always the case. If the same compact settlement was located in a hot–*humid* region, the close proximity of the buildings will likely block the free circulation of air in between and around dwellings thus exacerbating the high temperature and humidity. In this case, spreading out the buildings on the landscape will be the best option as this will ease the circulation of breezes throughout the development thus lowering temperature and humidity. This certainly represents an urban form that adapts well to extreme high temperatures and humidity in the summer months. Yet, although the energy inputs to cool down the dwellings in this second case will still be low, when such arrangement of buildings is extended to the entire city, the resulting density will be so low that attempting to collect and move people with a mass-transit system will be nearly impossible or, at the very least, extremely costly. This implies that people in this city will likely have to use their own cars to go about their daily activities, thus increasing the demand for petrol (fossil fuel) and its concomitant release of carbon dioxide into the atmosphere. The urban form for this kind of city is the worst for the mitigation of climate change.

This contradiction of an urban form that adapts well to global warming but performs poorly to mitigate it is exemplified in some of the cities founded by the Spaniards in the Caribbean Basin in the 16th century which follow the urban model of the medieval city. One of these cities, Cartagena in the Caribbean coast of Colombia, retains the compact urban form legacy from the Spanish times (Figure 3.1) with all its advantageous savings in infrastructure and transportation.

But, although the urban form is nicely dense and compact, and the urban space is shaded by the narrow streets and the proximity of the buildings, the region's high 33°C average temperature and 95 per cent humidity demand a more open spatial structure with buildings separated from one another to facilitate the movement of air. Thus, new suburban residential developments in the outskirts of the city (Figure 3.2), representing an urban form highly criticized for exacerbating climate change, have

Photo: By Author

Figure 3.1 Compact urban form in the town centre of Cartagena, Colombia

much more pleasant temperature and humidity than the compact counterpart in the urban core (Salmon, 1999; Pizarro, 2005).

So, within the same city we see examples of urban form that helps mitigate climate change but that will not adapt to the expected high temperatures and humidity for this part of the world under global warming, and vice versa; we see an urban form that will do well in adapting to future climatic conditions but that implies a heavy reliance on private automobiles for transportation (and thus on the burning of fossil fuels). This Colombian example, however, only touches on the complexity of elaborating a normative theory of urban form in the times of climate change, as we will see in the remainder of the chapter.

In sum, when it comes to mitigation, the compact city, advocated by the literature on sustainable development, is indeed the most appropriate urban form to reduce GHG emissions while the sprawling suburb is not (Gonzales, 2005). The sprawling suburb:

- is heavily reliant on private automobiles for transportation, with its concomitant high use of fossil fuel;
- is energy-intensive due to the high energy inputs needed to build the extensive infrastructure to distribute water, electricity, information, parcels and services, and to pipe sewage and collect solid waste across spread-out neighbourhoods;
- reduces carbon sink areas through the extensive clearing of land for development.

But, unlike *mitigation, adaptation* to climate change is more complex because adapting settlements to the negative effects of global warming depends entirely on the geographical location of the settlement, not on its attributes to minimize energy inputs. In adaptation, a compact or a

Photo: By Author

Figure 3.2 New suburban developments in the outskirts of Cartagena, Colombia

spread-out urban form does not have any intrinsic value to adapt to climatic conditions. It is the geographical location and its associated climatologic phenomena (e.g. high temperatures, high humidity, winds or floods) that determine what urban form is the most appropriate to adapt to those phenomena.

Adaptation strategies for climatic conditions

The literature on climate comfort and human settlements sets the standards for urban forms appropriate to deal with climate in a given geographical area (Olgyay, 1963; Konya, 1980; Kay et al, 1982; Givoni, 1998; Salmon, 1999; Koch-Nielsen, 2002; Emmanuel, 2005; Roaf, 2005). This literature tells us that there are too many factors that may affect the resilience of a development pattern to withstand heightened climatic conditions. For example, different densities in a built-up area will affect the local climate differently in different parts of that urban area. Overall density also affects how urbanization modifies the regional climate (wind conditions, air temperature near ground level, radiation balance, natural lighting, and the duration of fog and cloudiness) (Givoni, 1998, p282; Salmon, 1999; Emmanuel, 2005). Density of the built-up areas and the types of buildings affect the amount of solar radiation reaching the

ground level and the nocturnal radiant loss (Givoni, 1998; Koch-Nielsen, 2002). Urban ventilation depends greatly on the arrangement of buildings with different heights in the urban area. The distances between buildings, either across streets or within an urban block, greatly affect outdoor and indoor ventilation conditions. The distances between buildings along the north–south axis will affect the solar exposure of the buildings. Orientation of the streets with respect to wind direction affects the wind speed near the ground (Givoni, 1998; Salmon, 1999; Koch-Nielsen, 2002; Emmanuel, 2005). But even some architectural details of a building, such as its colour, for example, will affect reflectivity and the building's resilience to adapt to high temperatures (Givoni, 1998, p282). The actual shape, orientation and elevation of the dwelling, and its relationship to nearby hills and valleys may also worsen or ameliorate the adverse effects of the climate. The proximity to bodies of water, for example, and whether the development is on the lee or the windward side of water, will also have a direct effect on air temperature as developments on the lee side of water will have warmer temperatures in the winter and cooler in the summer (Konya, 1980; Emmanuel, 2005).

Furthermore, the orientation of streets can affect the climate of an urban area in several ways: it affects the wind conditions in the urban area as a whole; it affects exposure to the sun and creation of shade on the street and footpaths; it affects the solar exposure of buildings along that street; and it affects the potential for ventilation of houses along the street.

In extreme climatic conditions, houses built with different materials and construction techniques, but set in the same settlement, will perform differently under the same conditions. Other factors are the topography of the area, the tree-coverage in the neighbourhood, the amount and distribution of vegetation, the materials used in the paving of streets, and so on.

In sum, the development pattern alone is not the only factor to take into consideration when assessing the level of risk confronted by a neighbourhood under extreme climatic events. After all, 'the urban climate and the indoor climate are both parts of a climatological continuum, differing in scale, which starts with the regional natural climate and is modified at the urban scale, by the structure of the town, and at the site scale by the individual buildings' (Givoni, 1989, pxiii). The combination of factors to determine the resilience of an urban pattern to confront global warming, then, is too complex to assert with any degree of certainty whether a particular development pattern is negative or positive to confront the effects of climate change. In the following section I explain all the different factors that must be taken into consideration before deciding on the appropriate urban form to adapt to heightened climatic effects triggered by global warming.

Heat and urban form

One of the predictions in global warming is that temperatures will rise well above yearly averages to the point where the overuse of air conditioners may shut down the power generation systems in a city, thus leaving neighbourhoods literally to their own devices. It is under these conditions that subtle differences in the settlement pattern (including building heights, density, and organization of structures in space) will have a significant effect on how the area responds to this climatic effect. In the case of the spread-out development pattern, one general problem is its

low heights (as most houses are one or two stories high) because temperatures are generally higher as one is closer to the ground.

> *The portion of solar radiation which reaches the air raises the temperature of the ground [...] and during the daytime the highest temperature is always found at the boundary between the ground and the air. The temperature, in other words, increases considerably as one approaches the ground. [...] A peculiarity of microclimate, therefore, is that the closer one approaches the ground the more extreme it becomes.*
>
> (Konya, 1980, p34)

To this end, the large areas of dark paved surfaces, that generally characterize the streets of suburbia, are detrimental to keeping temperatures down in warm climates. They significantly absorb, radiate and keep the heat longer than unpaved or grassy surfaces (Kay et al, 1982; Givoni, 1998; Emmanuel, 2005).

In humid zones, air circulation is essential to lower high temperatures. Suburban neighbourhoods with long stretches of houses set up in rigid rows are bound to cast wind 'shadows' on the adjacent houses, thus impeding air flow. Checkerboard layout with the buildings staggered, however, will permit a great exposure to the winds when there is need for them. Solar orientation of lots, houses and streets have a great effect on the temperature of the houses (as much as the orientation of the house itself), but given that such orientations varied from block to block in suburban neighbourhoods everywhere, they can hardly be taken into consideration when assessing the performance of suburbia in relation to increased temperatures.

Hot–dry climates and urban form

The increased temperatures in hot–dry climates will further increase the necessity of providing shade to people when moving about the city and to obtain ventilation of buildings during the evenings (Givoni, 1998). The literature on architecture and climate (Olgyay, 1963; Konya, 1980;

Givoni, 1998; Salmon, 1999; Hyde, 2000; Koch-Nielsen, 2002; Emmanuel, 2005) tells us that, in these climates, cities must provide shade at all times to its residents. 'Neighbourhoods should be planned so that distances for walking people and playing children are short. Sidewalks should be shaded as much as possible, either by trees or by the buildings along them' (Givoni, 1998, p366). Also, in such climates, if a large area of the ground is covered by buildings with white roofs and walls, most of the solar radiation will be reflected back to the sky. This includes at least all the radiation impinging on the roofs and at least half the radiation impinging on the walls (Salmon, 1999; Koch-Nielsen, 2002; Emmanuel, 2005). This urban configuration can effectively lower the air temperature near the ground to a point where it is actually lower than the temperature in the surrounding open country. So, from the perspective of density and intensity of use the predominantly low-density single-family home residential neighbourhood is counterproductive in the prospect of higher temperatures in hot–dry regions (Salmon, 1999).

The orientation of streets, as explained in the above section, is also critical to lower temperatures in hot–dry climates. But given that the orientation of streets in a given neighbourhood can vary from block to block, there is no difference between the spread out pattern and one that is more compact. In general, in hot–dry climates the detached single family home typical of the spread out settlement pattern is not the best architectural option under conditions of increasing temperatures.

> *Single-family detached houses have the highest envelope surface area among the various buildings types. When they are built around an internal courtyard the envelope surface area is further enlarged. Consequently, in hot-dry regions, the rate of temperature rise during the daytime hours … is the fastest, for a given thermal conductance and mass of the walls and the roof. From this viewpoint this building type may exhibit the highest indoor discomfort and cooling requirement in summer and the highest heating load in winter, as compared with buildings of other types, properly oriented and ventilated.*
>
> (Givoni, 1998, p363)

Notwithstanding the above considerations, and to undergird the complexity of designing for adaptation to global warming, the detached single-family home in a spread-out suburb is the least sensitive to orientation and the one that can best use surrounding vegetation for climatic control (Salmon 1999; Hyde, 2000; Emmanuel, 2005). The extensive plant cover, shrubbery and trees so common in suburban neighbourhoods are beneficial to alleviate the discomfort generated by soaring temperatures. Plant cover, shrubbery and trees are beneficial to ameliorate climatic effects because they:

1 shade the roof, walls, and windows in a building;
2 shade the play and rest areas outside the houses;
3 elevate the humidity level in too-dry climates (although this only works in confined spaces);
4 reduce the temperature in the vicinity of the houses;
5 reduce the wind speed where it is desired;
6 concentrate airflow and increase airspeed were it is desired (Givoni, 1998, p356).

One of the problems in hot–dry regions, however, is that the scarcity of water drives municipalities to mandate xeroscopic landscapes to conserve this precious resource thus contributing to increased temperatures around the houses.

The provision of shade along the streets of a settlement in a hot–dry climate is critical.

> *Protection from sun … for pedestrians on the sidewalks can be provided by buildings with overhanging roofs, or colonnades, in which the ground floor is set back from the edge of the road, with the upper floors jutting out, supported by pillars (or other means). Such protection can create more pleasant climatic conditions for the urban pedestrian.*
>
> (Givoni, 1998, pxiii)

However, in *hot–dry* climates, the spread-out pattern of development is the absolute worst because there is little shade provided between structures. The compact dense neighbourhood, on the other hand, is more conducive to provide the shading needs suggested by Givoni above. With its narrow streets and mid- to high-rise buildings it provides shade between structures thus reducing the need for excessive use of air conditioning.

Hot–humid climates and urban form

One of the greatest challenges for urban design under global warming is prescribing urban form for hot–humid climates. First, in addition to high temperatures, the high humidity typical of these regions can make living conditions extremely uncomfortable. And, second, some hot–humid areas, especially those on the eastern side of continents, are the ones most prone to strong and destructive storms (hurricanes in the Caribbean islands and in south-eastern United States and typhoons in south-east Asia and north-east Australia) and floods (Salmon, 1999; Hyde, 2000; Koch-Nielsen, 2002; Emmanuel, 2005). The challenge is how the development pattern can respond simultaneously to high temperatures and humidity, strong winds and the risk of floods. To withstand hurricane-force winds, for example, structures must be heavy and sturdy, preferably of reinforced concrete. Yet, this type of structure usually represents high-mass buildings, and to respond appropriately to high temperature and high humidity, buildings should have very low mass.

In these climates, the performance goal for dwellings is to minimize solar heating during daytime and evenings and to provide effective natural ventilation, even during rain. To this end, the spread- out suburb with detached houses is the most suitable pattern for hot–humid climates.

Detached houses are exposed to the outdoor air on all sides. This feature provides good potential for natural ventilation and is advantageous in a hot–humid climate. For a given thermal resistance of the envelope, the expected indoor temperatures and human comfort during the daytime hours, if the building is ventilated, would not be worse than in a more compact building type. Furthermore, during the evening and night hours [...] a detached house will cool down faster than other types of buildings.

(Givoni, 1998, p404)

Street orientation has less significance in suburban neighbourhoods. 'In low-density residential areas, where detached single-family houses with private open spaces around them are common, and the problem [of street orientation] is of minor significance [...] detached houses can be designed with any street layout without compromising ventilation' (Givoni, 1998, p404). In such areas, breezes can circulate between and around the buildings regardless of street layout (Salmon, 1999; Hyde, 2000; Koch-Nielsen, 2002; Emmanuel, 2005). Urban density, however, is one of the major factors determining urban ventilation and temperature. An urban area with buildings arranged in high density and too close to each other will experience not just poor ventilation but a strong 'heat island' effect (Salmon, 1999; Emmanuel, 2005; Gill et al, Chapter 19). Thus, the appropriate arrangement of buildings in the wet tropics would naturally tend towards the scattering of structures. The complexity of designing an appropriate pattern of development for hot–humid climates starts to surface when we take into consideration other aspects of humidity such as dampness. The expected increase in temperatures and rainfall in some hot–humid climates as a result of global warming will likely result in increased dampness in dwellings as a result of transfers of dampness from the ground below the house. Given that all houses in suburban neighbourhoods are sitting directly on the ground, it is likely that the humidity inside will become unbearable.

The ideal urban design for this climatic zone is that of narrow high-rise apartment buildings (towers), built of reinforced concrete, with one apartment per floor, and as far apart from each other as possible. The occupants of the high-rise buildings in particular, will enjoy lower temperature and lower humidity (as vapour is generated by evaporation from vegetation and moist soil at, or near, ground level) (Lyons, 1984). But, then again, this design strategy may work well to respond to high heat and humidity (i.e. to adapt to a threat of global warming), but the unavoidable use of elevators in each tower will increase energy consumption, hence working against efforts to mitigate climate change.

Powerful winds and urban form

The threat of more frequent hurricane-force winds in certain geographical locations is certainly one of the major threats of the changing climate. Although no urban form would offer enough protection to human beings for such events, an appropriate development pattern may make living conditions less unpleasant under strong winds. At about 5 metres per second, wind becomes an annoyance by causing clothes to flap and disturbance to the air. At 10 metres/second, wind becomes disagreeable with dust and trash being picked up. But at 20 metres/second, wind most surely becomes dangerous (Roaf et al, 2005, p253). A compact pattern of development with medium height buildings (five to eight storeys) may shield downwind structures from damage. An area with skyscraper buildings may offer the worst exposure due to the accelerating effect very high buildings have on strong winds. But, a spread-out pattern of development will expose every structure to similar forces, thus adding to the wind's destructive effects.

Furthermore, detached low-rise single-family home neighbourhoods are unpredictable in relation to wind protection. The placement of houses in relation to others may create channelling or funnelling effects, doubling wind speeds, and thus create strong turbulence and eddies (Konya, 1980). Older suburbs, however, with mature large trees spread out evenly in the neighbourhood may 'feel' the force of the wind less because thick 'forests' have the effect of decelerating wind velocity and strength (Konya, 1980). The downside of the 'urban forest' strategy

is that trees themselves can become a hazard in violent winds. When such winds acquire hurricane force, uprooting and smashing of trees against dwellings and other structures is common. In general, however, 'tree belts produce a large reduction in wind velocities extending much farther to the lee than solid wind barriers such as walls and fences' (Kay et al, 1982, p62).

Density of the settlement is another factor in the effect of winds in a city. Although high building density reduces air flow in an urban area thus reducing the strength of powerful winds, orientation of the streets and the buildings will affect the final wind force.

> *The principal factors which determine the urban density effect on the urban wind speed are the average height of buildings and the distance between them. However, the most important factor with respect to a building's height, from the urban ventilation aspect, is the* difference in heights *of neighbouring buildings. While buildings reduce the speed of the 'regional wind' near ground level, individual buildings rising above those around them create strong air currents in the area.*

> (Givoni, 1998, p285)

The denser the urban area and the higher the buildings along the streets, the stronger the effects of buildings on the local wind patterns and wind strength. High-rise buildings create zones of low and high pressures above the built up areas generating vertical currents that stir the urban air mass, accelerating winds near building corners, reversing the flow of air in front of the building, generating turbulent air flow in the wakes behind and at the sides of tall buildings, and accelerating wind speeds through constricted areas such as passages and arcades (Isyumov and Davenport, 1978).

From the above considerations, the difficulty to prescribe urban form to withstand the force of strong winds becomes evident. Yet, it seems likely that an urban form made out of mid-rise buildings (five to eight storeys) is the most appropriate to adapt to increased wind forces in the geographical locations expected to be most affected by this effect of global warming. In the case of flooding, however, we will see in the next section that urban density and building height have little effect on protecting a settlement from destructive impacts.

The threat of flash/river/sea floods and urban form

Floods occur because urban soils cannot absorb or discharge fast enough the excess water that results from heavy rains or overflowed rivers and seas. In the case of flooding, urban form can do little to prevent flooding of a city. Whether it is a spread-out or a compact pattern of development, the advice is not to locate settlements, of any type, within areas prone to seasonal flooding (Davis 1984). 'No urban design details can prevent […] floods except simply avoiding locations prone to them' (Givoni, 1998, p409). To avoid flooding, settlements must simply increase soils' absorption rate, preserve land features of natural drainage such as interconnected valley systems, and collect excess runoff in urban reservoirs. Dunne (1984) recommends that the 'most obvious method of reducing runoff is to maintain as much as possible natural vegetation and permeable soils […] the planting of covers that are effective in maintaining high infiltration capacities […]' (in Givoni, 1998, p410). But, then again, these are features rather irrelevant to the form of the settlement.

The goal of a city layout must be to stop water flowing in from areas beyond the city limits and the rapid disposal of excess rainwater resulting from urbanization. As with the other urban design aspects of suburbia, the extensive paved surfaces of suburban neighbourhoods offer both advantages and disadvantages in relation to flooding. In areas prone to flooding by increased rain activity and sea surges, the flat impervious surfaces of the spread-out suburb may work as drainage channels to distribute runoff gradually and evenly towards retention basins and swales. However, in more dense and compact urban areas streets may flood more easily and runoff drainage may be retarded, possibly resulting in deep flooding.

Conclusion: Towards a framework to research appropriate urban forms to mitigate and adapt cities to climate change

Prescribing urban form to mitigate and adapt to climate change simultaneously is extremely challenging, if not impossible. Although mitigating climate change is best achieved by designing settlement patterns that do not contribute to more GHG releases due to the excessive use of energy (to move people, food and things, and to heat or cool down buildings), adapting to climate change involve a more complex set of issues. Two key words to address the adaptation challenge are context and compromise. Context because deciding which urban form and design features are appropriate to adapt to global warming depends on geographical location. Depending on the location, the urban form should be designed to withstand either the effects of extreme heat, humidity, strong winds, the threat of flooding, or a combination of some of these (e.g. high heat and humidity). And compromise because even the best design cannot respond to all extreme climatic conditions in all seasons. Such design may have to be crafted to withstand only the worst type of weather events in its particular geographical location. Unfortunately, but understandably, in certain locations the need to adapt to extreme climatic events exceeds the need to mitigate. Human lives may well depend on the former.

It is worth also keeping in mind that mitigating climate change depends as much on urban form as on lifestyle (especially consumption habits). Although debatable, it is argued that the best strategy to mitigate GHG emissions is to curb society's consumption habits to decrease the energy needs to produce goods. But assuming that the global society shifts towards a more frugal way of existence (although sadly the evidence points otherwise) and that planning and design professionals decide to address climate change through urban form, the strategy to *mitigate* is rather clear: an urban form that encourages walking, bicycling and mass-transit systems; a compact urban from that saves on

Table 3.1 Author's sample matrix to frame design strategies addressing four potential climate change-related risks in a hot–humid region

Aspects of urban form	Climate type and associated risks: hot–humid zone (or hot–dry, etc.*)			
	High temp	High humidity	Strong winds	Flooding
Land coverage				
Distance between buildings				
Average height of buildings				
Street layout and street orientation				
Vegetation and tree coverage				
Location and size of parks				
Space between buildings				
Topography				
Relation to other nearby topographic features (hills, valleys)				
Soils				
Proximity to large bodies of water				

Notes: * There are five main climatic zones: 1. hot–dry; 2. hot–humid; 3. temperate (mild winters and mild summers); 4. very cold winters with hot–humid summers; 5. very cold winters with mild summers.

infrastructure costs and energy requirements (i.e. less mileage of pipes and cables, less travel distance for fire trucks, police, delivery of parcels, and so on); the provision of combined heat and power (CHP) for urban districts as well as energy systems reliant on solar energy; and rooftop and community gardens to save on the transportation of food.

The strategies to adapt to the effects of global warming, however, are a different story and it underscores the complexity of the issue. As this chapter shows, many factors have to be taken into consideration before deciding on the appropriate urban form strategy to adapt to a specific global warming risk or, in some cases, to a combination of two or more of them. The number and type of risks will depend on the specific geographical location (including terrain inclination, yearly precipitation, whether there are nearby mountains, large bodies of water,

etc.). The factors need to be addressed within the framework of a matrix containing the following information (as illustrated in Table 3.1 above): on the top row, the climate types and their associated risks (for the sake of the example, I am using a hypothetical geographical location that would have a hot–humid climate and the main four risks associated with global warming – it is worth noticing, however, that it is rare to have the four types of risk present in one single location); on the left column, the different aspects of urban form that relate to those risks; and in each of the blank boxes, and corresponding to each of the aspects of urban form, the design strategy to address the specific risk associated with the specific climatic zone and geographical location. Filling out each box would require detailed studies of the relationship between the specific risk and the element of urban form in the left column.

Note

1   Even in works that seriously question the connection between sustainability and compactness, such as the edited piece by Jenks, Burton and Williams (Jenks et al, 2002), the underlying message is that, at least from the stand point of energy savings (which is the IPCC's main concern) the compact city, even in its 'decentralized concentration' variant, is the most appropriate model for sustainable development.

References

Beatley, T. (2000) *Green Urbanism: Learning from European Cities*, Island Press, Washington DC

Beatley, T. and Wheeler. S. (2004) *The Sustainable Urban Development Reader*, Routledge, London

Berke, P. (1995) 'Natural hazards reduction and sustainable development: A global assessment', Working Paper Number S95-02, Centre for Urban and Regional Studies, University of North Carolina at Chapel Hill, Chapel Hill, NC

Bulkeley, H. and Betsill, M. (2005) *Cities and Climate Change: Urban Sustainability and Global Environmental Governance*, Routledge, London

Bury, R. (ed) (1998) *Cooperating with Nature: Confronting Natural Hazards with Land-Use Planning for Sustainable Communities*, R. J. Joseph Henry Press, Washington DC

Davis, I. R. (1984) 'The planning and maintenance of urban settlements to resist extreme climatic forces', in World Meteorological Organization, *Urban Climatology and its Applications with Special Reference to Tropical Areas*, Proceedings of Technical Conference held in Mexico City, November, pp277–312

Dunne, T. (1984) 'Urban hydrology in the tropics: Problems, solutions, data collection and analysis', in World Meteorological Organization, *Urban Climatology and its Applications with Special Reference to Tropical Areas*, Proceedings of Technical Conference held in Mexico City, November

Elkin, T., Duncan, M. and Hilman, M. (1991) *Reviving the City: Towards Sustainable Urban Development*, Friends of the Earth, London

El-Masri, S. and Tapple, G. (2002) 'Natural disasters, mitigation and sustainability: The case of developing countries', *International Planning Studies*, vol 7, no 2, pp157–175

Emmanuel, R. (2005) *An Urban Approach to Climate Sensitive Design: Strategies for the Tropics*, Spon Press, New York

Frey, H. (1999) *Designing the City: Towards a More Sustainable Urban Form*, E & F N Spon, London

Girling, C. and Kellet, R. (2005) *Skinny Streets and Green Neighborhoods: Design for Urban Environment and Community*, Island Press, London

Gonzalez, G. A. (2005) 'Urban sprawl, global warming and the limits of ecological modernisation', *Environmental Politics*, vol 14, no 3, pp344–362

Givoni, B. (1998) *Climate Considerations in Building and Urban Design*, Van Nostrand Reinhold, New York

Hyde, R. (2000) *Climate Responsive Design: A Study of Buildings in Moderate and Hot Humid Climates*, E & F N Spon, New York

IPCC (2007) 'Climate Change 2007: Synthesis Report, Summary for Policymakers', 4th Assessment Report, Intergovernmental Panel on Climate Change, Geneva

Isyumov, N. and Davenport, A. G. (1978) 'Evaluation of the effects of tall buildings on pedestrian level wind environment', *Proceedings of the Annual Convention of American Society of Civil Engineers (ASCE)*, Chicago, IL

Jabareen, R. Y. (2006) 'Sustainable urban forms: Their typologies, models, and concepts', *Journal of Planning Education and Research*, vol 26, no 1, pp38–52

Jenks, M., Burton, E. and Williams, K. (eds) 2002) *The Compact City: A Sustainable Urban Form?*, Spon Press, London

Jenks, M. and Dempsey, N. (eds) (2005) *Future Forms for Sustainable Cities*, Architectural Press, Oxford

Kay, M., Hora, U., Ballinger, J. and Harris, S. (1982) *Energy-Efficient Site Planning Handbook*, The Housing Commission of New South Wales, Sydney, Australia

Koch-Nielsen, H. (2002) *Stay Cool: A Design Guide for the Built Environment in Hot Climates*, James & James, London

Konya, A. (1980) *Design Primer for Hot Climates*, The Architectural Press, London

Lyons, T. J. (1984) 'Climatic factors in the siting of new towns and specialized urban facilities', in World Meteorological Organization, *Urban Climatology and its Applications with Special Reference to Tropical Areas*, Proceedings of Technical Conference held in Mexico City

Olgyay, V. (1963) *Design with Climate*, Princeton University Press, Princeton NJ

Pizarro, R. E. (2005) 'The suburbanization of the mind: The Hollywood urban imaginarium and the rise of American suburbia in the Colombian Caribbean', Doctoral Dissertation (unpublished), The University of Southern California, School of Policy, Planning, and Development, Los Angeles

Roaf, S., Crichton, D. and Nicole, F. (2005) *Adapting Buildings and Cities for Climate Change: A 21st Century Survival Guide*, Architectural Press, Oxford

Rudlin, D. and Falk, N. (1999) *Building the 21st Century Home: The Sustainable Urban Neighbourhood*, Butterworth-Heinemann, Oxford

Salmon, C. (1999) *Architectural Design for Tropical Regions*, John Wiley & Sons, New York

Tyndall Centre for Climate Change Research (2004) *A Briefing on Climate Change and Cities: Briefing Sheet 30*, The British Council, www.britishcouncil.org/science-briefing-sheet-30-climate-and-cities-dec04.doc, accessed 25 January 2007

Van der Ryn, S. and Calthorpe, P. (1986) *Sustainable Communities: A New Design Synthesis for Cities, Suburbs, and Towns*, Sierra Club Books, San Francisco

Williams, K., Burton, E. and Jenks, M. (eds) (2000) *Achieving Sustainable Urban Form*, E & F N Spon, London

4

Patterns of Settlement Compared

Nick Green and John Handley

The growth of accommodation

'If you look at populations in southern England, everyone still lives within four miles of churches which had been planted by the 15th century' (Batty, 2001, p636). Thus wrote Mike Batty a few years ago, pointing out that some things don't change much. Things do change of course, often unhurriedly and imperceptibly, but spread over decades and centuries, these changes become profound. This 'deeper continuity' as Batty called it (he borrowed the phrase from George Holmes's *The Oxford History of Medieval Europe*) is crucial to understanding why settlement patterns are the way they are.

Batty made his point at the beginning of a paper which sought using computer simulations to show how, as a consequence of relentless positive feedback operating over timescales measured in centuries, a polynucleated urban landscape could be expected to arise spontaneously (Batty, 2001). Michael Breheny also highlighted the importance of relatively deep-seated changes in live–work patterns, arguing for a subtle approach to policy that aims not to reverse the trend, but simply to bend it (Breheny, 1997; Breheny and Hall, 1999). Other research has come to the same sort of conclusion: our

patterns of living and working are showing a consistent tendency to become more dispersed over time and the population of Britain continues to spread itself more evenly (Champion, 1989; Owens, 1992; Hall and Ward, 1998; Rogers and Power, 2000; Hall and Pain, 2006; Parkinson et al, 2006).

In the last half-century or so, one very obvious change has been the rise in car ownership: motoring has long since ceased to be the prerogative of the wealthy, and some have argued that socio-economic factors do in fact correlate more closely to travel patterns (and habits) than land-use characteristics (Stead, 2001). So people are travelling more than they used to, and by different means: trips by car, van and taxi increased more than ten-fold from 58 billion passenger kilometres in 1952 to 686 billion passenger kilometres in 2006; over the same period, the percentage of trips by rail almost halved, from 18 per cent to 7 per cent. The decline of bus and coach travel was more dramatic still, from 42 per cent of trips in 1952 to just 6 per cent in 2006 (National Statistics, 2007a). Crucially, this general trend of dispersal is happening within a general context of rapid population growth and a similarly hefty increase in the number of households: the UK's population is expected to reach over 70 million by 2031 (National Statistics, 2007b). Lastly, as the title of this book makes clear, the issue of climate change frames many considerations and, it might be argued, lays the emphasis on one of sustainability's 'three pillars' in particular: the environment.

Cities, though, are social and economic entities, as well as interventions that have an environmental impact. And the question of how to build cities that meet the needs of sustainability's 'three pillars' – society, economy, environment – is hardly a new one. Its modern origins lie in the 19th century slum 'city of dreadful night' (Hall, 1988) but if overly high densities were one of the problems then, the lower densities that come as a result of the modern trend to counter-urbanization are surely of vital importance now. Answers won't come easy, for as Ravetz has pointed out, 'a "sustainable" urban form for any city is a complex balance of many needs and goals, at larger and smaller scales' (Ravetz, 2000, p67).

A prejudgement

Invidious though it may be to prejudge the issue, at least within the narrative structure of this chapter, there is a consensus in the literature about what form a sustainable city would probably take, and it is Ebenezer Howard's original model of the social city (Hall, 1997; Hall and Ward, 1998). This is not to say that this is the only model that can work, and in 1993, Breheny, Gent and Lock opened their report *Alternative Development Patterns: New Settlements* with these words: 'This work was commissioned because of increasing political, public and professional concerns about how best to accommodate new development: its scale, location and consequences' (Breheny et al, 1993).

The literature since the mid-1990s contains little suggestion that much has changed in the decade and a half or so since the publication of this work, which continues to be much-cited in far more recent literature. This is not simply due to a lack of research, although it is certainly the case that in the UK at least, there has been relatively little research into many, although not all of these issues. One exception is the compact city, which has been the topic of considerable debate. But this review reflects the variation in coverage.

The lack of change is also because many of the arguments and issues raised by Breheny, Gent and Lock remain every bit as apposite in 2009 as they were in 1993: the concerns about housing and settlement type and location remain, and if anything have probably become greater; the phrase 'housing crisis' is no longer decried as unnecessary fear-mongering, but can be heard regularly in the mass media; environmental concerns are now in the political mainstream. An initiative by the UK government to develop a series of 'eco-towns' suggests the political will to try to deal with these issues, although overshadowed by global economic crisis. So while settlement patterns may be old ground, intellectually speaking, it behoves us to retread it; for things do change, even if unhurriedly.

Old ground though it may be, a simple route map will doubtless prove helpful. The chapter as a whole compares the three main types of new settlement pattern, and the next section, *The accommodation of growth*, comprises the bulk of it. It takes the form of a literature review which looks at some of the advantages and disadvantages – social, economic and environmental – that come with urban infill, urban extensions and entirely new settlements. The reader will quickly notice that scant mention is made of 'climate change' per se; much of the assessment in the literature tends to deal with 'sustainability', but we can reasonably infer that a more environmentally sustainable approach to doing things (and not just urban planning) will tend to be less damaging in terms of climate change. So a reduction in energy use, or in material use, will reduce carbon emissions, so helping to mitigate climate change. But the other way in which we will have to deal with climate change is by adapting to it, and here, different forms of settlement pattern can make a difference; one example would be in terms of the urban heat island generated by a particular settlement (see Gill et al, Chapter 19). But in the closing section, we shall return to the prejudgement above, in the light of the evidence reviewed, and with a view to the broad implications for climate change of both different patterns of settlement, and, albeit briefly, systems of settlements.

The accommodation of growth

Broadly speaking, there are three ways in which new households can be accommodated in a particular geographical area: new homes can be built on a site within an existing settlement; new homes can be built on a site that is connected to the edge of an existing settlement; and new homes can be built on a site that is not connected to an existing settlement. Where choices must be made, they will not be purely technocratic, or scientific, or economic, or social, or environmental. They will also be political. Let us look at each approach in turn.

Plugging the holes and filling the gaps

One way of accommodating more households is simply to fill in the holes in existing settlements, a process known as urban infill, or 'intensification'. Since it occurs within the boundaries of existing settlements it tends to vary in scale ranging from the large 'urban village' constructed on an old industrial estate, to the development of large back gardens, or of derelict gaps in the urban fabric (Breheny et al, 1993). It has its advantages and disadvantages. The advantages have the distinctive political overtones of broad acceptability: those who wish to prioritize urban regeneration find it acceptable; those seeking to preserve the countryside also find it acceptable; seemingly derelict land is put to obvious (and acceptable) use. The disadvantages by contrast are more practical in nature, but harder to articulate, based as they are on what may happen: urban areas have a limited capacity to absorb more homes; urban green space may come under increasing pressure to be developed; seemingly derelict land which may in fact harbour a diverse ecosystem comes under threat; town cramming, and the consequent decline in the quality of urban life is a threat; possibilities

for mitigating the urban heat island effect may be narrowed; adaptability to climate change may be compromised. The first thing to note, then, is that the arguments are as much political as practical; but there is a seed around which they have crystallized: the notion of the 'compact city'.

Richard Rogers and Anne Power, the chief proponents of the compact city in the UK, claim that 'people gravitate to compact cities because they like its energy, opportunity, diversity and excitement' (Rogers and Power, 2000), but there is plenty of evidence to suggest that this is not always the case. The trend to counter-urbanization identified by Champion (Champion, 1989, 2001) suggests that in Britain, the suburbs remain the favoured form of living environment; an observation, incidentally, made seven decades ago by Rasmussen (Rasmussen, 1982). And while the city centre 'loft living' identified nearly three decades ago by Zukin (Zukin, 1982) does indeed remain a preference for a significant minority, several authors have observed that many of the merits of urban intensification and the compact city have been based on assertion and theory rather than empirical evidence (Breheny, 1992a; Breheny, 1992b; Jenks et al, 1996; Williams, 2000; Williams et al, 2000; Vallance et al, 2005). For example, a study comparing a number of compact city scenarios with a 'trend' scenario in the UK found that shifting to a compact city strategy alone will not necessarily change car use (Simmonds and Coombe, 2000), a finding also noted by Banister (2005). However a compact city strategy need not worsen travel problems such as congestion unless densities are particularly high (Simmonds and Coombe, 2000). David Lock has pointed out that to advocate high-density development specifically to render public transport financially 'viable' is to place above all other considerations the profitability of private transport operators (Lock, 2006).

This is not to say that the compact city is a hopelessly weak idea. Its merits are such that Geurs and van Wee conclude that without it, urban sprawl and the concomitant car use in the Netherlands would be far greater than is currently the case (Geurs and van Wee, 2006). Over and above the fact that urban infill schemes can help limit urban sprawl, they are likely to bring good access to social facilities in general, and shops in particular (Breheny et al, 1993; Williams, 2000). Of course, an urban infill scheme may well come to an existing residential area, and the question of whether or not existing residents actually regard urban infill as a 'good thing' is not that simple. If the original residents perceive their quality of life to have benefited directly, they are likely to adjudge the urban infill itself as beneficial. Likewise, those who feel that their quality of life has suffered as a consequence of urban infill will project that negative perception onto the principle of urban infill in general (Williams, 2000).

Existing social networks might contribute to a good sense of community (Breheny et al, 1993), but a more circumspect reading of circumstance may be necessary in certain cases (Vallance et al, 2005). In Christchurch, New Zealand, a city whose suburban citizens guard their privacy closely, urban infill was often not well received by nearby local residents who resented the loss of privacy due to new residential buildings which sometimes gave a clear view into their houses. These same residents also felt that their own community was in danger of being damaged by the incomers (Vallance et al, 2005).

Breheny and colleagues were generally optimistic with regard to the potential social mix, which they felt was likely to be good. They also argued that urban infill creates relatively little disruption and this may make it more acceptable to local residents (Breheny et al, 1993). But in the end it all comes down to context, and the evidence suggests a need for sensitivity. In three suburban and predominantly residential areas in London, the social changes wrought by urban infill schemes actually had a negative effect, being perceived as damaging to both the sense of community and to local identity (Williams, 2000). Increasing the breadth of social mix might not be perceived as a 'good thing' in the abstract, but rather as a real threat to the existing community. Intriguingly, suburbanites feel the most threatened by such changes, so considerable sensitivity to the local context is required if such urban infill schemes are not to

cause resentment and distrust among existing residents (Williams, 2000; Vallance et al, 2005).

When it comes to reducing the use of the private motor car, urban intensification is just one part of the solution; cultural issues have historically played a strong role too, something that does not look like changing in the foreseeable future (Breheny, 1995; Williams, 2000; Banister, 2005). For while urban infill has high development costs relative to other forms of development, infrastructure costs are relatively low since much of the infrastructure already exists. Consequently, the maintenance costs of urban infill schemes are also likely to be low, since the infill 'plugs in' to existing systems (Breheny, 1992b). Access to employment for urban infill schemes is also likely to be good.

The occasionally confused nature of where the advantages lie came through in a study by Williams of three London boroughs (Camden, an inner-city borough; and Harrow and Bromley, both suburban), all three of which had undoubtedly seen improvements in the local economy during the study period and all three of which were happy enough to attribute these improvements to their urban intensification policies. The problem was that evidence to tie the improvements directly to the policy was actually rather sparse, leading to the disheartening conclusion that 'Determining the extent to which these benefits are a direct result of urban intensification, and how much they are the result of broader economic trends is almost impossible' (Williams, 2000, p44). The potential to regenerate depressed areas must therefore be seen as heavily dependent on context.

Even the environmental impacts come with a health warning. By definition urban infill does not result in loss of land, with the consequent expectation that the loss of natural habitats and the impacts on biodiversity might be expected to be low. But urban areas turn out to be surprisingly rich in wildlife habitats, especially on previously developed land where disturbance followed by neglect initiates natural succession: so urban infill can actually be problematic when it comes to safeguarding urban biodiversity. Indeed, loss of natural habitat as a consequence of urban infill is a distinct possibility: previously developed sites may have developed their own, possibly fragile ecosystems in the time since they were abandoned to nature (Breheny et al, 1993). Such sites will also play a role in ameliorating the effects of climate change.

When done properly, however, urban infill is for the most part benign. Its Achilles' heel is its potential to mutate into its malign variant, 'town cramming', a clearly undesirable development at a time when 'green-blue infrastructure' is seen as an increasingly important means of mitigating the effects of climate change in the city. For while the higher densities that come with urban infill may make community energy schemes more viable, which would have positive effects with regard to climate change, they also risk exacerbating the urban heat island effect (Shaw et al, 2007).

Furthermore, while urban infill has the advantage that it does not encroach on undeveloped land, the increase in suburb to suburb commuting, noted in Breheny et al (1993), and explored in greater depth by Breheny (1997) and Breheny and Hall (1999), does raise the possibility that the job–housing location balance may be 'wrong for many households'. As noted above, there are also tensions between urban infill and the need to enable cities to adapt to climate change through the provision of urban green space. What this all adds up to is that the potential impacts of urban infill need to be sensitively handled, and the social, economic and environmental contexts of any proposals carefully understood if urban infill schemes are not to do more harm than good.

Not edge city?

Rather than filling in the gaps in an existing settlement, one can of course extend it. Urban extensions, as their name suggests, comprise development that takes place at the edge of an existing settlement, usually on a green field site or other open land. Stimulated by improvements in transit systems over the previous century or so, this has been the favoured form of urban growth (Breheny et al, 1993). It has most commonly found its physical expression in the form of the low-density suburban development, long since identified by Rasmussen as a popular residential environment in the UK, and still in demand to this day (Rasmussen, 1982; Breheny et al, 1993).

Clearly, the extent to which urban extensions can offer access to social facilities will depend on the location and size of the development. There remains the possibility that new residents will have access to existing facilities in the more mature suburbs, but so too is the development of a sense of community dependent on the size and location of the extension. Notions of local identity, in particular, will be tied to overall scale of the development, and a larger development may be better placed than a smaller extension to take on its own identity; the smaller extension is more likely to take its identity from the older adjoining development (Breheny et al, 1993).

Villages, for example, may benefit from the fact that there is a pre-existing community, but as we saw above with urban infill, there is also the risk that this community may resent the intrusion of the incomers: this may be especially the case when the village in question is relatively small. Size will also have a bearing on access to social facilities. Again like urban infill schemes, urban extensions are among the least costly in terms of the provision and use of infrastructure, since they are well placed to take advantage of the existing infrastructure (Breheny et al, 1993).

Urban extensions can 'plug in' to existing amenities to an extent – schools for example – which means that their requirements for land will be less than those of a new settlement designed for a population of similar size and demographic profile. However, very large urban extensions can be expected to require new amenities, and can therefore be expected to use as much land as a new settlement, but the new (off-site) infrastructure required by an urban extension will be substantially less than that required for a new settlement (Breheny et al, 1993).

The potential impact on biodiversity is a subset of a wider range of impacts centred on the formerly rural landscape, although sometimes much modified by proximity to the town (Shoard, 2002). Clearly, decisions about peripheral expansion of settlements need to be made in a landscape context and then landscape character assessment is well placed to make an effective contribution (Swanwick and Land Use Consultants, 2002). Interestingly, a study by Ravetz and McEvoy, *Sustainable Development of the Countryside around Towns*, found that local authorities had more confidence in applying measures simply to control development, such as green belts, than they did in granting urban extensions (Ravetz and McEvoy, 2002).

Urban extensions also sidestep the unwanted possibilities of either town cramming or loss of urban green space, and may even provide positive opportunities to develop an effective, multifunctional green network, a topic explored in Chapter 19 in this book. For example, the lower density of urban extensions leaves room for green infrastructure which may assist in flood control (Shaw et al, 2007).

Urban extensions can also be expected to provide relatively good access to employment opportunities, offering the choice of ready access to both urban centre and urban hinterland. However, given that commuting from edge to edge of cities remains a growing trend, and that commuting patterns are becoming more dispersed (Breheny, 1997; Breheny and Hall, 1999; Green, 2008), the lack of employment provision in such developments does little to discourage long journeys to work (Breheny et al, 1993), and could be expected to continue the trend of 'edge-to-edge' commuting. Both urban extensions or new settlements that function as 'dormitory' suburbs or towns are obviously not energy efficient, since they encourage rather than discourage travel (Breheny et al, 1993).

Starting from scratch

The most extreme means of accommodating new homes is simply to start from scratch and build a new settlement which will provide a new geographical focus for development. An unambiguous definition of a new settlement is actually rather tricky, but Michael Breheny and his colleagues offered these approximate guidelines as being appropriate at the time they were writing (Breheny et al, 1993, p9):

- A new settlement may or may not incorporate a small settlement that already exists.
- Developers did not typically see a development of less than 350 dwellings as being a 'new settlement'.
- A 'new wave' new settlement could be expected to have between 350 and 5500 dwellings, although there is no reason in principle why it should not be larger.
- The criterion of 'free-standing' must be loosely applied, although some degree of functional separation from other settlements is a requirement.

Having laid down these basic rules, they defined a new settlement as:

> *A free standing settlement, promoted by private or public sector interests, where the completed new development − of whatever size − constitutes 50% or more of the total size of settlement, measured in terms of population or dwellings*
>
> (Breheny et al, 1993, p9).

> *New settlements share much in common with urban extensions in terms of environmental criteria, particularly in terms of: loss of land (inevitable); loss of natural habitats (likely); energy consumption due to transport (inefficient if a dormitory town); contribution to 'greening' the existing urban environment (no effect, by definition, although does no harm); and town cramming (again no effect, but does no harm)*
>
> (Breheny et al, 1993).

> *Intriguingly though, the definition set out above leaves open the possibility that in principle a new settlement could have the potential to regenerate depressed areas, although in practice this area would be the pre-existing settlement around which the new settlement is developed and there is little evidence one way or the other to support such an assertion. However, they might offer significant access to employment, as observed in a minority of new settlements in the mid-1990s*
>
> (Breheny et al, 1993).

The most difficult thing to achieve is perhaps the elusive 'sense of community', and although it is certainly the case that access to social facilities can be 'designed in' to a new settlement, the inconvenient fact remains that a 'sense of community' is built up over the longer term, since its successful nurture depends on trust bred through familiarity. Some observers have suggested that a period measured out in years (not months, not weeks) is required for such a sense of community to develop; the point is that a sense of community cannot be instilled overnight (Breheny et al, 1993).

Economically speaking, new settlements, along with urban infill, were likely to have the lowest cost of the end product. However, in terms of provision and use of infrastructure, new settlements were likely to be the most expensive (Breheny et al, 1993). Banister notes that 'evidence from Great Britain shows that large metropolitan settlements tend to be associated with low distance travel and energy consumption' (Banister, 2005, p105). He suggests that this may be because higher population densities widen the range of opportunities for personal contacts and activities that do not require motorized transport. He cautions, however, that diseconomies of scale may occur with very large settlement sizes, when travel distances between home and the urban centre increase. In short, the relationship between settlement size and travel patterns is complex.

If they are big enough, and built in the right place, new settlements can provide all that is required: Breheny and colleagues suggest a minimum of 3000–5000 dwellings, with around 10,000 dwellings being preferable (Breheny et al, 1993). They have considerable potential for reduced energy consumption in terms of space heating and lighting, not least because, being designed from scratch, such things can be built in rather than awkwardly retrofitted. Equally, the urban form itself can be designed in such a way as to mitigate the more onerous effects of climate change; that is, they can be designed with the future in mind (Shaw et al, 2007). But while they may be able to offer good energy efficiency at larger scales, they do require the use of rural land, raising the spectre of the loss of productive agricultural land at a time when food security is becoming an increasingly pressing issue. However, the historical precedents to demonstrate the efficacy of this approach can be found easily enough: Markelius's scheme for the post-war expansion of Stockholm is a classic example (Hall, 1988; Cervero, 1995), while Hall and Ward (1998) offer a blueprint for how the balanced regional growth originally advocated by Howard (1898) may be updated for the present day.

A prejudgement revisited

At the beginning of this chapter, we suggested that Ebenezer Howard's model of the social city was reckoned, generally speaking, to be a sustainable urban form. This need not always be the case, of course. We have seen that in certain instances simply filling in the holes in the urban fabric can do much to reinvigorate a tired metropolis. In rural areas, expanding villages may make the most sense, so that the increase in population can support a wider range of services for both the original and new inhabitants. Other times and places may leave no sensible alternative but to start from scratch on a greenfield site, and to bear stoically the slings and arrows of outraged nimbyism.

A study of energy use in transport in English towns with populations of approximately 100,000, found that as density increases, so energy use tends to decrease (Rickaby et al, 1992), reinforcing the findings of other research on this topic, although it has been pointed out that in theoretical models, the most efficient urban forms tended to include urban concentration plus nearby villages, in a polycentric regional structure (Rickaby, 1987; Breheny, 1992b).

There is an important point here, although it is beyond the scope of this chapter to explore it in the depth it deserves. It is not just the pattern of new settlement in terms of whether it is urban infill, or urban extension, or a completely new and free-standing settlement (even an eco-town) that matters. It is also to do with the resulting system of settlements cast net-like across a region and beyond. How individual settlements interact with one another with regard, for example, to where people live and work and shop and play, matters to how sustainable a system of settlements is. The edge-to-edge commuting identified by Breheny and Hall (1999) is as much a product of the prevailing economic system, and of the technology that enables it (in this case the automobile) as it is of spatial planning.

New settlements, for all their political and even environmental disadvantages, do have the support of a number of different studies which show that some form of 'decentralized concentration' is 'relatively efficient', since different models suggest that constraints on mobility will tend to encourage people to use those jobs and services that are nearest to them (Owens, 1992). And as Shaw and colleagues have pointed out, new settlements lend themselves to being designed in such a way that they can, as much as anything ever can, be future proofed (Shaw et al, 2007); the necessary green/blue infrastructure and the orientation of streets and buildings that is difficult or impossible to retrofit to an existing settlement can be an integral part of the whole in a new settlement, and to some extent in an urban extension.

The form of a new settlement itself is one part of the solution, then. But patterns of behaviour are every bit as important as patterns of building when it comes to dealing with climate change; some would argue that they are more so. As we have seen, any form of new settlement, be it urban infill, urban extension or a new, free-standing settlement will need compromises, and as has ever been the case, the context is all important. Accommodating new development in a village may be the best solution with regard to climate change, but the original residents may well be most concerned for the survival of the village's social structure when faced with such rapid change. Marshall has argued that cities have tended to grow organically, to evolve (Marshall, 2008) and one might add that they evolve in a particular environment that is physical, social and economic, developing in the process an underlying form and infrastructure that reflects that context. The current context is one of climate change in a post-industrial society, and if the right infrastructure is put in place we can be more optimistic that new settlements might grow and evolve in an environmentally sustainable way. The trick is to achieve this in tandem with social and economic benefits.

Banister (2005, p246) is unequivocal about how to proceed. 'The most sustainable urban form is the city', he says, and

> ... it should have over 25,000 population (preferably over 50,000), with medium densities (over 40 persons per hectare) with mixed use developments in public transport accessible corridors and near to highly accessible public transport interchanges... Settlements of this scale would be linked together to form agglomerations of polycentric cities, with clear hierarchies that would allow close proximity of everyday facilities and accessibility to higher order activities.

A prescription of urban concentration plus smaller surrounding settlements in a polycentric system all interlinked by high speed public transport may seem thoroughly modern; it isn't of course, as Orrskog and Snickars (1992) pointed out nearly two decades ago. It is actually very close to that set out by Howard over a century ago in his concept of the 'social city' (Howard, 1898). Ebenezer Howard himself would doubtless be gratified to find that he basically got it right. But one wonders if he would be as pleased that people are still arguing about it.

References

Banister, D. (2005) *Unsustainable Transport: City Transport in the New Century*, Routledge, Abingdon, Oxfordshire

Batty, M. (2001) 'Polynucleated urban landscapes', *Urban Studies*, vol 38, no 4, pp635–655

Breheny, M. (1992a) 'The contradictions of the compact city: A review', in M. Breheny (ed) *Sustainable Development and Urban Form*, Pion, London

Breheny, M. (ed) (1992b) *Sustainable Development and Urban Form*, Pion, London

Breheny, M., Gent, T. and Lock, D. (1993) *Alternative Development Patterns: New Settlements*, HMSO, London

Breheny, M. (1995) 'Compact cities and transport energy consumption', *Transactions of the Institute of British Geographers*, vol 20, no 1, pp81–101

Breheny, M. (ed) (1997) *The People: Where Will They Work?*, Town & Country Planning Association, London

Breheny, M. and Hall, P. (1999) *The People: Where Will They Live?*, Town & Country Planning Association, London

Cervero, R. (1995) 'Stockholm's rail-served satellites', *Cities*, vol 12, no 1, pp41–51

Champion, A. G. (1989) 'Counterurbanization in Britain', *Geographical Journal*, vol 155, pp52–59

Champion, A. G. (2001) 'A changing demographic regime and evolving polycentric urban regions: Consequences for the size, composition and distribution of city populations', *Urban Studies*, vol 38, no 4, pp657–677

Geurs, K. and van Wee, B. (2006) 'Ex post evaluation of 30 years of compact urban development in the Netherlands', *Urban Studies* vol 43, no 1, pp139–160

Green, N. (2008) 'City-states and the spatial in-between', *Town & Country Planning*, vol 77, May 2008, pp224–231

Hall, P. (1988) *Cities of Tomorrow*, Blackwell, Oxford

Hall, P. (1997) 'The future of the metropolis and its form', *Regional Studies*, vol 31, no 3, pp211–220

Hall, P. and Ward, C. (1998) *Sociable Cities; the Legacy of Ebenezer Howard*, John Wiley, Chichester

Hall, P. and Pain, K. (eds) (2006) *The Polycentric Metropolis: Learning from the Mega-city Regions in Europe*, Earthscan, London

Howard, E. (1898) *To-morrow: A Path to Real Reform*, Swan Sonnenschein, London

Jenks, M., Burton, E. and Williams, K. (eds) (1996) *The Compact City: A Sustainable Urban Form?*, E & F N Spon, London

Lock, D. (2006) 'Transport serves the town, not vice versa', *Town & Country Planning*, vol 75, July/August 2006, p203

Marshall, S. (2008) *Cities, Design and Evolution*, Routledge, London

National Statistics (2007) *Transport Statistics Great Britain*, Department for Transport, London

National Statistics (2007) 'National projections: UK population', www.statistics.gov.uk/cci/nugget.asp?id=1352, accessed 7 December 2007

Orrskog, L. and Snickars, F. (1992) 'On the sustainability of urban and regional structures', in M. Breheny (ed) *Sustainable Development and Urban Form*, Pion, London

Owens, S. E. (1992) 'Land-use planning for energy efficiency', *Applied Energy*, vol 43, pp81–114

Parkinson, M., Champion, A. G., Simmie, J., Turok, I.,

Crookston, M. and Park, A. (2006) *The State of the English Cities* (2 vols), ODPM Publications, London

Rasmussen, S. E. (1982) *London, the Unique City*, revised edition, The MIT Press, Cambridge, MA

Ravetz, J. (2000) *City Region 2020: Integrated Planning for a Sustainable Environment*, Earthscan, London

Ravetz, J. and McEvoy, D. (2002) *Sustainable Development of the Countryside around Towns*, Centre for Urban & Regional Ecology, University of Manchester

Rickaby, P. A. (1987) 'Six settlement patterns compared', *Environment and Planning B: Planning and Design*, vol 14, no 2, pp193–223

Rickaby, P. A., Steadman, J. P. and Barrett, M. (1992) 'Patterns of land use in English towns: Implications for energy use and carbon dioxide emissions', in M. Breheny (ed) *Sustainable Development and Urban Form*, Pion, London

Rogers, R. and Power, A. (2000) *Cities for a Small Country*, Faber and Faber, London

Shaw, R., Colley, M. and Connell, R. (2007) *Climate Change – Adaptation by Design: A Guide for Sustainable Communities*, Town & Country Planning Association, London

Shoard, M. (2002) 'Edgelands', in J. Jenkins (ed) *Remaking the Landscape: The Changing Face of Britain*, Profile Books Ltd, London

Simmonds, D. and Coombe, D. (2000) 'The transport implications of alternative urban forms', in K. Williams, E. Burton and M. Jenks (eds) *Achieving Sustainable Urban Form?*, E & F N Spon, London

Stead, D. (2001) 'Relationship between land use, socioeconomic factors, and travel patterns in Britain', *Environment and Planning B: Planning and Design*, vol 28, no 4, pp499–528

Swanwick, C. and Land Use Consultants (2002) *Landscape Character Assessment: Guidance for England and Scotland*, The Countryside Agency and Scottish National Heritage

Vallance, S., Perkins, H. C. and Moore, K. (2005) 'The results of making a city more compact', *Environment and Planning B: Planning and Design*, vol 32, no 5, pp715–733

Williams, K. (2000) 'Does intensifying cities make them more sustainable?', in K. Williams, E. Burton and M. Jenks (eds) *Achieving Sustainable Urban Form*, E & F N Spon, London

Williams, K., Burton, E. and Jenks, M. (eds) (2000) *Achieving Sustainable Urban Form?*, E & F N Spon, London

Zukin, S. (1982) *Loft Living: Culture and Capital in Urban Change*, John Hopkins University Press, New York

5

Transport Policies and Climate Change

David Banister and Jillian Anable

Introduction

There is a lively debate on the need to address the challenges of climate change, and of the role that transport should play, as it accounted for about 25 per cent of UK energy consumption, 27 per cent of greenhouse gases (GHG) and 29 per cent of CO_2 emissions in 2005 (Department for the Environment, Food and Rural Affairs (Defra), 2006, p61). The UK government has been positive about being able to meet its share of the EU Kyoto Protocol targets for CO_2 reductions (12.5 per cent from 1990 levels by 2008–2012), and it has gone further in a Climate Change Bill to propose legally binding reductions of 26–32 per cent by 2020.

Provisional emissions figures for 2007 show a small reduction, even in transport, with overall CO_2 emissions for the UK falling by 8.2 per cent since 1990 to 148.3MtC (total GHG emissions 639.4MtCO$_2$e). The transport figures have at last stabilized at about 41.8MtC (Department for Business Enterprise and Regulatory Reform (BERR), 2008). Even with the policies in the climate change programme, the transport figures are only expected to stabilize at current levels by 2020, and not decrease (BERR, 2008). It should be noted, however, that

these figures do not include international emissions from aviation and shipping which add another 11.5MtC (Department for Transport (DfT), 2007). The UK does seem to be on target for more than meeting its 12 per cent reduction target as there has been a reduction of 17 per cent in all GHGs from 1990 to 2007 (Table 5.1). But most of this reduction has taken place in other GHGs (not CO_2), and transport has remained stubbornly resistant to making any contribution.

Two main market based measures have been discussed within the transport sector to change behaviour. Firstly, the fuel duty escalator, introduced in 1993, was an annual increase in duty above the rate of inflation, initially set at 3 per cent and raised to 5 per cent (later in 1993) and to 7 per cent (July, 1997). The price of a litre of fuel was increased from 56 pence to 85 pence (1994–2000), of which about 64 pence was tax and duty. The escalator was removed in 2000, after pressure from industry and other interests, particularly those in rural areas, just when it seemed to be having an effect on distance travelled and new vehicle purchasing patterns. Real increases in fuel duty were increased in October 2007 (+2 pence), with further planned increases in 2008 (+2 pence, now postponed)

Table 5.1 Carbon dioxide emissions by end user in the UK

End user category	1990	1995	2000	2005	2006	2007
Transport – MtC	38.6	37.5	39.0	41.6	42.1	41.6
All emissions – MtC	161.5	149.8	149.5	150.5	151.1	148.1
Transport's share of carbon emissions	23.9%	25.0%	26.1%	27.6%	27.9%	28.1%
Levels of road traffic – Billions Veh km	410.8	429.7	467.1	499.4	506.4	509.3
Percentage change on 1990	0	+4.6%	+13.7%	+21.6%	+23.3%	+24.0%

Note: The figures for 2006 are actual and 2007 is provisional. End use emissions include an estimated share of upstream emissions from power stations and refineries allocated to the sectors responsible for using this fuel.

Source: DfT (2007) and BERR (2008).

and in 2009 (+1.84 pence) – these are the first increases in fuel duty above inflation since April 2004. This, together with the increases in the costs of oil, have all raised pump prices in the UK by about 20 per cent for petrol and 30 per cent for diesel over the last year to historically high levels.

Secondly, there have also been several documents produced by government and think tanks about the necessity for a national system for road pricing in the UK (Commission for Integrated Transport (CfIT), 2002; DfT, 2004, 2006b), but the only schemes that have been implemented have been in London and Durham. Even here, the motivation has not been to reduce CO_2 emissions, but to reduce traffic congestion.[1] There has however been a substantial improvement in local air quality in central London resulting from the congestion charge, and CO_2 emissions levels are down by 15 per cent, mainly due to fewer cars, higher speeds and less stop–start driving (Banister, 2008a). The 2007 Energy White Paper (Department of Trade and Industry (DTI), 2007) only makes one mention of pricing, and this in the context of the demand management and the Transport Innovation Fund (DfT, 2006b).

The trends briefly outlined above illustrate the substantial scale and continuing growth of carbon emissions from the UK transport sector, with little being done to reduce the rate of growth, let alone make a contribution to national and international targets. The question then arises as to why it has been so difficult to take effective action. This chapter examines the

options available in the four transport markets – city travel, long distance travel, freight and aviation – summarizing the progress that has been made on the technologies (vehicles and fuels), pricing and regulation, smarter choices,[2] land use and planning, and aviation (pricing and the EU Emissions-Trading Scheme (EU ETS)). It is argued that substantial behavioural change is required for both firms and individuals, including commitment to change and the need for effective implementation. It is here that the spatial planning system can have an important role through: location and density policies on housing, the development of accessible public transport hubs, providing good quality local services and facilities, engaging all major employers in reducing the use of single occupancy cars, providing high quality and safe facilities for walking and cycling, and the design of local neighbourhoods. It is only through a participatory process with all major stakeholders that encourages involvement, empowerment and ownership that effective action can be introduced. This in turn requires leadership and commitment to change that is consistent over time.

The four transport markets

To illustrate the different transport markets, we have divided all travel into four mutually exclusive forms of travel, one covering short distance city passenger travel (<80km), two for long distance passenger and freight travel (>80km)

Table 5.2 Travel in the UK 2005

Type of travel	Proportion of all travel distance	Distance per person per year	Modal split
City	37%	7380km	Car 65%, Walk 25% Cycle 1^1/$_2$%, Public transport 8^1/$_2$%
Long distance	20%	4035km	Car 83%, Rail 9%, Bus and coach 5%, Air 1%
Freight	21%	4232 ton km	Lorry 66%, Coastal shipping 24%
Air	22%	4383km	

Notes: Long distance is over 80 km and all distances here are given as averages per person in the UK. The freight figures are tonne km per person per year – light vans (under 3.5 tonnes) account for 14 per cent of all road traffic and heavy goods vehicles account for 6 per cent of all road traffic (2007).

Source: Based on own calculations and data from DfT (2006a, 2006b).

and one for air travel. Much of the research on transport and climate change has concentrated on the city (e.g. Banister, 2005), but this market is relatively stable and there are good opportunities here for change. It is in the other three markets that substantial increases in travel are taking place, and it is also where there are far fewer opportunities available for reductions in carbon based energy use. Figures for the UK (2005) illustrate the scale of this problem (Table 5.2). City travel accounts for 98 per cent of land based trips, but 65 per cent of distance, with long distance travel making up 35 per cent of distance, and only 2 per cent of trips. It is in the long distance, the freight and the air markets that growth in travel is taking place. But it is these non urban travel markets that the low-carbon alternatives are much harder to envisage, let alone implement.

In some cases long distance car journeys can be replaced by rail or bus, but overseas journeys by air present a real problem. Road freight traffic is one of the fastest contributors to carbon emissions from transport in the UK. The use of vans has resulted in steep growth in emissions by nearly half since 1990, and lorries by a third (Defra, 2007a).

Since 1990, domestic air traffic has grown nearly 100 per cent and international trips by 125 per cent, with the largest growth coming on scheduled flights offered by 'no frills' carriers and people flying more for leisure (Cairns et al, 2006). Although the airlines have improved their efficiency in terms of carbon emissions per seat kilometre,[3] the growth in air travel has more

than outweighed these gains, as passenger numbers are doubling every 10–15 years. One long distance return air journey (say to New York from London or Frankfurt) produces as much carbon as the annual use of the car by the average driver (Box 5.1).

As well as in aircraft, there has already been a significant improvement in engine technology. Average new petrol car efficiency, for example, has improved by around a quarter since the late 1970s (CfIT, 2007). Nevertheless, there has not been a corresponding reduction in emissions, as increased travel distance and the purchase of higher performance cars have offset the gains. There is a need to travel less, but the 'rebound' effect[5] can also erode the benefits of freeing up road or air capacity as it simply fills up with latent travel demand. Thus, for every policy, be it

---

Box 5.1 Travel, energy and carbon use

The average person in the UK (2005) travels 11,000km per year and produces about 1168kg CO_2 – about 90 per cent of this is from car travel. This is equivalent to one long haul flight (10,593km) to New York from London, or a series of short haul flights (6500km in total). Note that in addition to personal travel, there is an energy cost associated with freight distribution (Table 5.2). The total average travel energy budget in the UK is personal plus air and freight giving a total of 2300kg CO_2 per person (2005),[4] about 27 per cent of all CO_2 emitted per person.

(Based on authors' calculations and Defra, 2007a)

technological or demand management, it would seem that complementary policies to 'lock-in' any benefits are necessary (CfIT, 2007). Perhaps it is here that the debate needs to take place on the role that transport should play in any carbon reduction strategy.

Spatial planning can be most effective through reducing the levels of car-based city transport, and enabling the most efficient public transport modes to be used for longer distance travel. It can also 'lock-in' the benefits through energy efficient city urban form and organizational structures, as there are many aspects of travel behaviour that spatial planning can influence. Firstly, the number of trips made to undertake a given set of activities is important. By organizing land uses to enable travel to take place in 'tours', single link journeys can be reduced by combining several destinations and activities in one trip chain. The increased use of telecommunications, including the internet, can reduce the number of trips made through e-commerce activities such as online shopping. This can also extend to reductions in air travel through the use of teleconferencing and video-conferencing. Secondly, travel distance can be reduced through selecting nearer destinations rather than those that are further away. Thirdly, the most efficient forms of transport can be used, for example, walk, cycle and public transport, so that energy use and carbon emissions are reduced, and finally each form of transport should be fully loaded so that efficiency per vehicle is maximized and energy use per person (or tonne of freight) is minimized. In each of these situations, the same role is being assigned to the planning system, namely that cities, towns and even small villages need to be designed so that the opportunity is given to use local services and facilities. Distance reduction is essential for the transport system to become more sustainable, and it is here that the interface between planning and transport is strongest (Banister, 2008b). Such a strategy would also target social objectives to improve accessibility to needed services and facilities for those people without access to a car. In terms of sustainable development, it directly addresses the environmental and social imperatives.

Spatial planning and transport

A substantial amount of research has tried to establish links between travel, land use and urban form. This ranges from simple analyses of trip generation and attraction characteristics of particular land uses (e.g. residential and shopping) to more detailed analyses of travel (and energy use) in locations with distinctly different characteristics. The verdict on this empirical work is mixed. For example, Anderson et al (1996) concluded that the current level of understanding of the influence of urban form on the generation of emissions and the use of energy is weak. But others (e.g. Stead, 2001; Hickman, 2007) have found far more significant relationships between land use and transport (Box 5.2). But even here, the socio-economic variables explain substantially more of the variation in trip making activities than the land use factors.

Note that the levels of explanation of the relationships between socio-economic variables and land use variables and travel or energy use fall as the level of disaggregation increases, so

---

Box 5.2 Explanation of travel from land use factors

Stead (2001): The most extensive UK study used regression analysis on National Travel Survey data. Here, it was concluded that socio-economic factors are more important than land-use factors, explaining 23–55 per cent of the variation in the amount of travel by wards (there are some 8400 wards in England) at the aggregate level. The most important socio-economic factors are car ownership, socio-economic group and employment. Land-use characteristics explain up to 27 per cent of the variation in trip making – this includes density, settlement size and public transport accessibility.

Hickman (2007) and Hickman and Banister (2007): Household data were collected from new housing developments in Surrey (1998). Land use and socio-economic variables together explain 60 per cent of the variation in the travel patterns of households, and individually the levels were 9 per cent for land use and 28 per cent for socio-economic variables.

higher levels of explanation are found at the regional and city wide levels – see Newman and Kenworthy's analysis below.

Three main elements encapsulate the planning and transport interface:

1 Density of development
2 Proximity and quality
3 Local neighbourhood and design

1 Density of development

Density and development has an important effect on the distances travelled, the modes used and the energy profiles. The most cited research here has been carried out over the last 15 years by Newman and Kenworthy (1989a, 1989b, 1999) in their comparison of the transport energy profiles of 84 cities. The powerful conclusion reached was that when urban density in the 58 wealthier cities was correlated with car passenger kilometres, urban density explained 84 per cent of the variance (Kenworthy and Laube, 1999; Kenworthy, 2007). When energy use was correlated with activity intensity (persons and jobs per hectare), 77 per cent of the variance was explained. Despite concerns over the methods used and the quality of the data, clear relationships have been established at the city level. A general conclusion is that an increase of 10 per cent in local density results in a 0.5 per cent decrease in vehicle trips and vehicle miles travelled (Ewing and Cervero, 2002; Table 5.5).

In Hong Kong, the role of land use in mode choice is clear due to the density of the built environment. Empirical modelling confirmed that the role of land use in influencing travel was independent from travel time and monetary costs. Elasticity estimates show that the composite effect of land use on driving could be comparable in magnitude to that of driving cost. Land use strategies influence travel more effectively when complemented by pricing policies (Zhang, 2004).

Settlement size is also important in influencing both modal shares and the distance travelled, as use of public transport and walking increases with population size (Dargay and Hanly, 2004).

Diseconomies of scale may feed in with the largest cities, which have a complexity of movement that is substantially greater than the smaller monocentric cities, as circumferential trips become as important as radial trips (Banister, 1997).

The US literature is also variable in its findings. Ewing (1997) estimated that a doubling of density resulted in a 25–30 per cent lower level of vehicle miles travelled (VMT), whilst Holtzclaw (1994) concluded that the difference between 50 dwellings/hectare (urban densities) and 12.5 dwellings/hectare (suburban densities) was a 40 per cent increase in travel. Overall, the US evidence seems empirically powerful, suggesting that higher density developments can reduce VMT by at least 10–20 per cent as compared with urban sprawl (Litman, 2007).

2 Proximity and quality

Land use patterns in post industrial cities are changing as greater mixed use is the dominant feature. This means that journey lengths can be reduced through the use of local facilities and services. Considerable effort is now being placed in transport development areas (or the similar transit oriented developments in the US), where high quality public transport accessibility can be combined with office development, residential, leisure and retail activities, all in close proximity to each other. The importance of quality is paramount as these accessible locations become the centre of activity giving possible implications for public transport use. This is a concentration of activity that has beneficial impacts on modal split and the use of local facilities, but it needs to be balanced against the counter trend of dispersal (and sprawl) that has an opposite effect on trip lengths and a greater level of car dependence.

Cervero and Duncan (2006) examined the degree to which job accessibility is associated with reduced work travel, and how closely retail and service accessibility is correlated with miles and hours logged getting to shopping destinations. Based on data from the San Francisco Bay Area, they found that jobs–housing balance reduces travel more, by a substantial margin, than accessibility to shopping. But they also concluded that it is important to look at access to public transport at both ends of the journey. Concentrating 'housing near rail stops will do little to lure commuters to trains and buses unless the other end of the trip – the workplace – is similarly convenient to and conducive to using transit' (Cervero and Duncan, 2006, p53).

3 Local neighbourhood and design

The new urbanism debate encourages more local activity through more walking, direct routing for slow modes of transport, and quieter and narrower streets (Duany et al, 1992; Calthorpe, 1993). People travel shorter distances when they move into neighbourhoods with higher accessibility (Krizek, 2003), with median distance increasing from 3.2km in the more accessible neighbourhoods to 8.1km in less accessible neighbourhoods. Street connectivity is also important here as it can reduce distances for slow modes, but cul de sacs are also popular with residents, even though they tend to extend travel distances. Main Street programmes in the US (and more recently in the UK) are intended to revitalize town centres by restricting access at certain times and to create vibrant communities day and night (Handy, 2004). Other initiatives to encourage urban living include extensive pedestrianization, the closure of residential streets, gated communities and even the removal of freeways (e.g. the Embarcadero Freeway in San Francisco). The issue of parking management is central here, and this is one decision that is still under the direct control of planners, who can determine the number of spaces, the prices and

the time limitations, at least for the publicly controlled stock. They also now have the possibility of charging for private work place parking, as recently implemented in Nottingham.

One of the few detailed empirical studies has been carried out in Toronto (Norman et al, 2006) for city centre apartments (net residential density 150 dwellings/hectare) and suburban detached housing (net residential density 19 dwellings/hectare). Although the GHG emissions and energy density were similar per unit of living space (m^2) for construction materials, building operations and transport, the figures per person are very different (Table 5.3). This is due to the additional space available per person in the suburban detached housing. The GHG emissions are 2.5 times higher in the suburban than the urban housing. For transport, the figures are stark, with GHG emissions (and energy use) being more than 3.5 times as high in the low density housing for car and 6.5 times as great for public transport. The densities used in the Toronto study are different to those used in UK cities, where gross densities average about 20–40 dwellings/hectare (net densities 80–160 dwellings/hectare).[6] For example, the average Inner London (20 per cent of area) gross density is about 45 dwellings/hectare, and that for Outer London (80 per cent of area) is about 15 dwellings/hectare, a 3 to 1 ratio (Banister, 2007).

A large sample of the Great Britain National Travel Survey was taken by Dargay and Hanly (2004) for 1989–1991 and for 1999–2001 to test for the impact of land use characteristics on the level of mobility and the use of cars. They concluded that land use characteristics (population density, settlement size, local access to shopping and other facilities and accessibility of

Table 5.3 GHG emissions for different housing types in Toronto

Annual GHG emissions – kg CO_2e/person/year in 1996	Suburban Detached	%	Urban Apartments	%
Construction	597	7	391	12
Building operations	2730	32	1510	45
Car travel	5180	60	1420	43
Bus transport	130	1	20	–

Source: Based on Table 4 in Norman et al (2006).

public transport) play a significant role on car ownership and use of the car. Density has a greater impact than settlement size, and proximity to local facilities encourages walking instead of car travel.

Cumulative effects

Land use effects on travel behaviour tend to be cumulative and mutually reinforcing (Hickman, 2007; Litman, 2007). This effect can be illustrated in two ways. Ewing and Cervero (2002) calculated the elasticity of vehicle trips and travel per capita with respect to four land use variables (Table 5.4). Their estimates suggest that a doubling of local density reduces car trips by 5 per cent per capita and travel by about the same amount. Although the elasticities are low, Ewing and Cervero (2002) concluded that the land use effects were cumulative, thus giving the potential for 13 per cent and 33 per cent decreases in trips and trip distance respectively.

The second study was by Lawton (2001) using data from Portland Oregon to examine the impact of land use density, mix, and road network connectivity on personal travel. As urbanization increases, per capita vehicle travel declines significantly from about 20 average daily travel miles per adult (32kms) to just over 6 miles (10kms). The main conclusions with respect to the impacts of the land use factors on travel distance can be summarized as follows (Hickman and Banister, 2005):

1   At the regional level, the location of new development, particularly housing, should be of a substantial size and located near to or within existing settlements (see also Chapter 4) so that the total population is at least 25,000 and probably nearer to 50,000. The provision of local facilities and services should be phased so as to encourage the development of local travel patterns.

2   Density is important and average journey lengths by car are relatively constant (around 12km) at densities over 15 persons per hectare, but at lower densities car journey lengths increase by up to 35 per cent. Similarly, as density increases, the number of trips by car decreases from 72 per cent of all journeys to 51 per cent. Car use in the high density locations is half that in the lowest density locations.

3   Mixed use developments should reduce trip lengths and car dependence. Although research here is limited and concentrates on the work journey, there is considerable potential for enhancing the proximity of housing to all types of facilities and services.

4   As settlement size increases, the trips become shorter and the proportion of trips by public transport increases. Diseconomies of size appear for the largest conurbations as trip lengths increase to accommodate the complex structures of these cities.

5   Development should be located near to public transport interchanges and corridors so that high levels of accessibility can be provided. But this may also encourage long-distance public transport commuting. Free flowing strategic highway networks are likely to encourage the dispersal and sprawl of development and stretch commuting.

Table 5.4 Elasticities of trips and travel by land use factors

Factor	Description	Trips	Travel (VMT)
Local density	Residents and employees divided by land area	-0.05	-0.05
Local diversity	Jobs/residential population	-0.03	-0.05
Local design	Sidewalk completeness/route directness and street network density	-0.05	-0.03
Regional accessibility	Distance to other activity centres in the region	–	-0.20

Note: VMT = vehicle miles travelled.

Source: Ewing and Cervero (2002).

6   The availability of parking is a key determinant of whether a car is used or not, and further research is required to determine appropriate standards linked to accessibility levels.

These points are well summarized by Litman (2007), who concludes that in the US a 10–20 per cent cumulative total saving in VMT is possible through density and mixed design, and a further 20–40 per cent is possible from regional decisions on the location of new development. The figures in the UK are likely to be less, as the trip distances travelled are lower and there is already a much greater use of land use and development controls than in the US. The McKinsey Report (2007, p42) set a carbon abatement cost at \$50 per tonne CO_2e, and concluded that the US can reduce its emissions by between 3.0 and 4.5Gt CO_2e by 2030 (31 per cent to 49 per cent reduction). About a third of this figure would come from action on the built environment (buildings) and transport, but it was assumed (McKinsey and Company, 2007, p42) that there was no change in consumer utility and urban design; denser and more transport efficient communities were not assessed. It was also expected that there would be significant increases in distances travelled in the US over the period 2005–2030. The Stern Report on the economics of climate change also concluded that the carbon abatement in the transport sector would be more expensive to achieve than in other sectors due to the cost of technology and negative impacts on welfare (Stern, 2007). The evidence cited here suggests that behavioural change and land use and development decisions can all have a substantial influence on travel and energy use, and can contribute to reductions in CO_2 emissions.

Spatial planning and freight transport

Two major land use factors influence the efficiency of freight operations, namely handling factors and the average length of haul. They both relate to the distribution networks, as they look at cutting the number of separate journeys made from source to consumption, which in turn reflects the amount of outsourcing and vertical disintegration that has taken place in this sector. McKinnon (2007, p21) concludes that transport cost increases would have to be 'very large to induce such a structural change'. The use of consolidation points to assemble loads into larger units to save vehicle mileage and improve load factors (including reducing empty running) would be the aim of any such reorganization, but there is little information that links these individual movements to overall supply chains.

The empirical evidence shows a substantial increase in the UK freight haulage lengths, with the centralization of production and the widening of supply chains, but this trend has stabilized recently (1953–2004: 72km to 117km for all freight; 35km to 87km for road freight). The extension of supply chains means cheaper production (in locations with cheaper labour costs), but less energy efficiency and longer journeys (so more CO_2/tonne km). McKinnon (2007) concludes that reconfiguration of production and distribution systems would require transport costs to be more than doubled. Such a change may now be taking place as market forces have substantially raised the price of diesel in the UK (by 30 per cent from May 2007 to May 2008). More local sourcing is also counter to global trends, but there is a need for full Life Cycle Analysis (LCA) of CO_2 emissions. Spatial factors do affect the structure of the UK freight distribution system, with pricing and location decisions having a limited effect on reducing distances and improving load factors, but the scale of any intervention would have to be substantial to have a real effect. There does seem to be potential for action, but pricing alone will not resolve the problem. There also needs to be clear statutory guidance on determining where new freight distribution centres are located to ensure that haul distances are minimized.

Table 5.5 Income and travel

Trips per person per year by main mode	Lowest household income quintile		Average household income quintile		Highest household income quintile	
Walk	307	(34)	249	(24)	208	(18)
Car	406	(46)	658	(63)	818	(71)
Bus and coach	114	(13)	65	(6)	32	(3)
Rail	14	(2)	24	(2)	61	(5)
Other	41	(5)	41	(4)	39	(3)
Total	**882**		**1037**		**1158**	
Distance per person per year by main mode (miles)						
Walk	225	(5)	201	(3)	200	(2)
Car	2838	(69)	5693	(80)	9213	(79)
Bus and coach	530	(13)	359	(5)	212	(2)
Rail	276	(7)	541	(7)	1297	(11)
Other	255	(6)	339	(5)	666	(6)
Total	**4124**		**7133**		**11,588**	
Total CO_2 emissions	690kg		1250 kg		2030 kg	

Notes: Percentages in brackets. 'Other' includes bicycle, other private transport, taxi and minicab, and other public transport. Air travel information is not collected in the National Travel Survey. Total CO_2 emissions calculated from Defra (2007b) Guidelines to Defra's GHG conversion factors and uplifted by 15 per cent to allow for real world conditions as presented in CfIT (2007, Figure 2.6).

Source: Based on DfT (2006c) and Table 5.3.

Social justice and travel

The evidence and discussion presented here gives an average picture for the UK, but this conceals huge variations in the amount of travel and the carbon intensity of that travel. If the differences by income are taken, those in the highest income quintile travel nearly three times further (2.81) than those in the lowest income quintile, even though the number of trips made is only about 30 per cent higher. But as distances increase, so does the use of more energy consumptive forms of transport, and car and rail dominate the travel modes. The poor only make more use of the bus, whilst the rich make much more use of car and rail (Table 5.5), with very steeply rising curves at the highest income levels.

The conclusions on the unequal distribution of GHG emissions from personal travel in the UK have been brought into clear focus by Brand and Boardman (2008), where their empirical analysis clearly identified car and air travel as the dominant factors in energy use and carbon emissions from travel. They concluded that there is a highly unequal distribution of emissions, independent of mode, location and social characteristics. The top 10 per cent of emitters are responsible for 43 per cent of emissions and the bottom 10 per cent for only 1 per cent. Such a wide distribution would suggest that the market mechanisms have failed to reduce carbon emissions in transport, both overall and for those responsible for most of the emissions. Prices have risen recently, but mainly driven by world oil prices rather than direct government intervention, and where fuel duties are increased it is largely to raise revenue rather than to reduce emissions.

It would therefore seem that seriously tackling carbon emissions from transport requires tougher economic instruments, including road pricing, car purchase taxes, vehicle excise duty and parking charges. These also need to be designed to specifically reduce carbon emissions (i.e. targeting the most polluting vehicles) and to be combined with efficiency improvements, regulation and information. Strong action is needed on all available policy instruments, including the regulation of emissions standards for new technology, smarter choice instruments, speed enforcement and eco-driving and information communication technology. Technical measures such as agreed efficiency standards with motor manufacturers and biofuels are in place, but so far have been too lenient and designed without enough attention to the broader sustainability issues and rebound effects.

Smarter choice measures, including travel plans at schools and workplaces, individualized marketing and car clubs, have been proven to be very successful at securing a modal shift from the car to 'soft modes' and public transport at specific locations. If intensified and grossed up to the national level, these measures are estimated to have the potential to make a substantial (ca. 11 per cent) cut in UK national road traffic, after about a decade, if the benefits are 'locked-in' by supportive policy (Cairns et al, 2004). Enforcing the 70mph speed limit just on trunk roads could save at least 1 per cent of road traffic emissions immediately, and if combined with other efficient driving techniques (eco-driving), about twice as much as this (Anable and Bristow, 2007; CfIT, 2007).

Information and communications technology should help to optimize traffic management systems, facilitate more sophisticated road user charging as well as substituting for some travel, including air travel. Alongside these options, it may be that spatial planning may only have a limited role to play in the short term. However, it facilitates many of these other policy areas, fosters more sustainable location choices in the longer term, and enables efficiency to be maximized within the network as a whole.

Just as current patterns of travel and emissions are not even across society as a whole,

the effectiveness of policy instruments will also differ across groups of people. We can see from the Brand and Boardman (2008) analysis that the wealthier segments may be insulated from even the toughest taxes and charges. However, these people may be most likely to adopt more efficient technologies. Other population segments may be more likely to optimize their fuel with eco-driving and be open to information about alternative transport modes (Anable, 2005). All these segments differ according to their attitudes to climate change and belief that individual actions can make a difference (Stradling et al, 2008 and Box 5.3).

The question here is whether the high mileage drivers and air travellers identified by Brand and Boardman (2008) are the same as those hard core drivers (25 per cent) who think that they have a right to use their cars irrespective of the consequences to the environment (Box 5.3). If so, it may be that even those policies listed above will not be enough to sufficiently reduce emissions from these segments and that the only guaranteed way to do this might be to give each an equal personal carbon allowance which is tradable within an overall cap. However, there is not the political will to introduce such a system. A recent Government study concluded that:

> while personal carbon trading remains a potentially important way to engage individuals and there are no insurmountable technical obstacles to its introduction, it would nonetheless seem that it is an idea currently ahead of its time in terms of public acceptability and the technology to bring down its costs.
>
> (Defra, 2008, p4)

Nevertheless, within such a system the role of planning will be vital to provide local services and facilities, and the means of accessing them by modes other than the car to enable people to be able to live within their carbon allowance.

Box 5.3 Attitudes towards climate change and transport

Overall

80 per cent of all respondents think that current levels of car use have a serious effect on climate change.

66 per cent of all respondents agree with the view that, for the sake of the environment, everyone should reduce how much they use their cars.

59 per cent of all respondents disagree with the defeatist statement 'anyone who thinks that reducing their own car use will help the environment is wrong – one person doesn't make a difference.'

Drivers are concerned about these issues as well. 82 per cent think that current levels of car use have a serious effect on climate change, and 66 per cent think that everyone should reduce their use of the car.

Use of car for short journeys

About 45 per cent of drivers are both willing and able to do so. More encouragement and support is needed to get them to switch away from their cars.

About 12 per cent of drivers are able to reduce car use, but are unsure about whether they are willing to do so. They would also benefit from more encouragement and support to switch their modes.

About 18 per cent of drivers are willing to reduce their car use, but are unable to do so. This group might benefit from improved public transport and enhanced walking and cycling facilities.

About 25 per cent of drivers think that they should be able to use their cars as much as they like, even if it causes damage to the environment. High mileage drivers are likely to be in this group

Source: Stradling et al (2008).

Implications for strategy

Many of the strategies implemented in transport for decades which have set out to solve other societal issues such as accessibility, congestion and safety, also have the potential to tackle climate change. In this sense, the transport sector is in a strong position. This cannot mean business as usual, as transport's relative share of emissions is growing in all regions of the world. There is now very little time left to stabilize atmospheric concentrations of GHG emissions and this means early and strong action is needed to reverse the relentless growth in all transport markets identified in this chapter. Any successes must be maintained over the longer term with complementary 'lock-in' mechanisms so that the benefits are not quickly eroded away.

Within the debate about sustainable communities, decisions need to be taken at all levels, each with the appropriate scale of gover-nance, policy formulation and implementation powers to deliver targeted and deep cuts in carbon emissions (Anable and Shaw, 2007). The EU is giving greater guidance on the principles of sustainable development through their recent statement on urban mobility (Commission of the European Communities, 2007). The national government in the UK is responsible for overall planning policy, and this is changing with the new Planning Act (2008), which is designed to speed up the approval process for certain types of major projects. The key issue here is the location of new housing and other development in the UK, as this will have substantial implications for the levels of demand on the transport system, journey distances and the use of the different modes of transport over the next 20–30 years.

At the regional and city levels, there are questions about density of development, the availability of land for infill or reuse, the extent of mixed use development, the shape and size of different settlements, and concentration and distribution of services and facilities. Local issues include neighbourhood design and quality decisions, including the layout of developments and the role for slow modes of transport.

In all cases, there is a need for all actors at all levels to work together across the different sectors so that sustainable development becomes a reality. Too often in the past decisions have been made in isolation, and it is often the transport system that has had to accommodate the additional demand for movement. There may now be an increasing realization of this in decisions that people are now making in terms of where they choose to live. New lifestyle decisions mean that an urban location with shorter distances, good public transport and good accessibility to services and facilities become much more attractive. People do not like spending large amounts of time stuck in traffic.

At the strategic level, crucial decisions are now being made on the location of new housing and other forms of development that generate and attract substantial traffic. Eco towns and zero carbon settlements only work if transport is seen as being part of the design to encourage shorter journeys, less use of the car and greater use of public transport, walk and cycle. This must include clear guidance on density, settlement size, provision of local services and facilities, mixed land uses, proximity to public transport accessible developments, and limited availability of parking. Best practice, benchmarking and strategic guidance would all help here.

At the neighbourhood level (and in city centres), design standards should be used to encourage ownership of the local environment by residents and other stakeholders, so that its quality is maintained and improved. Residential, shopping and even commercial space should be very clearly designated for different priority uses (perhaps varying by time of day for schools, or day of week for markets), so that people and slow traffic have priority. Similarly, local facilities and services raise levels of accessibility so that travel distances to shops, schools and health centres can be reduced. Through design it is possible to lock in the benefits of lower transport emissions levels.

This is the basic dilemma facing society in terms of climate change and transport. We all like travelling and we are doing much more of it. Yet we are also aware of the environmental costs of travelling and our responsibilities both locally and globally. Our social networks are increasingly international and the global economy is also dependent on long supply chains. To some extent individual behaviour can be modified and we can substitute travel with technological communication. But in many cases there is no substitute for face-to-face communication, and we want to see the world. It presents a classic case of the conflict between individual preferences and choices, as opposed to the wider needs of society to protect the environment and future generations.

At present the scale and nature of the changes necessary in the transport sector to address climate change have not been seriously debated. Pricing for the external costs of transport would help, as would regulations on emissions and heavy investment in clean technology. But even here, the price rises necessary to create real change are not politically acceptable, as both industry and the electorate are powerful pro-travel lobbies. How can individual preferences be matched to societal responsibilities? Travel is a major and increasing contributor to climate change, yet there are few signs that we are prepared to make substantial behavioural change. The real challenge confronting society is greater than this, namely the expected growth in travel from all countries and the desire for long distance travel. Serious debate and action on these issues has not even started, and all the time the climate change clock is ticking.

Notes

1   Note that within the Congestion Charging zone the CO_2 levels are calculated from vehicle-km driven and fuel consumed. The traffic and speed changes observed in the charging zone are estimated to have led to 15.7 per cent savings, half

from reduced traffic and half from more efficient driving conditions (less congestion) (Transport for London (TfL), 2007).

2 Smarter choices (sometimes called 'soft' measures) include travel planning, individualized marketing, car clubs, car sharing, teleworking, videoconferencing, travel awareness campaigns and other measures designed to encourage voluntary travel behaviour change (Cairns et al, 2004).

3 Note that this measure assumes that all planes are full and so there are considerable gains to be made by using larger planes.

4 These calculations are based on the UK Office of National Statistics Environmental Accounts, which measure GHG emissions on a UK resident basis and includes emissions generated by UK households and companies in the UK. They also include emissions from UK residents transport and travel activities abroad.

5 The 'rebound' effect refers to compensatory increases in travel as efficiency increases, and it is sometimes called the Jeavons Paradox. For example, a fuel economy rebound effect of 20 per cent would mean that an increase of 10 per cent in fuel economy would only result in an 8 per cent saving, as the 2 per cent would be 'lost' through travelling further or faster than before.

6 The relationship between gross and net density is not a simple one, as net density excludes open space (public and private), roads, parking and footpaths. Buildings normally occupy 20–30 per cent of the total site area, so the ratio would be between 3 and 5 to 1.

References

Anable, J. (2005) 'Complacent car addicts or aspiring environmentalists? Identifying travel behaviour segments using attitude theory', *Transport Policy*, vol 12, no 1, pp65–78

Anable, J. and Bristow, A. L. (2007) 'Transport and climate change: Supporting document to the CfIT report', The Commission for Integrated Transport, London

Anable, J. and Shaw, J. (2007) 'Priorities, policies and timescales: Geography and the delivery of emissions reductions in the UK transport sector', *Area*, vol 39, no 4, pp443–457

Anderson, W., Kanaroglou, P. and Miller, E. (1996) 'Urban form, energy and the environment: A review of issues, evidence and policy', *Urban Studies*, vol 33, no 10, pp7–35

Banister, D. (1997) 'Reducing the need to travel', *Environment and Planning B*, vol 24, no 3, pp437–449

Banister, D. (2005) *Unsustainable Transport: City Transport in the New Century*, Routledge, London

Banister, D. (2007) 'Cities, urban form and sprawl: A European perspective', European Conference of Ministers of Transport Round Table 137, pp113–142, and paper presented at the OECD/ECMT Regional Conference Workshop, Berkeley, CA, March 2006

Banister, D. (2008a) 'The big smoke: Congestion charging and the environment', in H. W. Richardson and C.-H. C. Bae (eds) *Road Congestion Pricing in Europe: Implications for the United States*, Edward Elgar, Cheltenham, pp176–197

Banister, D. (2008b) 'The sustainable mobility paradigm', *Transport Policy*, vol 15, no 1, pp73–80

Brand, C. and Boardman, B. (2008) 'Taming the few: The unequal distribution of greenhouse gas emissions from personal travel in the UK', *Energy Policy*, vol 36, no 2, pp224–238

Cairns, S., Newson, C., Boardman, B. and Anable, J. (2006) *Predict and Decide. Aviation, Climate Change and UK Policy*, Environmental Change Institute, University of Oxford

Cairns, S., Sloman, L., Newson, C., Anable, J., Kirkbride, A. and Goodwin, P. (2004) 'Smarter choices – Changing the way we travel', Research Project 'The Influence of Soft Factor Interventions on Travel Demand', Final Report to the Department for Transport, London

Calthorpe, P. (1993) *The Next American Metropolis: Ecology, Community and the American Dream*, Princeton Architectural Press, New York

Cervero, R. and Duncan, M. (2006) 'Which reduces vehicle travel more: Jobs–housing balance or retail-housing mixing?', *Journal of the American Planning Association*, vol 74, no 4, pp475–490

Commission for Integrated Transport (CfIT) (2002) *Paying for Road Use*, Commission for Integrated Transport, London

CfIT (2007) 'Transport and climate change', *Advice to Government from the CfIT*, London, October

Commission of the European Communities (2007) 'Green Paper: Towards a new culture for urban mobility', COM(2007) 551 Final, Brussels, September

Dargay, J. and Hanly, M. (2004) 'Land use and mobility', paper presented at the World Conference on Transport Research, Istanbul, July

Department for Business Enterprise and Regulatory Reform (BERR) (2008) 'Updated energy and carbon emissions projections', *The Energy White Paper*, Department for Business Enterprise and Regulatory Reform, London

Department for Environment, Food and Rural Affairs (Defra) (2006) *Climate Change: The UK Programme 2006*, The Stationery Office, Cm 6764, London, www.defra.gov.uk/ENVIRONMENT/ climatechange/uk/ukccp/index.htm, accessed 30 June 2008

Defra (2007a) 'Passenger transport emissions factors: Methodology paper', Department for the Environment, Food and Rural Affairs, London, www.defra.gov.uk/environment/business/envrp/ pdf/passenger-transport.pdf, accessed 30 June 2008

Defra (2007b) 'Estimated total emissions of UK "basket" greenhouse gases on an IPCC basis weighted by global warming potential (Table 4)', Department for the Environment, Food and Rural Affairs, London, www.defra.gov.uk/environment/ statistics/globatmos/gagccukem.htm#gatb4, accessed 30 June 2008

Defra (2008) 'Synthesis report on the findings from Defra's pre-feasibility study into personal carbon trading', Department for the Environment, Food and Rural Affairs, London

Department for Transport (DfT) (2004) 'Feasibility study of road pricing in the UK', Department for Transport, London, www.dft.gov.uk/pgr/roads/ introtoroads/roadcongestion/feasibilitystudy/study report/, accessed 31 July 2008

DfT (2006a) *Transport Statistics Great Britain 2006*, 32nd Edition, The Stationery Office, London

DfT (2006b) 'Transport Innovation Fund: Guidance', Department for Transport, www.dft.gov.uk/ pgr/regional/tif/, accessed 31 July 2008

DfT (2007a) 'National Travel Survey 2007', *Transport Statistics Bulletin*, The Stationery Office, London

Department of Trade and Industry (2007) *Meeting the Energy Challenge: A White Paper on Energy*, Cm 7124, The Stationery Office, London, www.dtistats.net/ewp/ewp_full.pdf, accessed 30 June 2008

Duany, A., Plater-Zyberk, E. and Speck, J. (1992) *Suburban Nation: The Rise of Sprawl and the Decline of the American Dream*, North Point Press, New York, NY

Ewing, R. (1997) 'Is Los Angeles style sprawl desirable?', *Journal of the American Planning Association*, vol 63, no 1, pp107–126

Ewing, R. and Cervero, R. (2002) 'Travel and the built environment: Synthesis', *Transportation Research Record 1780*, TRB, Washington DC

Handy, S. (2004) 'Accessibility v mobility-enhancing strategies for addressing automobile dependence in the US', paper presented at the European Conference of Ministers of Transport Round Table on Transport and Spatial Policies: 'The Role of Regulatory and Fiscal Incentives', RT124, Paris, November 2002, pp49–85

Hickman, R. (2007) 'Reducing travel by design: A micro analysis of new household location and the commute to work in Surrey', unpublished PhD Thesis, Bartlett School of Planning, University College London, London

Hickman, R. and Banister, D. (2005) 'Reducing travel by design', in K. Williams (ed) *Spatial Planning, Urban Form and Sustainable Transport*, Ashgate, Aldershot, pp102–122

Hickman, R. and Banister, D. (2007) 'Transport and reduced energy consumption: What role can urban planning play?', paper presented at the Transport Planning and Management Conference, Manchester, July

Holtzclaw, J. (1994) *Using Residential Patterns and Transit to Decrease Auto Dependence and Costs*, National Resources Defense Council, Washington DC

Kenworthy, J. (2007) 'Urban planning and transport paradigm shifts for cities of the post-petroleum age', *Journal of Urban Technology*, vol 14, no 2, pp47–60

Kenworthy, J. and Laube, F. (1999) *An International Sourcebook of Automobile Dependence in Cities 1960–1990*, University of Colorado Press, Boulder

Krizek, K. J. (2003) 'Residential relocation and changes in urban travel: Does neighbourhood scale urban form matter', *Journal of the American Planning Association*, vol 69, no 3, pp265–281

Lawton, K. T. (2001) 'The urban structure and personal travel: An analysis of Portland, Oregon data and some national and international data', E-Vision Conference, www.rand.org/scitech/stpi/ Evision/Supplement/lawton.pdf, accessed 30 June 2008

Litman, T. A. (2007) *Land Use Impacts on Transport: How Land Use Factors Affect Travel Behaviour*, Victoria Transport Policy Institute, Canada

McKinnon, A. (2007) 'CO_2 emissions from freight transport in the UK', report prepared for the Climate Change Working Group of the Commission for Integrated Transport, Edinburgh, www.cfit.gov.uk/docs/2007/climatechange/pdf/ 2007climatechange-freight.pdf, accessed 30 June 2008

McKinsey and Company (2007) 'Reducing US greenhouse gas emissions: How much at what cost?', US Greenhouse Gas Abatement Mapping Initiative

Newman, P. W. G. and Kenworthy, J. R. (1989a) *Cities and Automobile Dependence: An International Sourcebook*, Gower, Aldershot

Newman, P. W. G. and Kenworthy, J. R. (1989b) 'Gasoline consumption and cities: A comparison of US cities with a global survey', *Journal of the American Planning Association*, vol 5, no 1, pp24–37

Newman, P. W. G. and Kenworthy, J. R. (1999) *Sustainability and Cities: Overcoming Automobile Dependence*, Island Press, Washington DC

Norman, J., MacLean, H. L. and Kennedy, C. A. (2006) 'Comparing high and low residential density: Life-cycle analysis of energy use and greenhouse gas emissions', *Journal of Urban Planning and Development*, vol 132, no 1, pp10–21

Stead, D. (2001) 'Relationships between land use, socio-economic factors, and travel patterns in Britain', *Environment and Planning B*, vol 28, no 4, pp499–528

Stern, N. (2007) *The Economics of Climate Change: The Stern Review*, Cambridge University Press, Cambridge

Stradling, S., Anable, J., Anderson, T. and Cronberg, A. (2008) 'Car use and climate change: Do we practise what we preach?', in *British Social Attitudes: The 24th Report*, published by Sage for NatCen

Transport for London (TfL) (2007) 'Central London congestion charging: Impacts monitoring', Fifth annual report, London

Zhang, M. (2004) 'The role of land use in travel mode choice', *Journal of the American Planning Association*, vol 70, no 3, pp344–360

6

Transitioning Away from Oil:
A Transport Planning Case Study with Emphasis on US and Australian Cities

Peter Newman

The problem

Oil vulnerability has become a major focus of the world's cities in the early part of the 21st century. This is fundamentally because the world is peaking in oil production as many pundits have been predicting for the past two or three decades. Added to this is the climate change agenda which suggests that oil needs to be phased out anyway. Reducing oil use is thus a political necessity for many reasons. The waning of petroleum resources and the global climate change imperatives require all cities to act on their transport systems; if they don't their citizenry will not be impressed at the inevitable increase in prices and indeed many cities will face complete economic collapse unless they are rebuilt for a decarbonized economy. The $100-a-barrel oil barrier has been broken and some analysts are saying that it could go over $300 within five years, though the collapse in the price in late 2008 with the collapse in the economy may have given us a short period to prepare for this ultimate eventuality.[1] However even if the looming fuel shortage was not driving this issue we should be doing it anyway for the following reasons:

- Reducing oil use will reduce impacts on the environment. Oil use is responsible for approximately one-third of greenhouse gases (GHGs). Transport emissions are seen as the most worrying part of the climate change agenda as they continue to grow during a period when more renewable or efficiency options are available.
- Reducing oil use will reduce smog. Improvements in urban air quality from technological advances are being washed out by growing use of vehicles in 39 different air quality districts in the US that are over the required standards (this is 40 per cent of the United States). Developing cities desperately need to lower air emissions as they are often well above the health limits recommended by the World Health Organization (WHO).
- Reducing car dependence will improve human health, safety and equity. The inequities of heavily car dependent cities for the elderly, the young and the poor, will be reduced; the health impacts of car dependence such as poor air quality, obesity due to lack of activity, and depression will be reduced; the social issues such as noise, neighbourhood severance, road rage and loss of

public safety will be reduced; the economic costs from loss of productive agricultural land to sprawl and bitumen, the costs of accidents, pollution and congestion, all will be reduced.

- Reducing our dependence on petroleum fuels will make us less economically vulnerable. The next agenda for the global economy, sometimes called the Sixth Wave (outlined below) is about sustainability, about responding with technology and services for a new and cleverer kind of resource use. Cities will compete within this economic framework and those cities that get in first will likely do best. But the same economic competition is facing households depending on which city they live in and where they live in those cities. In US cities the proportion of household expenditure on transportation increased from 10 per cent in the 1960s to 19 per cent in 2005, before the 2006 oil price increases (which only reduced the percentage to 18 per cent), with very car dependent cities like Houston and Detroit having even higher percentages (Surface Transportation Policy Project (STPP), 2005). A more detailed study by the Centre for Housing Policy shows working families with household incomes between $20,000 and $50,000 spend almost 30 per cent on transportation (Lipman, 2006). In Atlanta within this income range the percentage is 32, and for families who have found cheap housing on the fringe it can be over 40 per cent of their income. So in car dependent cities there is an increase in household income spent on transportation especially in their urban fringe areas. Almost all of this is for car travel. Households on the fringes of car dependent cities are highly vulnerable as the cost of transport escalates. The 2007/8 dramatic increase in oil prices coincided with the subprime mortgage crisis, hitting many with a double whammy of increased transportation costs and a ballooning mortgage payment. It is now evident that cities, and parts of cities, are economically vulnerable to oil as it increases in cost and that financial institutions will not easily lend money to highly car dependent land development.

- Reducing dependence on foreign oil is likely to result in more resilient, peaceful cities. Cities that are able successfully to reduce their dependence on imported oil, especially from politically sensitive areas, will have greater energy security. Terrorism and war have many causes but one deep and underlying issue is the need of high oil-consuming cities to secure access to oil in foreign areas, whether they are friendly or not. As oil becomes more and more valuable the security of supply will become a more and more central part of geopolitics. Fear can drive us to make security decisions that are not going to help create resilient cities.

- Most importantly these more resilient cities will be better places to live (Newman et al, 2009). The many benefits of a resilient city include greater overall physical and emotional health; ease of movement from higher density, mixed-use communities that are walkable and have accessible transit options; better food that is produced locally and is therefore fresher; more energy efficient, affordable and healthy indoor environments; access to natural environments; and more awareness of the local urban area and its bioregion enabling us to have a greater sense of place and identity. Some of these factors are challenging to quantify but are nevertheless real factors.

The opportunity

The response to these challenges can often be one of panic — that it will have a severe impact on our economy. However, it is also possible to see that this is a real opportunity and that indeed the ability of cities to compete effectively in reducing oil will be a major part of their new economies. The next phase of innovation in our cities is seen by some to be based around sustainability innovations as set out by Hargraves and Smith (2004) in Figure 6.1.

Figure 6.1 shows how crashes in our industrial cities have occurred before, generally after a boom based on a particular set of technologies.

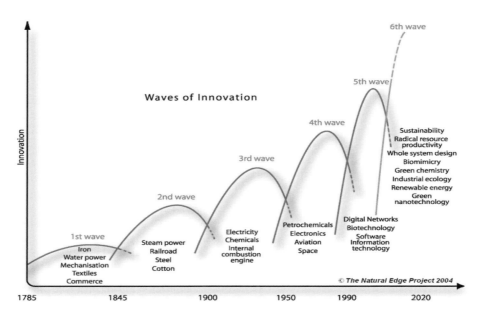

Source: Hargraves and Smith (2004)

Figure 6.1 The waves of economic innovation showing the booms and busts related to how cities have adopted technologies and built themselves around these

The First Wave began in traditional walking cities where new industries began to develop along rivers and canals using water power to manufacture textiles. The cities that resulted were dense and filled quickly with the new wave of urban immigrants. These cities were never more than five to eight kilometres across (the distance you can walk in one hour which is the universal average travel time budget in cities). Dominated by the smoke and waste of industry the new industrialism quickly overwhelmed the traditional walking cities which were not built for such activity. The crash of the 1830s and 1840s saw an end to this kind of urban development.

The Second Wave of industrialism then used the new technology of steam trains to build new cities which spread out along the railways of the steel and steam era. These cities had dense industrial and population activity in their main centres, built along these rail lines as people walked within each centre or sub-city. These also became limited by the wastes and human activity that could be accommodated so that by the 1890s the crash of investment provoked a new way to build cities to absorb the continually growing urban population.

This emerged in the Third Wave of electricity which saw lighting and power delivered without the immediate smoke of the old coal-fired boilers and enabled the transport system to be electrified as well. It saw electric railways and tramways built as the basis of most cities, enabling their residential estates and commercial activity to be spread along its streetcar and electrified rail systems. These cities followed the trains and trams which spread 30 kilometres or so. These all crashed in the 1930s as urban fringe speculation collapsed. The Great Depression meant cities had to invent technology that would enable them to expand further.

So the Fourth Wave was dominated by cheap oil and cars which enabled cities to spread and sprawl for 50 to 80 kilometres or so in every direction. Thus the automobile city was invented enabling houses to be built in successive rings, absorbing each new wave of immigration or urbanization. These cities could not have contained their growing populations if they had continued to be based around the industrialization of petrochemicals and manufacturing. Thus by the 1980s a new kind of downturn occurred and the cities of the modern western world moved to find a new basis for their economies.

The Fifth Wave of internet and digital technologies has replaced the old industrial manufacturing parts of central and inner areas with knowledge jobs, thus helping to minimize some of the sprawl and start the renewal of these older industrial sites from the previous Waves. However the Fifth Wave still had cheap oil, enabling cars to dominate the transport system, and suburbs to be built further and further out.

Travel time limits began to undermine these scattered developments, leaving them vulnerable to financial vagaries. The shift in oil prices exposed this underlying vulnerability of highly car and fuel dependent urban development from the Fourth and Fifth Waves. Once the fuel price increased, the loans which were used to form these scattered urban areas became toxic. At the same time a more global limit was reached with climate change; the cities of the world now faced a new limit whereby they must phase out all fossil fuels. Although not yet part of the main marketplace, the undermining of confidence in the long term future of heavily fossil fuel-dependent industry and land development is already underway. The crash of September 2008 signals the end to the urban economy based around oil in particular, but all heavily fossil fuel-dependent urban development as well.

Thus radical resource productivity and renewable energy can be seen to be linked to the digital networks of the previous innovation phase to produce a whole new set of economic opportunities. These opportunities now confront cities and their regions to overcome their oil addiction and move towards a much greater degree of resilience. The question from this perspective then becomes: what are these opportunities and how can we respond best to them? This chapter will present five areas, listed below, where transport opportunities appear to be presenting themselves that could help make resilient cities and regions. It suggests them within a context of cities and regions in Australia

and the US though many other cities and regions will be similar.

- New generation electric transit systems and their associated Transit-Oriented Development (TOD), Pedestrian-Oriented Development (POD), and Green-Oriented Development (GOD) structures.
- Renewable energy-based electric vehicles linked through Smart Grids.
- Natural gas and biofuels in freight and regional transport.
- Telepresence, high speed rail and airships.
- Indigenous settlements going diesel-free.

New generation electric transit systems and their associated TOD, POD and GOD structures

Cities need to have a combination of transportation and land use options that are favourable for green modes, and offer time savings when compared to car travel. This means transit needs to be faster than traffic down each major corridor. Those cities where transit is relatively fast are those with a reasonable level of support for it. The reason is simple – they can save time.

With fast rail systems, the best European and Asian cities with the highest ratio of transit to traffic speeds have achieved a transit option that is faster than the car down the main city corridor. Rail systems are faster in every city in the 84-city sample, studied by Kenworthy and Laube (2001), by 10–20 kilometres per hour (kph) over bus systems, as buses rarely average over 20–25kph. Busways with a designated lane can be quicker than traffic in car-saturated cities, but in lower-density car-dependent cities it is important to use the extra speed of rail to establish an advantage over cars in traffic. This is one of the key reasons why railways are being built in over 100 US cities.[2]

Rail has a density-inducing effect around stations, which can help to provide the focused centres so critical to overcoming car dependence. Thus transformative change of the kind that is needed to rebuild car-dependent cities

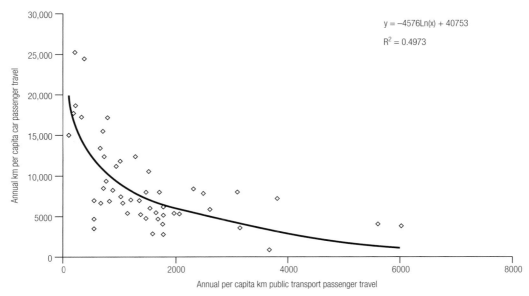

$y = -4576\text{Ln}(x) + 40753$

$R^2 = 0.4973$

Source: Newman and Kenworthy (1999)

Figure 6.2 City ratios for car travel and public transport use

comes from new electric rail systems as they provide a faster option than cars and can help build transit-oriented centres.

How much is it possible to change our cities? It is possible to imagine an exponential decline in car use in our cities that could lead to 50 per cent less passenger km driven in cars. The key mechanism is a quantitative leap in the quality of public transport whilst fuel prices continue to climb, accompanied by an associated change in land use patterns.

Figure 6.2 shows the relationship between car passenger km and public transport passenger km from the Global Cities Database (Newman and Kenworthy, 1999). The most important thing about this relationship is that as the use of public transport increases linearly the car passenger km decreases exponentially. This is due to a phenomenon called *transit leverage* whereby one passenger km of transit use replaces between five and seven passenger km in a car due to more direct travel (especially in trains), trip chaining (doing various other things like shopping or service visits associated with a commute), giving up one car in a household (a common occurrence that reduces many solo trips) and

eventually changes in where people live as they prefer to live or work nearer transit (Newman and Kenworthy, 1999).

The data on private transport use and public transport use in selected Australian cities for 1996 are given in Table 6.1. The values in Figure 6.2 show Australian cities are somewhat down the curve from the very high US cities, which have almost no transit (some around the 100 to 200 pass km per person) and very high private transport use of over 15,000 pass km per person. The data show that the highest Australian city Sydney had 12.3 per cent of its total motorized pass kms on transit and that the lowest was Perth with 4.5 per cent (this was before the remarkable increase in patronage associated with Perth's rail revival).

If Sydney doubled its transit use to 3018 pass kms per person, Figure 6.2 shows that it would have a per capita private transport use of 4088 passenger kms per capita which is a 61 per cent reduction in car passenger kms per person over the 1996 figure. If Perth was able to continue the rapid growth in transit patronage and triple its 1996 use to around 2000 pass km per person then it would reduce its private transport use per

Table 6.1 Car and public transport use per capita in four Australian cities, 1996

City	Private transport use (pass km/person)	Public transport use (pass km/person)
Sydney	10,506	1509
Melbourne	11,918	994
Brisbane	12,487	720
Perth	13,546	642

Source: Newman and Kenworthy (1999).

capita to 6000 car passenger km per capita, which is a reduction of 56 per cent over the 1996 level. Similar calculations can be done for the other Australian cities. Indeed it is feasible that each city could set a target of increases in passenger kms per capita for public transport in order to achieve certain target reductions in car use as part of their commitment to reaching the national goal of 80 per cent reduction in GHGs by 2050.

These remarkable reductions suddenly become imaginable. But are they real? Could it happen? The driving force would need to be a combination of *push* and *pull*. The *push* would come from fuel prices that rise inevitably as supply of oil declines and other alternative fuels just cannot fill the gap in supply. In the US in 2008 where fuel price rises were more severe (as Australia was shielded by the rising dollar), there was a reduction in vehicle kilometres of travel (VKT) of 4.3 per cent and a substantial rise in transit patronage.

This trend cannot continue unless there is a simultaneous *pull* from the provision of transit. Already capacity limits have been reached across Australian cities in their public transport. Therefore, for a start, substantial increases in trains, trams and buses are needed to fill the rapid growth in transit. There will also need to be new lines and new technology like Metros and light rail to increase the capacity and speed of transit to make it attractive to use.

At the same time the cities will need to develop rapidly around transit stations. This can be a significant source of funding for the required rail infrastructure through 'value transfer public private partnerships' (PPP) as in the very success-

ful Chatswood Transport Interchange PPP which has created a new railway station and bus interchange along with a retail and residential complex that makes a small city around and over the station (Dawson, 2008). It can be the main mechanism for replacing the development of car dependent suburbs which are already beginning to die as the price of fuel climbs. Significant new local transit options linking across the heavy rail corridors – especially with light rail systems – will also be needed.

How realistic is it to assume that public transport can increase as described, and what are the capacity implications of such an assumption for our public transport systems?

Table 6.2 shows the medium population projections for the five largest Australian cities to 2051. As can be seen, these reveal that:

- the five largest cities are expected to grow by around 20 per cent between 2004 and 2021, and by 45 per cent by 2051;
- they will increase their share of Australia's population slightly from 61 per cent to 63 per cent over that time;
- although all cities will grow, Brisbane is expected to grow the fastest (almost 90 per cent growth by 2051) and Adelaide the slowest.

Table 6.3 shows the implications in terms of per capita passenger-kilometres in those cities ranging from a doubling by 2051 for Sydney to a tripling for the small cities (Brisbane, Adelaide and Perth) as suggested in the analysis above. Thus, they suggest per capita public transport use in Melbourne in 2021 would be slightly above

Table 6.2 Medium population projections for Australia, 2004–2051

City	Thousands			Growth 2004–2021	Growth 2004– 2051
	2004	2021	2051		
Sydney	4225	4871	5608	15%	33%
Melbourne	3593	4252	5041	18%	40%
Brisbane	1778	2404	3355	35%	89%
Perth	1455	1875	2454	29%	69%
Adelaide	1123	1201	1203	7%	7%
Rest of Aust	7917	9268	10509	17%	33%
Australia	20,091	23,871	28,170	19%	40%
Five city sub-total	12,174	14,603	17,661	20%	45%
% in five largest cities	61%	61%	63%		

Source: Australian Bureau of Statistics: Population Projections 2004–2101; ABS 3222.0.

that achieved in Sydney in 2004, while Perth and Adelaide's use in 2051 would equal that of Sydney currently.

The total public transport travel task implied by these predictions is shown in Table 6.4, combining the derived per capita growth figures with the predicted population increases. This shows that across the five largest cities' total patronage would need to be lifted by 80 per cent by 2021, and more than trebled by 2051.

However the increase in patronage in peak periods would not need to be as large as in off-peak periods, given the much lower share achieved for non-work or education trips (such as social/recreation, shopping and business trips) which are largely made in off-peak periods. This is shown in Table 6.5 below to illustrate the task in terms of augmenting public transport capacity at peak periods in each of the cities to achieve the increase in public transport use in Table 6.4.

Hence to achieve major reductions in car use it would be necessary to increase capacity in Sydney by around 50 per cent by 2021, and by 120 per cent by 2051. For Brisbane the increases are more like a doubling in capacity by 2021 and a quadrupling by 2051. These are not difficult to imagine as they represent growth rates of around 2 per cent per year. With such growth the transformation of Australian cities to achieve significant reductions in car use can then happen. The first stage of making this happen has now occurred with the Federal Budget in 2009 providing $4.6 billion for urban public transport (mostly rail) which is an historic step after no Federal involvement in this area before. The biggest beneficiary was Melbourne but Infrastructure Australia's report to the Federal Government is based on at least doubling the capacity of urban rail in all major Australian cities (Infrastructure Australia, 2009).

Table 6.3 Assumed per-capita public transport use in major Australian cities (pass km per year)

City	1996	2004	2021	2051
Sydney	1509	1500	2100	3000
Melbourne	994	990	1600	2500
Brisbane	720	800	1300	2200
Perth	642	700	1200	2000
Adelaide	500	500	800	1500

Table 6.4 Implications for overall public transport use – estimated pass-km (billion)

City	2004	2021	2051	Growth 2004–2021	Growth 2004– 2051
Sydney	6.3	10.2	16.8	61%	165%
Melbourne	3.6	6.8	12.6	91%	254%
Brisbane	1.4	3.1	7.4	120%	419%
Perth	1.0	2.3	4.9	121%	382%
Adelaide	0.6	1.0	1.8	71%	221%
Total	12.9	23.4	43.5	81%	237%

The biggest challenge in an age of radical resource-efficiency requirements will be finding a way to build fast rail systems for the scattered car-dependent cities of the New World. How can a fast transit service be built back into these areas? The solution may well be provided by Perth and Portland which have both built fast rail systems down freeways. Freeways are public facilities that may well be in decline in the future as car traffic faces the double whammy of increasing fuel prices due to peak oil and carbon taxes due to climate change. To build fast electric rail down the middle of these roads is easier than anywhere else as the right of way is there and engineering in terms of gradients and bridges is compatible. They are not ideal in terms of ability to build TOD but it can still be done using high-rise buildings as sound walls. Linkages from buses, electric bikes and park and ride are all easily provided so that local travel to the system is short and convenient. The key is the speed of the transit system and in Perth the new Southern Railway has a maximum speed of 130kph (80mph) and an average speed of 90kph

(55mph), which is at least 30 per cent faster than traffic. The result is dramatic increases in patronage far beyond the expectations of planners who see such suburbs as too low in density to deserve a rail system. There is little else that can compete with this kind of option for creating a future in the car dependent suburbs of many cities.

Fast electric rail services are not cheap. However, they cost about the same per mile as most freeways and we have been able to find massive funding sources for these in the past 50 years. In the transition period it will require some creativity as the systems for funding rail are not as straightforward. In Perth the state government was able to find all the funds from Treasury due to a mining boom and was able to pay off the entire rail system, including the new Southern Railway even before it was opened. But for most cities this is not possible.

To solve this funding problem cities have had to find innovative solutions such as financing transit through the use of taxes or direct payments from land development, as in Copenhagen's new rail system, or through a

Table 6.5 Estimated increase in peak and off-peak capacity

City	Growth 2004–2021		Growth 2004–2051	
	Peak	Off-Peak	Peak	Off-Peak
Sydney	50%	70%	120%	200%
Melbourne	70%	110%	200%	300%
Brisbane	100%	140%	300%	500%
Perth	100%	140%	280%	480%
Adelaide	50%	90%	150%	300%
Total	65%	95%	160%	320%

congestion tax as in London. Funding of transit in congested cities can occur as it has in Hong Kong and Tokyo, where the intensive requirements around stations means that the transit can be funded almost entirely from land redevelopment. In poorer cities the use of development funds for mass transit can increasingly be justified through the transformation of their urban economy. Peak oil and climate change will increasingly be part of that rationale.

TOD has become a major technique for reducing automobile dependence and hence tackling peak oil. For the full agenda of sustainability and peak oil to be addressed TODs need to also be PODs and GODs.

The facilitation of TODs has been recognized by all Australian and many American cities in their metropolitan strategies, which have developed policies to reduce car dependence through centres along corridors of quality transit. The major need for TODs is not in the inner areas as these have many from previous eras of transit building. However the newer outlying suburbs, built in the past four or five decades, are heavily car dependent with high fuel consumption and almost no TOD options available. There are real equity issues here as the poor increasingly are trapped in the fringe with high expenditures on transport. A 2008 study by the Center for Transit Oriented Development shows that people in TODs drive 50 per cent less than those in conventional suburbs (CTODRA, 2004). In both Australia and the USA, homes that are located in TODs are holding their value the best or appreciated fastest under the pressure of rising fuel prices. The Urban Land Institute's report (2008) suggested that TODs would appreciate fastest in bouyant markets and hold value best in time of recession.

Thus TODs are an essential policy for responding to peak oil, especially when they incorporate affordable housing. The economics of this approach have been assessed by the Center for Transit Oriented Development and the NGO Reconnecting America (CTODRA, 2004). In a detailed survey across several states these NGOs assessed that the market for people wanting to live within half a mile of a TOD was 14.6 million households. This is more than double the number who currently live in TODs. The market is based on the fact that those living in TODs now (who were found to be smaller households, the same age and the same income on average as those not in a TOD) save some 20 per cent of their household income by not having to own so many cars – those in TODs owned 0.9 cars per household compared to 1.6 outside. This freed up on average $4000 to $5000 per year. In Australia a similar calculation showed this would save some $750,000 in superannuation over a lifetime. Most importantly, this extra income is spent locally on urban services, which means the TOD approach is a local economic development mechanism.

TODs must also be PODs, that is pedestrian-oriented developments, or they lose their key quality as a car-free environment in attracting business and households. This is not automatic but requires the close attention of urban designers. Jan Gehl's transformations of central areas such as Copenhagen and Melbourne are showing the principles of how to improve TOD spaces so they are more walkable, economically viable, socially attractive and environmentally significant (Gehl, 1987; Gehl and Gemzoe, 2000; Gehl et al 2006). It will be important for those green developers wanting to claim credibility that scattered urban developments, no matter how green in their buildings and renewable infrastructure, will be seen as failures in a post peak-oil world unless they are building pedestrian-friendly TODs.

At the same time TODs that have been well designed as PODs will also need to be GODs – green-oriented developments. TODs will need to ensure that they have full solar orientation, are renewably powered with Smart Grids, have water sensitive design, use recycled and low impact materials, and use innovations like green roofs.

Perhaps the best example of a TOD-POD-GOD is the redevelopment of Kogarah Town Square in Sydney. This inner city development is built upon a large City Council car park adjacent to the main train station where there was a collection of poorly performing businesses adjacent. The site is now a thriving mixed-use development consisting of 194 residences, 50,000 square feet of office and retail space and

35,000 square feet of community space including a public library and town square. The buildings are oriented for maximum use of the sun with solar shelves on each window (enabling shade in summer and deeper penetration of light into each room), photovoltaic (PV) collectors are on the roofs, all rain water is collected in an underground tank to be reused in toilet flushing and irrigation of the gardens, recycled and low impact materials were used in construction, and all residents, workers and visitors to the site have a short walk to the train station (hence reduced parking requirements enabled better and more productive use of the site). Compared to a conventional development, the Kogarah Town Square saves 42 per cent of the water and 385 tons of GHG – this does not include transport oil savings that are hard to estimate but are likely to be even more substantial.

While the demand for TODs is growing, creating TODs can still present significant challenges given the complexity of financing TODs and the number of private and public actors involved. TODs are in great demand, which often results in housing priced out of the range of middle- and lower-income households. Thus, along with the other green requirements for TODs there needs to be a requirement of a certain proportion of affordable housing. In Perth the 20 or so TODs being planned have been suggested to be progressed via a new TOD zoning that requires minimal amounts of parking, maximizes density and mix, includes green innovations and has a minimum of 15 per cent affordable housing to be purchased by social housing providers. Ellen Greenberg, director of policy and research for Congress of New Urbanism, suggests six steps for a planning and policy approach to implementing TODs: create customized zoning for projects integrating transit facilities; minimize customized planning and discretionary review for standardized projects; provide an explicit foundation in policy and politics; engage transit organization policy leadership; meet multiple objectives (e.g. affordable housing, commuter parking, transit transfer station, meeting carbon reduction goals); anticipate a lengthy timeline for customized projects (Dittmar and Ohland, 2004).

Renewable energy based electric vehicles linked through Smart Grids

Even if we manage to reduce car use by 50 per cent as suggested above, by a rather Herculean effort, we still have to reduce the oil and carbon in the other 50 per cent of vehicles being used. The question should therefore be asked: what is the next best transport technology for motor vehicles? The growing consensus seems to be: plug-in electric hybrid vehicles (PHEVs). Plug-in electric vehicles are now viable alternatives due to new batteries such as lithium ion, and with hybrid engines for extra flexibility they are likely to be attractive to the market. The key issue here is that plug-in electric vehicles not only reduce oil vulnerability but they are becoming a critical component in how renewable energy will become an important part of a city's electricity grid. The PHEVs will do this by enabling renewables to have a storage function.

After electric vehicles are recharged at night they can be a part of the peak power provision next day when they are not being used but are plugged in. Peak power is the expensive part of an electricity system and suddenly renewables is offering the best and most reliable option. Hence the Resilient City of the future is likely to have a significant integration between renewables and electric vehicles through a Smart Grid. Thus electric buses, electric scooters and gophers, and electric cars have an important role in the future Resilient City – both in helping to make its buildings renewably powered and in removing the need for oil in transport.

Electric rail can also be powered from the sun either through the grid powering the overhead wires or in the form of new light rail (with these new Li-ion batteries) which could be built down highways into new suburbs without requiring overhead wires. Signs that this transition to electric transport is underway are appearing in demonstration projects such as Google's 1.6MW solar campus in California (with 100 PHEVs) and by the fact that oil companies are acquiring electric utilities.[3]

What sort of impact could there be? According to one study the integration of hybrid cars with the electric power grid could reduce gasoline consumption by 85 billion gallons per year. That's equal to:

- 27 per cent reduction in total US greenhouse gases;
- 52 per cent reduction in oil imports;
- $270 billion not spent on gasoline (Kintner-Meyer et al, 2007).

The real test of a resilient city will be how it can simultaneously be reducing its global greenhouse and oil impact through these new technologies whilst reducing the need to travel by car through the policies outlined in the first strategy on transit and TODs.

Natural gas and biofuels in freight and regional transport

What do you do with freight transport and regional transport outside of cities where electric grids are not so easily used with vehicles?

There will almost certainly be a reduction in the amount of freight moving around as fuel prices eat into the transport economics of consumption. Containers will be reduced as their fuel costs move from being 10–15 per cent to over 50 per cent. Food miles will start to mean something to food prices when the cost of fuel triples. But trucks and trains and regional transport will still go on.

The next stage for larger vehicles and for regional transport would appear to be to switch to greater use of natural gas and biofuels. Trucks and trains and fishing boats can use CNG (compressed natural gas) or LNG (liquefied natural gas) in their diesel engines (with pay-off times of just a few years due to high diesel costs). Cars can be switched over as well (particularly if the manufacturer makes them standard as occurred in Sweden when the government committed to natural gas cars for their vehicle fleet). The attraction is that natural gas is already in place in terms of infrastructure although actual filling stations are not commonplace.

The conversion to natural gas is an obvious step in places like Australia where there is a good supply of natural gas available. However in Europe and in the US this is not the case. Europe is going to faraway places in the east, such as Russia, to bring their gas and already some signs of an OPEC-like protection of the resource are developing. In the US natural gas has already peaked and officials are now looking to import it using LNG tankers – starting an overseas dependence similar to oil.

Global natural gas production has had similar estimates on its peak as oil production and they range from 2010 to 2030 with a little less certainty than for oil. The peak in discoveries occurred in the late 1960s to early 1970s so the same pattern as oil seems to be evident. It is not surprising that oil and natural gas patterns are parallel as they have similar geological origins in marine sediment (unlike coal which comes from ancient forests). In addition, oil and natural gas prices are closely linked so as oil goes up in price the same occurs for natural gas. Natural gas can only be a small part of the transitional arrangements for oil; it cannot be seen as the long-term replacement as it is also peaking. Moreover its use will need to be eventually phased out as part of our response to climate change. The benefit of the transition to natural gas is that it enables the long-term transition to hydrogen to be facilitated.

Biofuels have promised a lot but since they began being delivered they have become rather tarnished due to their impact on food prices when used to convert fuel from grain, and when some estimates suggested they may be worse than oil when it comes to climate change. However they still have a potentially significant role in some areas where there is surplus sugar for example, and eventually when the technology improves to make them from cellulose materials (agricultural and forestry waste) and from blue green algae. It is likely that biofuels will be used as a do-it-yourself fuel on farms. Thus biofuels may have a role in agricultural regions as a fuel to assist farmers in their production but as a widespread fuel for cities it is not an option that can yet be taken seriously.

Telepresence, high speed rail and airships

Transport to meet people by long distance or even short distance trips within cities may not be needed once the use of broadband-based telepresence begins to make high quality imaging feasible on a large scale. There will always be a need to meet face-to-face in creative meetings in cities, but for many routine meetings the role of computer-based meetings will rapidly take off.

Aircraft are not going to easily cope with the rapid rise in fuel. At the height of the 2008 fuel crisis there was panic amongst airlines as the price of fuel went to more than 50 per cent of the price of a ticket (Demerjian, 2008; McCartney, 2008). Gilbert and Perl (2007) suggest a few ways that air travel will adapt but mostly they see little of potential other than regional high speed rail and a return to ship travel.

Perhaps the technology that could make a come-back is airships. These are able to fly at low levels at speeds of 150–200kph and carry large loads with one tenth of the fuel of aircraft technology. They are already being used to carry large mining loads to remote areas and to take groups of 200 or so on eco-tourism ventures similar to a cruise ship.

Indigenous settlements going diesel-free

Remote settlements in Australia are under the spotlight due to serious health and social problems. The obvious lack of governance to enable decent services to these areas is likely to be overcome through federal and state commitments. But they must also begin to show how they can become diesel free as these settlements are highly vulnerable to price rises. Renewable power can be used in these settlements but not much is there yet for transport. These settlements need to be provided with upgraded road access to enable weekly services by 'bush bus' that can enable them to have reasonable access to regional towns. Fewer vehicles are likely to be the main response, however, to the global fuel crisis on these areas.

Conclusions

There are not many guidelines to the future of our cities and regions that take account of what could happen to transport in response to climate change and peak oil. It is understandable therefore why some people get very upset about the possibilities of such oil-vulnerable cities collapsing. As Lankshear and Cameron (2005, p10) say:

> Peak oil has already become a magnet for post-apocalyptic survivalists who are convinced that western society is on the brink of collapse, and have stocked up tinned food and ammunition for that coming day.

The alternatives require substantial commitment to change in both how we live and the technologies we use in our cities and regions. The need to begin the changes is now as they will take decades to get in place and the time to respond to peak oil and climate change is of the same order, probably less. But at least by imagining some of the changes as suggested above it is possible to see how we can get started on the road to more resilience and sustainability in our settlement transport systems which are so bound up in their land use patterns.

Notes

1 See a summary of the peak oil issue in Newman (2007).
2 Data are from Kenworthy and Laube (2001), which was a study of 100 cities (16 were incomplete) and 27 parameters using highly controlled processes to ensure comparability of data. See also Kenworthy et al (1999).
3 www.google.org/recharge/; http://energy smart.wordpress.com/2007/06/22/rollerblading-to-a-phev-future/

References

CTODRA (2004) 'Hidden in plain sight: Capturing the demand for housing near transit', Center for Transit Oriented Development and Reconnecting America, www.reconnectingamerica.org

Dawson, B. (2008) 'The new world of Value Transfer PPPs', *Infrastructure: Policy, Finance and Investment*, May, pp12–13

Demerjian, D. (2008) 'As fuel costs rise airlines can't make the math work', *Autopia*, 10 June, http://blog.wired.com/cars/2008/06/prices-for-jet.html, accessed 25 June 2009

Dittmar, H. and Ohland, G. (eds) (2004) *The New Transit Town*, Island Press, Washington DC

Gehl, J. (1987) *Life Between Buildings: Using Public Space*, translated by Jo Koch, Van Nostrand Reinhold, New York

Gehl, J. and Gemzøe, L. (2000) *New City Spaces*, The Danish Architectural Press, Copenhagen

Gehl, J., Gemzøe, L., Kirknaes, S. and Sondergaard, B. S. (2006) *New City Life*, The Danish Architectural Press, Copenhagen

Gilbert, R. and Perl, A. (2007) *Transport Revolutions: Making the Movement of People and Freight Work for the 21st Century*, Earthscan, London

Hargraves, C. and Smith, M. (2004) *The Natural Advantage of Nations*, Earthscan, London

Infrastructure Australia (2009) 'National Infrastructure Priorities', *Infrastructure Australia*, May, Sydney

Kenworthy, J., Laube, F., Newman, P., Barter, P., Raad, T., Poboon, C. and Guia, B. (1999) *An International Sourcebook of Automobile Dependence in Cities, 1960–1990*, University Press of Colorado, Boulder, CO

Kenworthy, J. and Laube, F. (2001) *The Millennium Cities Database for Sustainable Transport*, UITP, Brussels

Kintner-Meyer, M., Schneider, K. and Pratt, R. (2007) 'Impacts assessment of plug-in hybrid vehicles on electric utilities and regional US power grids, Part 1: Technical analysis', Pacific Northwest National Laboratory, US DoE, DE-AC05-76RL01830

Lankshear, D. and Cameron, N. (2005) 'Peak oil: A Christian response', *Zadok Perspectives*, vol 88, pp9–11

Lipman, B. (2006) *A Heavy Load: The Combined Housing and Transportation Burdens of Working Families*, Center for Housing Policy, Washington DC

McCartney, S. (2008) 'Flying stinks', *Wall Street Journal*, 10 June, pD1

Newman, P. (2007) 'Beyond peak oil: Will our cities collapse?', *Journal of Urban Technology*, vol 14, no 2, pp15–30

Newman, P. and Kenworthy, J. (1999) *Sustainability and Cities: Overcoming Automobile Dependence*, Island Press, Washington DC

Newman, P., Beatley, T. and Boyer, H. (2009) *Resilient Cities: Responding to Peak Oil and Climate Change*, Island Press, Washington DC

STPP (2005) 'Driven to spend: The impact of sprawl on household transportation expenses', Surface Transportation Policy Project, Washington DC

Urban Land Institute (2008) 'Emerging trends in real estate', www.uli.org, accessed 10 January 2009

7

Climate Change Vulnerability:
A New Threat to Poverty Alleviation in Developing Countries

Kirsten Halsnæs and Nethe Veje Laursen

Development and climate change

Climate impacts will affect regions and countries unevenly – the poorest countries will suffer most from the negative consequences of climate change because they are in areas most exposed to disasters and they experience extreme climate events already. On top of that, the poorest countries lack resources to cope with the damages after being affected or even to gather information on what can be done beforehand (Intergovernmental Panel on Climate Change (IPCC), 2007a). Adaptation to climate change is a process that makes societies able to cope with the uncertainties that lie in the future through appropriate adjustments and well planned development. Throughout history the climate has changed and populations and planners have adjusted accordingly and adapted gradually to seasonal changes and changes in extreme weather events. However, changes are now happening faster and extreme events are becoming stronger and more frequent. The exact economic costs of an increase in the frequency of extreme climate events are uncertain, but the

magnitude is large enough to have the potential to threaten development in many countries.

The development stress resulting from current trends is magnified by the increasing population pressure on scarce land resources that, together with land degradation, is creating a vicious circle in which poverty increases and the pressure on land resources goes up. Climate change therefore threatens to undermine sustainable development in these countries through the additional burden it lays on poverty eradication and other development goals.

Table 7.1 illustrates that climate events are already a major stress to development. Overall, the table shows that, on average, more than 200 million people have been affected by weather/climate related disasters annually, from 1990 to 2008.

The large developing countries, India and China, are most vulnerable in terms of actual numbers of people affected. Third is Bangladesh, where 148 million people have been affected by weather/climate related disasters in 1990–2008. Flooding is the single most destructive event in the majority of the countries, followed by droughts and storms. Only two countries from the developed world figure on the list, namely

Table 7.1 Top 20 countries with the highest number of people affected by climate related events 1990–2008

		Grand total (millions)	Drought	Earth-quake	Epidemic	Extreme temperature	Flood	Mass movement (wet and dry)	Storm	Other
1	China	2207.46	311.41	62.24	0.01	7.90	1481.63	0.02	344.22	0.03
2	India	904.45	351.18	5.56	0.35	0.00	512.15	0.22	31.95	0.04
3	Bangladesh	147.95	0.00	0.00	2.21	0.19	114.76	0.00	30.77	0.03
4	Philippines	70.20	2.85	1.97	0.01	0.00	7.55	0.29	56.16	1.36
5	Thailand	49.07	23.50	0.06	0.00	0.00	22.57	0.01	2.93	0.01
6	Kenya	44.61	35.70	0.00	6.87	0.00	2.04	0.00	0.00	0.00
7	Iran	40.84	37.00	1.37	0.00	0.00	2.30	0.00	0.17	0.00
8	Viet Nam	32.24	6.11	0.00	0.02	0.00	16.84	0.00	9.27	0.00
9	Ethiopia	31.04	29.09	0.00	0.05	0.00	1.89	0.00	0.00	0.01
10	Pakistan	29.41	2.20	1.25	0.02	0.00	23.98	0.00	1.97	0.00
11	United States	22.81	0.00	0.03	0.41	0.00	0.92	0.00	20.67	0.78
12	Malawi	19.86	18.25	0.00	0.05	0.00	1.56	0.00	0.00	0.00
13	Korea, Dem. Rep.	18.72	0.00	0.00	0.00	0.00	10.22	0.00	0.49	8.00
14	Cambodia	17.13	6.55	0.00	0.42	0.00	9.26	0.00	0.00	0.90
15	Sudan	16.54	11.36	0.01	0.08	0.00	2.49	0.00	0.00	2.60
16	Australia	15.67	7.00	0.01	0.00	4.60	0.07	0.00	3.94	0.05
17	South Africa	15.64	15.30	0.00	0.10	0.00	0.12	0.00	0.11	0.00
18	Mozambique	15.51	6.04	0.00	0.31	0.00	6.76	0.00	2.40	0.00
19	Zimbabwe	13.94	13.16	0.00	0.52	0.00	0.27	0.00	0.00	0.00
20	Brazil	13.88	12.00	0.00	0.85	0.00	0.85	0.01	0.15	0.01

Source: Based on data from EM-dat (www.emdat.be).

US and Australia. Most affected in terms of number of people touched by weather events, are Asian and African countries, and among those, many of the world's poorest countries.

Asia is very disaster prone and thus highly affected by climate change. Capacity to adapt is, however, increasing. In Bangladesh, for example, there has been success with early warning systems. Capacity is still low, however, and is constrained by the poor resource base (United Nations Framework Convention on Climate Change (UNFCCC), 2007).

Africa has many challenges: the continent is not only the poorest continent but also has some of the most variable climate in the world, with dramatic changes over seasons and decades. Africa lacks the skills, technology and financial means to adapt, and suffers from weak institutional structures which further compound the

problem of resources. As can be seen in Table 7.1, drought affects the greatest numbers in Africa. With ongoing climate change a serious shortage of water can be expected to lead to loss of crops and other resources and to conflicts over the rights to use the water as a consequence (Africa Partnership Forum, 2008).

The Millennium Development Goals: A policy framework

The Millennium Development Goals (MDGs), as agreed by the international community, form one of the key international frameworks for assessing development policies. They comprise eight goals to be achieved by 2015 drawn from the actions and targets included in the Millennium Declaration. The Declaration was

Table 7.2 Millennium Development Goals

Goal no.	Description of goal
1	Eradicate extreme poverty and hunger
2	Achieve universal primary education
3	Promote gender equality and empower women
4	Reduce child mortality
5	Improve maternal health
6	Combat HIV/Aids, malaria and infectious diseases
7	Ensure environmental sustainability
8	Develop a global partnership for development

Source: Adapted from www.un.org/millenniumgoals.

adopted by 189 nations, and signed by 147 heads of state and governments during the UN Millennium Summit in September 2000. The 8 goals are supported by 18 more specific targets and 48 indicators (see www.un.org/millenniumgoals). The MDGs are shown in Table 7.2.

The overall MDG target is formulated as goal number one: to eradicate extreme poverty and hunger. All the subsequent goals are related to achieving this goal. Goal seven is the only goal that directly addresses environmental issues like climate change. However, several other goals, including those that are related to health issues, water and food provision, are closely linked. The task of including climate change concerns in the development planning required to achieve these goals should not be underestimated.

Social vulnerability and development goals

In the context of the MDGs, reducing the number of poor people affected by extreme climate events and climate change will depend on reducing the vulnerability of people, settlements, regions and countries through adaptations to climate change. Vulnerability is an important concept when looking at poverty and the capacity to adapt to a changing climate. The literature currently addresses climate change impacts on societies in terms of social vulnerability. One way to understand the concept is through the IPCC definition which states that 'vulnerability is the degree to which a system is susceptible to, and unable to cope with, adverse effects (of climate change) including climate variability and extremes' (IPCC, 2007c, p6). The 'system' includes not only humans but also our surroundings, and the capacity this system has to minimize negative effects from climate change.

The assessment and management of vulnerability is complex and multidisciplinary and includes studies in development and poverty, public health, climate, geography, political ecology, and disaster and risk management. The focus on reducing vulnerability is especially relevant in relation to poverty eradication and the other MDGs. The MDGs try to address not only poverty but also vulnerability by focusing on different aspects of poverty: illiteracy, health, malnutrition, mortality and so on (Birkmann, 2006). Measuring the level, or the magnitude, of vulnerability depends on levels of poverty and income and on human, natural and social capital in a geographical and social context. Large-scale climate modelling or macro-economic studies cannot reflect these aspects properly, and local assessments are needed to decide the right initiatives that can help to reduce vulnerability.

Generally, people in rural areas are vulnerable because they are directly dependent on natural resources as a source of income. Climate change will affect the main source of income, leading to hunger, malnutrition and even to migration. Three quarters of the world's poor live in rural areas, making these areas important when focusing on poverty alleviation (World Resources Institute, 2008). However, the urban poor are also vulnerable to climate change, and migration to big cities means that urban areas are growing all the time. The quality and robustness of infrastructure and buildings play a crucial role in relation to increasing frequency and intensity of climate events, and is important when determining vulnerability levels. Furthermore, populations exposed to pollution will have health problems, and health problems could be worse combined with contaminated water and poor sanitation. The large cities in

Table 7.3 Millennium Development Goals and climate change impacts

MDG	Examples of links to climate change
Eradicate extreme poverty and hunger (Goal 1)	• Climate change may reduce poor people's livelihood assets, for example, their homes, infrastructure, access to water and healthcare. • Climate change may alter the path and rate of economic growth due to changes in natural systems and resources, infrastructure and labour productivity. A reduction in economic growth will directly influence poverty through the reduced income opportunities. Reduced income opportunities in the formal sector and traditional agriculture forces people to over-exploit the natural resources available. • Climate change may also alter regional food security.
Achieve universal primary education (Goal 2)	• Links to climate change are less direct but loss of livelihood assets may reduce opportunities for full time education in numerous ways such as school desertion for work and increase in time to reach school. • Natural disasters might damage schools or lead to school closure, and cause damage to access roads.
Promote gender equality and empower women (Goal 3)	• Woman are particularly vulnerable to climate induced diseases such as malaria when they are pregnant. • Woman often carry a relatively large burden in relation to provision of water and wood fuels and this can be more time consuming with decreases in resources due to climate change.

Health related goals:
  Combat major diseases
  Reduce infant mortality
  Improve maternal health
  (Goals 4,5 & 6)
	• Increase in heat related mortality and illness associated with heat waves.
	• Increase in the prevalence of some vector borne diseases (e.g. malaria, dengue fever) and water borne diseases.
	• Decline in quantity and quality of drinking water – a prerequisite for good health – i.e. through reduced natural resources productivity.

Ensure environmental sustainability (Goal 7)	• Climate change could alter the quality and productivity of natural resources and ecosystems, some of which may be irreversibly damaged, and these changes may also decrease biological diversity and compound existing environmental degradation.
Global partnerships (Goal 8)	• Climate change is a global issue and responses require global cooperation, especially to help developing countries adapt to the adverse impacts of climate change.

Source: Sperling (2003) (selected parts of table).

coastal zones and the poor living in unplanned urban settlements in these cities are disproportionately affected by climate change. It is important to focus on both urban and rural development planning. To focus, for instance, solely on infrastructure or water drainage might not be the optimal choice for the rural population which may urgently need to adapt their agricultural base to climate change to ensure food security.

Linking development goals and climate change

An exploration of the links between the MDGs and climate change was carried out by a number of UN agencies in the report *Poverty and Climate Change: Reducing the Vulnerability of the Poor through Adaptation* (Sperling, 2003).

Table 7.3 illustrates how the MDGs may be affected by climate change impacts. Adaptation has therefore to be integrated into development planning in order to reach the development goals. In order to address the challenges that climate change places on MDG 1, the focus should be on the development of infrastructure, water, natural resources, agriculture and food security and human settlements. Development planning in these sectors should be prepared in accordance with climate predictions and potential climate change risks. Actions to be taken into account could include avoidance of human settlements in exposed areas or adequate protection measures. Access to water should be secured for all population groups and settlements, and agricultural planning and advice is essential if conditions for growing certain types of crop suddenly disappear. Generally, sound natural resources management is increasingly necessary in the face of growing pressures.

The second MDG relates to primary education which will play an important role in teaching school children and their families what to do in case of disaster or in response to more gradual climate change which modifies their living conditions. At the same time, schools can be built to withstand disasters, which is particularly important in areas where they are also used as shelters.

Gender issues (MDG 3) are especially important in the poor countries where women often have all responsibilities related to the household such as finding wood fuel, water and food. Climate change could complicate this work through, for example, pressure on the productivity of woods and crops and contamination of water. A focus on natural resources and land use management should aim to reduce the possible effects.

Three MDGs relate to health. The goal is to combat major diseases, and reduce infant mortality and improve maternal health. Responses will include climate-proofed drainage systems and sealed water systems to reduce water-based contamination. MDG 7 on environment highlights the need for sustainable natural resources management to maintain fertile and productive soils. Also education in sustainable land use practices to avoid degradation can promote reduction of environmental damage.

The last MDG deals with international, global partnerships. This will mainly benefit the poor through strengthening the role of developing countries at international negotiations.

Overall, to reach the MDGs, disaster preparedness and planning has to be included in general planning. Droughts may lead to drying of wells; storms may lead to destruction of houses, bridges, latrines, roads and wells. As a consequence, people could be forced to migrate to other areas, putting extra pressure on existing infrastructure. Droughts and floods may also lead to crop loss, shortage of fodder and food, low price on livestock, and cut off market access for remote areas. Disaster damage could be substantially reduced by 'climate proofing' of buildings and infrastructure and designing water and sanitation systems based on climate change data so that they can withstand weather extremes. Similarly, agricultural coping strategies related to crops, livestock, insurance, extension services and improved information can be developed. Establishment of disaster warning and management systems is particularly important.

Vulnerability to climate change: Case examples

This section considers the detailed assessment of climate change vulnerability based on experiences that have been gained through the implementation of the Danish Climate and Development Action Programme (Danida, 2005), where studies were carried out in Bangladesh, Uganda and Ghana in 2007 and 2008 (Danida, 2007a; Danida, 2007b; Danida, 2008). The aim of conducting a climate screening of Danida programmes was to assess whether the activities included in the programmes were vulnerable to climate change. This was used as a basis for recommendations to ensure activities are 'climate proof'. In this context, climate proof means that the activity design has taken climate risks into consideration – that is, it has considered whether some adaptation options should be included, depending on the size of the risk relative to the costs.

Uganda

Uganda is landlocked and lies in East Africa. The country has status as a Least Developed Country, and has a poverty level of almost 40 per cent measured by the national poverty line (United Nations Development Programme (UNDP), 2007) The northern part of the country is poorest and least developed, with some areas experiencing an increase in the poverty incidence recently despite a general downwards trend in the country as a whole. The climate sensitivity of Uganda is closely related to its abundant water resources, including Lake Victoria and the Nile. Water covers almost 15 per cent of the country, but the water levels in Lake Victoria and other lakes and rivers show a very large seasonal and inter-annual variation. Furthermore, water is very unevenly distributed in Uganda. The water levels in Lake Victoria and rivers have a large influence on many people's livelihoods in terms of fishery options and water supply. Water is also a key economic resource in terms of hydropower potential and tourism (Ben, 2005).

With only about 13 per cent of the population living in urban areas, agriculture and rural development is very important. The agricultural sector employs four out of five Ugandans; hence many households depend solely on natural resources, agriculture and subsistence farming as their source of income (UNDP, 2007; Central Intelligence Agency, 2008). Production in the agricultural sector has experienced increased revenues lately. However, this is not due to productivity gains but rather expansion of land cultivation. This means that it will be difficult to sustain positive growth rates due to increasing demand for land. The limited land is becoming a problem especially for the poorest groups who live off small scale farming (Ghana Poverty Reduction Strategy, 2005; Mugyenyi et al, 2005; Uganda NAPA, 2007). Agriculture in Uganda is almost exclusively rain-fed, which means that Uganda is extremely vulnerable to rainfall variability. Various adaptation measures could help farmers to minimize the threat of losing harvests (Uganda NAPA, 2007). One of Uganda's main export commodities is coffee. Coffee is particu-

larly vulnerable to changes in climate such as temperature increase. If Uganda wants to reduce the risk of relying on income from the coffee harvest, it will need to either diversify income sources and export earnings by including other cash crops in production, or find strains of coffee that are suitable for a different climates (UNFCCC, 2002; Ben, 2005).

The 'cattle corridor', running from the north-east to south-west, is one of the most vulnerable areas of the country. There is no formalized land use structure in the cattle corridor, and the land is used by pastoralists that drive their herds around according to the availability of grazing. This poses a threat to fragile ecosystems that already suffer from frequent droughts and floods. Existing practices in this area encourage land degradation, leading to deforested highlands that can create mudslides and flooding in the low-lying areas (UNFCCC, 2002). Population growth and increasing demand for land has created serious conflicts between settlers and pastoralists in the cattle corridors and its surroundings due to unsettled land allocations, and these conflicts can be expected to increase with larger water variability due to climate change.

The low current urbanization level in Uganda could be turned into an opportunity for integrating climate change into development planning, since many large infrastructure projects have yet to come. Some of the important sectors in this context are roads, water resources, sanitation and agriculture. Uganda is still in a planning phase for establishing a national road system. The majority of existing and planned roads are made from gravel without any top sealing. This reflects the priority placed by the government on low construction costs and a large share of local manual labour, but has a number of negative side effects including high dust pollution in the nearby environment as well as the low reliability and short lifetime of the roads. Many roads are likely to be washed away during floods or can be damaged during storms, and the gravel roads are thus very vulnerable to climate change (Poverty Eradication Action Plan (PEAP), 2004). The Danida climate screening (Danida, 2007b) recommended assessment of the costs of climate proofing gravel roads by proper planning of drainage systems and sealing with, for example, bitumen.

Changes in water availability are considered to be a serious threat to development in Uganda because of its structural dependence on very abundant water resources. Water is used for key economic sectors like fisheries, agriculture, energy production and human livelihoods and is also a major means of transportation. Several climate change adaptation options could help to reduce the vulnerability of the water sector. These include management of small watersheds and construction of new boreholes in areas with decreasing groundwater tables (Ben, 2005). The link between clean water supply and sanitation becomes even more important with climate change since more frequent flooding increases the risk of water contamination from open sewage systems and pit latrines. An adaption option is to dimension the systems so that they can tolerate more variation in water levels and flooding. There is also a special need to manage increased seasonal water variability in relation to rain-fed agriculture. One option is to use more drought resistant crops and to implement water storage facilities such as rainwater harvesting. A change in agriculture may also consider changing cropping strategies in relation to, for example, coffee plantations, where large investment in plants can be at risk if temperature and precipitation patterns change. The challenge here is to find alternative crops with high value added and export possibilities.

Ghana

Ghana is one of the 'wealthier' poor countries and is hoping to move into the group of middle-income countries by 2015. It is expected that it could be the first African country that actually reaches the MDG 1 of halving its national poverty rate (Breisinger et al, 2008). Its impressive per capita GDP growth in recent years has been due to rich natural resources and the existence of export commodities, including gold and cocoa, that have experienced increasing prices. The general poverty level is decreasing,

and is at the moment around 28 per cent (Department for International Development (DFID), 2008). The incidence of poverty, however, remains high in rural areas and in northern Ghana. Thus, even with decreasing poverty rates, some rural areas are left out of the positive economic development and will have difficulties in meeting the MDG 1. Ghana has a relatively high degree of urbanization with about 50 per cent of the population living in urban areas, putting extra pressure on urban infrastructure and proper development (UNDP, 2007).

Despite the many urban settlers, subsistence farming is still widespread in the poorest rural areas and many livelihoods depend on this as their main source of income. In this context, it is important to encourage sustainable development and land use management, and to avoid a vicious circle where human activities such as deforestation, erosion of river banks, and other land degradation activities worsen living conditions and make livelihoods more vulnerable to climate change. The poorest areas are located in the north, which already experience very bad climatic conditions both in terms of temporary droughts and flooding events originating from short but very intensive precipitation events (UNFCCC, 2001). Despite some uncertainty in the detailed regional climate projections for West Africa, it is expected that the poorest northern areas in Ghana will also be most seriously affected by climate change, and up to 10 per cent decrease in annual precipitation is possible. In contrast, the more prosperous southern Ghana is expected to get increasing precipitation (Hulme et al, 2001; IPCC, 2007b).

In addition to agriculture, the key economic sectors of infrastructure and water are also vulnerable to climate risks. The vulnerability of infrastructure was clearly indicated during the 2007 flooding events in northern Ghana, when bridges, highways and local roads were seriously damaged with very negative impacts on access to emergency relief and market access (e.g. Reliefweb, 2007). Based on this experience, the climate proofing of infrastructure projects is highly recommended. This can include new designs of culverts and other drainage systems, and adding top sealings on rural gravel roads as explained in the case of Uganda. Improved infrastructure designs in some cases can be very costly, so very detailed climate risk assessment is recommended for large projects with a long lifetime. Such a risk assessment can include probability density functions for climate variability parameters (precipitation and floods) coupled with technical and economic design data which, for example, can be included in standard design manuals.

With climate change, there will be a need for improved water resource management in order to manage larger inter-annual variations and the expected decreased precipitation in the north. Planning of water use involves a broad menu of activities, including licensing schemes, demand side management, the location of wells, and balancing water use across different sectors including irrigation, industrial production and hydro power (UNFCCC, 2001; The Netherlands Climate Assistance Programme (NCAP) et al, 2007).

Bangladesh

Bangladesh is one of the poorest countries in the world, and ranks low on almost all measures of economic development. Twenty per cent of GDP comes from the very climate sensitive agricultural sector but almost two thirds of the population relies on agriculture as its main source of income. It is extremely vulnerable to natural disasters and climate change due to its geographical location in a low-lying delta with a long coastline, downstream to three large river systems. Most of the country is less than ten metres above sea level, and cyclones and storms often result in disasters and inundations over large parts of the country. Storm surges from the sea and increasing sea levels also result in salinity intrusion in the coastal area. Flooding from the rivers is already a major risk and it is a regular phenomenon that 30–70 per cent of the country is flooded during the summer monsoon. Recently, there has been a tendency for more frequent flooding of relatively large areas and for longer periods (Agrawala et al, 2003; Bangladesh NAPA, 2005). The population is therefore

extremely vulnerable to natural disasters and climate events. Sea level rise of one metre, for example, would mean that 11 per cent of the population would be directly threatened by inundation (Agrawala et al, 2003). Other key climate change threats in Bangladesh include contamination of fresh water resources in flooding events from poor sanitary systems, cholera epidemics, reduced fresh water availability, loss of agricultural crops due to droughts and flooding, and destruction of infrastructure and human livelihoods (Rahman and Alam, 2003).

The urban areas are threatened by climate change, especially since many settlements are in low-lying areas. Given that urbanization in Bangladesh is increasing, and the urban share of population has increased to 25 per cent, compared to 10 per cent 30 years ago, urban settlement development planning is urgently needed (UNDP, 2007). High population density throughout the country exacerbates the exposure to and effect of climate events, and increases the levels of vulnerability, along with poverty and the poor institutional development of the country (Agrawala et al, 2003).

Despite being one of the most climate-vulnerable countries in the world, Bangladesh is also characterized by very high awareness about disasters and climate change. This has led to a decrease in the number of people affected by large flooding events over the past decades due to improved disaster warning systems. However, Bangladesh needs a stronger adaptive capacity, and this will include a wide range of institutional, technical and financial measures (UNFCCC, 2007).

Not only is more than 40 per cent of Bangladesh's population settled in low elevation coastal zones but the entire country is dependent on activities taking place in the coastal zones (McGranahan et al, 2007). The most exposed and vulnerable settlements are often occupied by the poorest households. Coastal zones face numerous risks, including sea level rise, floods and storms, which can destroy livelihoods and income possibilities, erode banks and cause salinity intrusion. The vulnerability is enhanced by poor planning and land degradation related to settlements, fish ponds and agricultural activities.

Implementation of proper planning measures and land use regulations is a key component of climate change adaptation. This would include climate risk assessment for roads and other infrastructure, river bank strengthening and anti-erosion measures for beaches. Adaptation options for agriculture include the introduction of crops that are more tolerant to salinity intrusion, drought resistant rice, rainwater harvesting measures, and improved tube well designs that are adjusted to decreasing groundwater table and the risk of intrusion with contaminated water through flooding events. The training of officers facilitates the implementation of climate change coping measures by local farmers. This can include training in timing of planting and location of fields due to short term weather prognosis, improved crops, harvesting schedules and so on.

Conclusions

Climate change and natural disasters are emerging as serious additional stresses on development in some of the poorest parts of the world. Alleviating the negative impacts will require improved planning and various adaptation measures. The new climate reality requires the inclusion of climate change risk assessments when making important and costly decisions. This is especially so for investments with long time horizons. Many developing countries are in a process of building key infrastructure such as roads, railways and energy supply in both rural and urban settlements due to economic growth and population increase. This creates important opportunities for integrating climate adjusted risk in decision making and investments. However, many national planning activities are currently characterized by weaknesses in terms of lack of data, limited coordination between different agencies, and various governance and institutional issues related to property rights and enforcement. This will be complicated further by adding climate change as yet another planning issue with all its inherent uncertainties and its long term perspective. The challenge is both to include some short term response measures that

can mitigate the consequences of climate related disasters and increased variability, and to make more detailed climate risk assessment for larger investments with a long lifetime.

Climate vulnerability both reflects natural factors and social issues. Some developing countries, like Bangladesh, are physically vulnerable to climate change but on top of this, poverty related issues, such as land degradation, poor health conditions, weak sanitary systems, widespread subsistence agriculture and low education levels, increase climate vulnerability. The importance of similar social vulnerability factors can also be seen in African countries such as Uganda and Ghana. The physical climate impacts in the latter may be less than in Bangladesh, but their economies are even more dependent on climate sensitive sectors, like agriculture, than in the case of Bangladesh, and this could result in very serious climate change impacts.

As discussed earlier, there are many examples of potential appropriate adaptation measures. Given that there are similarities between the appropriate measures that can be used in different countries, there are important opportunities to share experiences and develop some common standards and guidelines (e.g. related to infrastructure and water supply systems) (Danida, 2005).

References

Agrawala, S., Ota, T., Ahmed, A. U., Smith, J. and van Aalst, M. (2003) 'Development and climate change in Bangladesh: Focus on coastal flooding and the Sundarbans', working paper, Organisation for Economic Co-operation and Development, Paris

Africa Partnership Forum (2008) 'Climate challenges to Africa: A call for action', www.oecd.org/dataoecd/46/47/40333574.pdf, accessed 1 October 2008

Bangladesh NAPA (2005) 'National Adaptation Programme of Action (NAPA)', Ministry of Environment and Forest, Government of the People's Republic of Bangladesh, http://unfccc.int/resource/docs/napa/ban01.pdf, accessed 1 October 2008

Ben, T. (2005) 'A content analysis report on climate change impacts, vulnerability and adaptation in Uganda', www.iied.org/pubs/pdfs/10011IIED.pdf, accessed 1 October 2008

Birkmann, J. (ed) (2006) *Measuring Vulnerability to Natural Hazards: Towards Disaster Resilient Societies*, United Nations University Press, Tokyo

Breisinger, C., Diao, X., Thurlow, J., Yu, B. and Kolavalli, S. (2008) 'Accelerating growth and structural transformation: Ghana's options for reaching middle-income country status', International Food Policy Research Institute Discussion Paper 00750

Central Intelligence Agency (2008) *CIA World Factbook*, www.cia.gov

Danida (2005) 'Danish Climate and Development Action Programme', Ministry of Foreign Affairs of Denmark, Danida, www.netpublikationer.dk/um/5736/pdf/samlet.pdf, accessed 1 October 2008

Danida (2007a) 'Climate screening Bangladesh', Ministry of Foreign Affairs of Denmark (not publicly available)

Danida (2007b) 'Climate screening Uganda' Ministry of Foreign Affairs of Denmark (not publicly available)

Danida (2008) 'Climate change screening of Danish cooperation with Ghana' Ministry of Foreign Affairs of Denmark (not publicly available)

DFID (2008) 'Ghana Factsheet', www.dfid.gov.uk/pubs/files/ghana-factsheet.pdf, accessed 1 October 2008

Ghana Poverty Reduction Strategy (2005) 'The Republic of Ghana: Growth And Poverty Reduction Strategy – GPRSII 2006–2009', http://siteresources.worldbank.org/INTPRS1/Resources/GhanaCostingofGPRS_2(Nov-2005).pdf, accessed 1 October 2008

Hulme, M., Doherty, R., Ngara, T., New, M. and Lister, D. (2001) 'African climate change: 1900–2100', *Climate Research*, vol 17, pp145–168

IPCC (2007a) 'Climate Change 2007: Synthesis report', contribution of Working Groups I, II and III to the Fourth Assessment Report of the Intergovernmental Panel on Climate Change, core writing team, R. K. Pachauri and A. Reisinger (eds), IPCC, Geneva, Switzerland, www.ipcc.ch/ipccreports/ar4-syr.htm, accessed 1 October 2008

IPCC (2007b) 'Climate Change 2007: The physical science basis', contribution of Working Group I to the Fourth Assessment Report of the Intergovernmental Panel on Climate Change, Chapter 11, S. Solomon, D. Qin, M. Manning, Z. Chen, M. Marquis, K. B. Averyt, M. Tignor and H.

L. Miller (eds), Cambridge University Press, Cambridge, UK and New York, USA. www.ipcc.ch/pdf/assessment-report/ar4/wg1/ar4-wg1-chapter11.pdf, accessed 22 January 2009

IPCC (2007c) 'Climate Change 2007: Impacts, adaptation and vulnerability', contribution of Working Group II to the Fourth Assessment Report of the Intergovernmental Panel on Climate Change, M. L. Parry, O. F. Canziani, J. P. Palutikof, P. J. van der Linden and C. E. Hanson (eds), Cambridge University Press, Cambridge, UK, www.ipcc.ch/ipccreports/ar4-wg2.htm, accessed 22 January 2009

McGranahan, G., Balk, D. and Anderson, B. (2007) 'The rising tide: Assessing the risks of climate change and human settlements in low elevation coastal zones', *Environment and Urbanization*, vol 19, no 1, pp17–37

Mugyenyi, O., Tumushabe, G. and Waldman, L. (2005) 'My voice is also there: The integration of environmental and natural resources into the Uganda Poverty Eradication and Action Plan', study initiated under the Poverty and Environment Partnership (PEP), and jointly funded and managed by Canadian International Development Agency (CIDA), DFID and German Society for Technical Cooperation (GTZ)

NCAP, Environmental Development Action, United Nations Organization for Education, Science, Culture and Communications and Stockholm Environment Institute (2007), 'Climate change adaptation and water resources management in West Africa', Synthesis report – writeshop 21–24 February 2007, available at www.nlcap.net/fileadmin/NCAP/News/032135.070920.NCAP_West_Africa_Climate_Change_Writeshop_Report_English_Version.v1.pdf

PEAP (2004) Poverty Eradication Action Plan 2004/05–2007/08, Ministry of Finance, Planning and Economic Development, Uganda

Rahman, A. and Alam, M. (2003) 'Mainstreaming data adaptation to climate change in least developed countries (LDCs)', Working Paper 2: Bangladesh Country Case Study, Bangladesh Centre for Advanced Studies, www.iied.org/pubs/pdfs/10003IIED.pdf, accessed 15 November 2008

Sperling, F. (ed) (2003) 'Poverty and climate change: Reducing the vulnerability of the poor through adaptation', report for AfDB, ADB, DFID, EC DG Development, BMZ, DGIS, OECD, UNDP, UNEP and the World Bank, Washington DC

Uganda NAPA (2007) 'Climate change: Uganda National Adaptation Programme of Action',

Ministry of Tourism, Environment and Natural Resources, http://unfccc.int/resource/docs/ napa/uga01.pdf, accessed 1 October 2008

UNDP (2007) 'Fighting climate change: Human solidarity in a divided world', Human Development Report 2007/2008, United Nations Development Programme, available at http://hdr.undp.org/en/media/HDR_20072008_ EN_Complete.pdf

UNFCCC (2001) Ghana's National Communication to UNFCCC, available at http://unfccc.int/resource/docs/natc/ghanc1.pdf

UNFCCC (2002) Uganda's National Communication to UNFCCC, available at http://unfccc.int/resource/docs/natc/uganc1.pdf

UNFCCC (2007) 'Climate change: Impacts, vulnerabilities and adaptation in developing countries', available at http://unfccc.int/files/essential_ background/background_publications_htmlpdf/ application/txt/pub_07_impacts.pdf

World Resources Institute (WRI) in collaboration with United Nations Development Programme, United Nations Environment Programme and World Bank (2008) *World Resources 2008: Roots of Resilience – Growing the Wealth of the Poor*, WRI, Washington, DC, http://pdf.wri.org/world_ resources_2008_roots_of_resilience.pdf, accessed 22 January 2009

8

Climate Change Vulnerability:
Planning Challenges for Small Islands

Thanasis Kizos, Ioannis Spilanis and Abid Mehmood

Introduction

What will happen to the small islands under the climate change scenarios currently envisaged? Despite the fact that island societies are small contributors to climate change due to their restricted populations and their production and consumption patterns, 'they will suffer dispro-portionately from the damaging impacts of climate change' (United Nations Framework Convention on Climate Change (UNFCCC), 2007, p7). This vulnerability is largely the result of insularity or 'islandness' (Baldacchino, 2004, p272), their small size, remoteness and low acces-sibility, combined with unique and fragile natural and cultural environments. Due to their small size and the subsequent lack of natural resources (Tompkins et al, 2005), economies of scale are unattainable and therefore competitiveness in the world markets very low. At the same time, their economies – just like other peripheral and coastal areas – are usually characterized by mono-activity based on the exploitation of natural resources (e.g. agriculture, fishery, mining and tourism) that are excessively dependent on international trade. The issues related to their remoteness, adaptive capacity and accessibility increase the operational cost for enterprises,

households and governance (administration and infrastructure) as well as for mitigation and adaptation measures. However, with their well-preserved local assets, customs and practices, these small islands have the potential to provide the exemplars of sustainability and endurance in the wake of a changing climate.

In this chapter, we discuss the various charac-teristics of small islands, especially those features that render them amongst the areas most suscep-tible to climate change. These vulnerability issues are discussed first within the context of small islands. The specific case of the Aegean Islands in Greece is then set out in detail to demonstrate the urgent need for proactive spatial planning, along with suggestions for further action.

The global situation of small islands

Islands face a number of difficulties within contemporary global and national relations. Their socio-economic and political status is mixed: some form independent states of one or more islands (also termed as Small Island Developing States (SIDS)); others are autonomous or administrative regions; and a third group comprises the parts of nearby or far away continental states. In all cases, they may appear to have a limited role in the global social, economic, cultural and political arena, as discussed below.

A large body of literature has been developed for the SIDS, which are explicitly identified in Agenda 21 chapter 17 (UN, 1992) as particularly vulnerable areas that have to be managed in an integrated way in order to achieve global sustainability goals. The Barbados Conference in 1994, and subsequently the Mauritius Declaration (2005), followed by the Programme of Action for the Sustainable Development of Small Island Developing States (www.sidsnet.org/) have highlighted the potential significance of climate change impacts. Data collection systems and methods are also developed in order to assess impacts and propose effective policies (Tompkins et al, 2005; Gilman et al, 2006; UNFCCC, 2007). Programmes such as the South Pacific Sea Level and Climate Monitoring project (SPSLCM) and Caribbean Planning for Adaptation to Climate Change (CPACC) have created monitoring and observation networks for Pacific SIDS and Caribbean SIDS, respectively.

As regards the second group of the autonomous or administrative island regions, Chapter 16 of the Intergovernmental Panel on Climate Change (IPCC) Working Group II's Report on 'Impacts, adaptation and vulnerability' (see Mimura et al, 2007) focuses on their vulnerabilities to climate change impacts. Particular focus in the report is on policy implications and adaptive measures to sea-level rise in the 'autonomous small islands predominantly located in the tropical and sub-tropical regions' (Mimura et al, 2007, p690). Major vulnerabilities identified in the report include: sea-level rise and its effects on infrastructure; lower precipitation leading to limited fresh-water resources, as forecast by the IPCC Special Report on Emissions Scenarios (IPCC, 2000); varying degrees of effects on natural systems (flora and fauna) and displacement of species; effects on local agriculture (food security), tourism and human health. The report acknowledges the low adaptive capacity of island systems, discusses

opportunities and constraints from the examples of SIDS, and recommends a number of integrated measures for adaptation and capacity building by means of public engagement and traditional local knowledge.

For small islands, archipelagos and regions that form parts of the EU member states, there have been efforts under way by various think tanks and regional networks (e.g. Islands Commission of the Conference of the Peripheral and Maritime Regions of Europe (CPMR)) to raise policy-level recognition of the specific attributes of small islands that should be taken into account at the national and European scales. Today, climate change is considered as one of the main external factors (along with globalization) in discussion of European islands' policy. However, these discussions primarily relate to consideration of impacts on territorial cohesion and balanced development in the insular regions, rather than specifically focusing on appropriate mitigation and adaptation measures.

In the Mediterranean region, the Mediterranean Action Plan (MAP), sponsored by the United Nations Environmental Programme (UNEP), has put climate change as one of the seven essential issues in its Strategy for Sustainable Development, endorsed by the 21 Contracting Parties in 2005 (MAP, 2005). While sea level rise is viewed as a major threat, the rise in temperature will exacerbate problems such as lack of water, reduction of wetlands area, invasion of new species and migration or extinction of existing ones, desertification and loss of agricultural productivity. These impacts, combined with the growing population pressure from both sides of the Mediterranean Sea, lead to continued degradation of the environment (Benoit and Comeau, 2005). Different policy measures have been proposed, promoting Integrated Coastal Zone Management and efficient use of energy, water and renewable resources, in order to reduce growing environmental pressures. All these measures focus on the issues of high vulnerability of small islands and the low adaptive capacities of their resources.

Vulnerability and adaptive capacity of small islands

As mentioned in Chapter 1, vulnerability is a function of both exposure and sensitivity. Islands are among the most vulnerable places, and hence, have to develop 'mitigation-friendly' adaptive measures to become resilient to the impacts of climate change. Furthermore, as Halsnæs and Laursen (Chapter 7) argue, vulnerability is both a social and a development issue. As regards small islands, this relates to the peripherality and marginality from the mainland areas. In this respect, vulnerability does not remain invariable for all islands: size, morphology and geographical location differentiate the impacts. For example, under all projections of sea-level rise scenarios (Chapter 18), small and low-lying islands could see large parts of their coastlines submerged by sea and lose a significant part of their resources and coastal developments. Islands located in tropical zones are more likely to suffer from frequent and vigorous tropical cyclones and hurricanes, droughts and desertification, threatening human and ecosystem safety and making sustainable development difficult. Tompkins et al

(2005) have provided a number of vulnerability assessment indicators and tools (referring to agriculture, biodiversity, economy, natural resources and public health) along with examples and suggestions for adaptive measures on small islands (see Table 8.3). However, their proposed adaptation strategies give relatively less attention to mitigation, probably on the grounds that not only do greenhouse gas (GHG) emissions from small islands have relatively minor impacts on climate change but that mitigation may also mean cutting energy use in construction and transport, the two sectors that underpin mass tourism.

UNEP has recorded some characteristics of the 2000 most important islands of the world (islands.unep.ch). This figure is smaller than the actual number, but since national definitions also vary there is no definitive number of islands and the criteria against which the relative 'importance' is measured (e.g. in Greece UNEP records a total of 36 islands, while the number of inhabited islands in 2001 was 112). Fifty-two per cent of the islands that were recorded by UNEP are found in the Pacific Ocean, where the smaller ones in size are also located (median size of 136.9km^2). Average altitude is lower for Arctic

Table 8.1 Altitude classes for islands according to the UNEP islands' database

	Total N	Total with altitude data		Altitude classes %				
		N	%	< 50m (N = 127; 10.1% of total)	50–100m (N=72; 5.8% of total)	100–500m (N=397; 31.7% of total)	500–1000m (N=358; 28.6 of total)	> 1000m (N=298; 23.8 of total)
Pacific	1038	639	51.0	52.8	37.5	46.1	54.2	56.4
Atlantic	378	225	18.0	22.8	19.4	16.6	15.9	19.8
Indian	218	115	9.2	9.4	18.1	12.1	7.5	5.0
Arctic	170	135	10.8	10.2	16.7	14.9	8.9	6.4
Mediterranean	88	71	5.7	0.8	0.0	5.0	8.9	6.0
Southern Antarctic	79	45	3.6	0.0	0.0	3.3	4.5	5.4
Baltic	21	18	1.4	3.9	8.3	1.8	0.0	0.0
Rest*	8	4	0.3	0.0	0.0	0.3	0.0	1.0
Total	2000	1252	100	100	100	100	100	100

Note: * The category 'Rest' includes 8 islands that are classified from UNEP as being parts of the 'borders' between oceans (Atlantic/Pacific, Atlantic/Arctic, Indian/Pacific and Pacific/Arctic).

Source: islands.unpe.ch, processed by the authors.

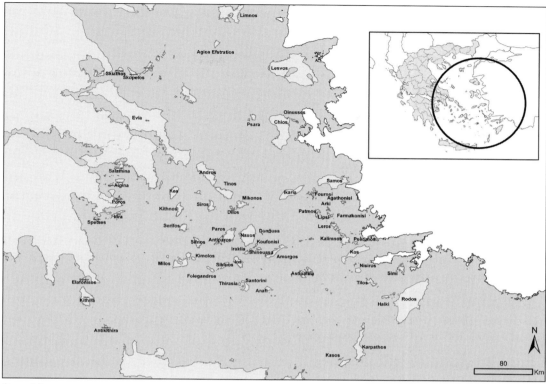

Source: Authors.

Figure 8.1 Islands of the North and South Aegean Regions, Greece

Islands followed by the ones in the Atlantic and the Pacific. A sea level rise would put many islands at direct risk (e.g. 15.9 per cent are lower than 100m, of which 10.1 per cent are lower than 50m, with a median size of 4km² in the Pacific, Table 8.1).

In a study of the European Islands System of Links and Exchanges (EURISLES) network it is explicitly mentioned that Greek islands will be particularly exposed to risk from sea level rise (EURISLES, 2002, p50). The following section presents an illustration of planning challenges for mitigation and adaptation measures in the Aegean archipelago in Greece. These islands are of different sizes (from the very small to relatively large) with varying degrees of accessibility and development. They serve as a remarkable opportunity to examine planning responses at the national, regional and local levels for mitigating and adapting to climate change in island contexts.

The Aegean Islands: An overview

Geographically, the Aegean Islands in Greece are a complex of 2800 islands (with a further 253 in Turkey) in a space defined by the Island of Crete in the south, continental Greece in the north and west and continental Turkey in the east, in total 210,240km². Administratively, the 112 inhabited Greek islands are in 4 insular (i.e. including only islands) regions: the Crete, North and South Aegean regions and the Ionian Islands, while some more islands are parts of continental Regions. The importance of islands within Greece (18 per cent of the territory and 13 per cent of the population), their diversity concerning size and level of economic development, the limited availability of data at the island level, and the administrative complexity indicate the difficulty of elaborating and implementing effective sectoral and territorial policy addressing economic, social and environmental issues. The 53 inhabited islands in the North and South Aegean Regions form the case study in this chapter.

The islands' climate is typical Mediterranean, characterized by dry and hot summers and short rainy winters, with major differences in seasonal precipitation among localities, namely more arid ecosystems, less forest and more savannas from north to south (Grove and Rackham, 2002). Vegetation is also Mediterranean and consists of sclerophyllous, evergreen flora forming mixed forests of maquis, phrygana and pine–oak forests (Allen, 2001).

The population of these islands (507,393 people in 2001) had expanded until the 1950s (978,339 people in 1951), but dropped significantly until the 1990s as a result of economic decline, with 41 of the 53 islands losing population (−25 per cent on average with 11 cases over −50 per cent). In the 1990s population either remained stable at 1991 levels, or slightly increased (only six islands lost population again), but this increase offered only partial compensation for the losses of the previous 40 years for most islands. At present, the island societies are ageing (18.8 per cent of population were over 65 in 2001 compared with the national average of 16 per cent), natural growth is negative and immigration trends are positive. Therefore, population stability can be largely attributed to immigration (especially foreign workers and Greek pensioners).

Despite common perceptions, the Aegean economy is based more on tourism and agriculture than on fisheries. Although agriculture has declined in the last decades, it still remains important.[1] However, the decline has significantly affected the overall land use patterns (see Kizos et al, 2007 for a more detailed analysis). Until the 1950s to 1960s, production was principally oriented towards self-sufficiency, with diversification of production and land uses, storage of raw or processed products and distribution to markets lowering risks and ensuring strong connection with markets in the dense communication networks of the area (Horden and Purcell, 2000). These features have generated

characteristic landscape elements such as terraces, drystone walls, footpaths, traditional storehouses, windmills and water mills in the Aegean Islands. In general, the islands have limited fertile and flat arable areas as well as resource availability (especially irrigation water).

Organic agriculture (especially for permanent crops such as olives and lately for vegetables) and animal husbandry are increasing. Organic production does not involve less irrigation or lower grazing densities and therefore does not resolve vulnerability issues of water scarcity, soil degradation and erosion. Aquaculture has developed in the last two decades and is now a very dynamic and exporting sector for the whole Greek economy, with the Aegean Islands accounting for about 20 per cent of national activity.

Tourism is the most important activity in the majority of islands, balancing economic decline and population loss after the 1950s in some islands and affecting almost all with development pressures. In some, tourism (including related activities in commerce, restaurants, entertainment and transportation services) represent more than 50 per cent of GDP and employment, as well as of energy and water consumption. A rich variety of localities, settings, accessibility and tourism development levels form the basis of the regional tourism industry. However, tourism is unequally developed both temporally and spatially (Spilanis and Vayanni, 2004). Tourists (approximately 3.5 million per annum) travel mainly if not exclusively in summer − most of the times with chartered flights (67 per cent in 2001). This increases vulnerability in terms of socio-economic activity with intense seasonal changes in transportation frequency and environmental pressure. Spatially, most of the hotel beds (250,000 in total) are found on a small number of islands: 44 per cent are found on Rhodes and Kos, and more than 65 per cent are found on six islands. The numbers of nights that tourists stay (more than 25 million in total) are even more unequally distributed, with 49 per cent on Rhodes and 23 per cent on Kos.

In addition to tourists, the presence of 'vacationers' (i.e. people who own houses in the islands but do not live there all year round) is

Table 8.2 Number of houses, new houses and their changes for Aegean Islands Prefectures

	Number of houses change %		Number of houses (2001)	New houses change %		New houses (2002)
	1961 to 1981	1981 to 2001		1996 to 2002	2000 to 2002	
Greece	72.5	37	5,476,162	47.9	43.5	128,297
Athens	140.7	26.5	1,529,998	35.1	20	26,177
Aegean Islands	21.5	40.1	330,697	51.3	56.6	8,980

Source: Greek National Statistics Service, processed by the authors.

very important in economic and social terms, but generates some of the most intense land use and landscape changes and environmental pressures (water and energy consumption, waste production). The local economy has benefited greatly from construction and associated activities.

Another key concern in these islands is housing developments. Land tenure and speculations can result in economic and social vulnerability on small islands. Data on new houses built on the Aegean Islands reveal that, with the exception of the Prefecture of Lesvos, the last 20 years have been a period of rapid house construction, at a greater rate than in Athens (Table 8.2). This has been accompanied by constant rise of land and house prices (e.g. prices for houses on Mykonos and Santorini are currently the highest in Greece). It is worth mentioning that local taxes are based on the size of area/land covered by the house rather than the value of the building itself, while construction and value-added taxes (VAT) are collected by the central government. Such tax regime encourages sprawl, and the resulting new constructions threaten landscape character in the areas, raise land prices and turn all pieces of land into potential building plots. This simultaneously raises concerns about the vulnerable local natural resources and habitats. However, it remains such a powerful driving force that all restrictions to individual building permits largely end up in illegal construction, which is becoming a major problem throughout Greece, but mainly in the coastal zones.

The most important difference between tourists and vacationers arises from the demand

for buildings, as vacation development requires more infrastructure and space. This fact puts additional pressure on the resources of the area, particularly fresh-water resources.

The consumption patterns of visitors have a double impact on the islands: a direct one from their own behaviour (high mobility; preference for fast ships and short stays; use of airplane, private car and air-conditioning; consumption of imported food and beverages; use of swimming pools; high water consumption; demand for big houses etc.) and an indirect one as they transform the perceptions, expectations and behaviour of the local population.

Conservation of natural resources on the Aegean Islands is based around the NATURA 2000 network that has been slowly developing since the late 1990s. Many rare and endemic species and specific habitats are found on Aegean Islands in a significant number of sites (15 per cent of the total, 28 sites of roughly 50,000ha with 9 more sites in a second catalogue, www.minenv.gr). Although the actual management plans are not realized yet, it is used as a means of pressure for the protection of the environment. Many locals see protection as barriers to 'development', especially in relation to building permits. So far, only two organized institutionalized efforts have been developed on islands in marine protected areas: one on the Ionian Islands (Zakynthos) for the protection of the sea turtle (*Caretta caretta*) and another on the Aegean Islands (Alonissos) for the protection of the monk seal (*Monachus monachus*). Their implementation was not without problems as the competencies of the Authorities of Natural Parks are not clear enough and financing from central

Table 8.3 Type of climate change phenomena and their expected impacts for the Aegean Islands

Type of climate changes	Impacts
Increase of frequency of extreme events such as heat stress, drought or flood conditions Changes in precipitation and storminess Increased evaporation rates	Diminution of fresh water availability Increase of fires, runoff and soil erosion Increase of energy demand Destruction of man-made capital Changes in habitats and species Changes in agriculture and tourism activities
Sea level rise	Coastal erosion Change in shoreline Loss of beaches Coastal lands inundated Inundation of wetlands Destruction of human settlements, tourism investments and infrastructures Salinization of coastal aquifers and diminution of fresh water
Rise of sea temperature	Increase of sea's acidity Loss of sea-grass beds Changes in marine habitats and species Structural changes in the fisheries and aquaculture sector

Source: Based on Tompkins et al (2005).

government is not secured. Efforts of individuals and NGOs to protect certain areas have met fierce resistance and limited results. Such efforts often aim to block harmful practices of different public (national and local) and private actors, rather than planning initiatives for the protection or conservation of resources.

Other important environmental issues on the islands include water and waste management. Seasonal tourism demand makes the problems worse; as it is in the summer when water availability is naturally low. In many islands, water has to be delivered by special ships in the summer. In some islands desalinization plants have been operating with conventional energy sources. A recent pilot project by the University of the Aegean desalinates sea water on an off-shore (floating) platform using wind power; wind power is also used in a recently constructed installation on Milos Island producing 2000m^3 per day.

Energy is another major issue as demand increases annually. Growing tourism and second home activity as well as air-conditioning use are the main reasons for the increase in demand. Electricity is produced in 27 small or medium size inefficient generation units using fossil fuels, producing power of high cost that is subsidized in order to keep the same price as on the mainland. Few islands are connected to the continental network and some islands have interconnections. The constant rise in demand repeatedly raises the issue of building new units or increasing the production potential of existing ones. Permanent connection via underground cables is facing economic, social and environmental problems. Renewable energy production is for the moment restricted to solar domestic water heating and small wind farms. Talk of solar and wind power developments have been delayed as the national plan for the spatial allocation of the units that will produce renewable energy production is still under discussion. Many locals and most NGOs react against the proposed development of wind farms with huge turbines (150m high) relative to the scale of the islands.

The question of a low carbon development trajectory for the Aegean Islands is still open to debate and tangled up with national energy plans: on the one hand, it seems that the renewable or 'clean' energy sources (solar, wave and wind power) are ideal choices for most of the islands, especially smaller ones, as they can easily be developed in small scale. On the other hand, if islands are not linked with the national network, 'conventional' energy power plants are necessary to complement all types of autonomous systems. The situation in Greece in general does not provide a basis for much optimism, as in recent years the energy supply has been based on imported power during the summer when demand is at peak. Moreover, the power plants planned for the near future would still be using lignite (low grade coal) as fuel.

Climatic vulnerabilities for the Aegean Islands

Three questions emerge: (a) What kind of vulnerabilities have been observed on the Aegean Islands? (b) What impacts are expected as a result of these vulnerabilities? (c) What kind of planning measures have to be implemented?

The main phenomena related to climate change on Aegean Islands (a mixture of slow onset changes and sudden extreme events, as described by Tompkins et al, 2005) and their expected impacts are summarized in Table 8.3. Although these phenomena have not been the subject of specific assessment at the level of Greece or that of the Aegean Islands, we will try to provide some evidence below.

As evidence of climate change, water scarcity is already an important issue as demand is growing and heat stress and droughts are becoming more frequent. In some islands precipitation has decreased as much as 25 per cent in the last ten years compared to the last century's average, while the salinization of local underground aquifers is becoming intense. Also, exotic species

of algae and fishes are migrating into the Aegean from warmer seas. These vulnerabilities, along with the impacts identified in Table 8.3, will not only affect the capacity of islands to achieve sustainability and development goals (such as higher GDP, lower unemployment, and population well-being), but also endanger their viability.[3] Since in most of the islands the majority of tourism infrastructure and activities are settled in coastal zones it is not hard to envisage a situation of islands that will have difficulty pursuing tourism activity if scenarios of sea level rise materialize.[4]

What would the content of mitigation and adaptation strategies be in this context? In terms of general mitigation measures, even if the small islands by themselves may not appear to be able to significantly diminish overall carbon emissions (the economic activity of Aegean Islands is less than 4 per cent of the Greek GDP), they can contribute with their own (limited) forces to achieve this goal, mainly by reducing the intensity of energy consumption per product unit (e.g. per night spent) and by replacing fossil fuel uses by renewable sources. Although availing of the public/private investments in such projects remains a challenge for small islands. The main potential for reducing GHGs lies in:

- changing consumption patterns for both tourists and permanent inhabitants (e.g. countering the increases in travel frequencies, transport activity and energy consumption per km by faster sea vessels);
- increasing energy efficiency of houses, both for tourism and private use. More and more hotels are now investing in reducing energy consumption (e.g. use of low energy lamps, interruption of electricity in a vacant room, interruption of air-conditioning when a window is open, etc.) but there is space for improvement;
- substituting conventional with renewable energy sources (such as wave-action and wind-power, etc.) via local and private initiatives.

There are also general options for the adaptation measures, such as to take action in order to reduce stress on the resources identified in Table 8.3 that are going to be most affected (i.e. fresh water, beaches, habitats and soil). Islands can invest in vulnerability reduction, as this is the main option to reduce the damage caused by environmental hazards. The causes of vulnerability are closely linked to an island's social, economic and geophysical characteristics (Tompkins et al, 2005) and to their development pattern.

The weakness of planning policy at the national level appears to have turned almost every piece of land in the Aegean region into prospective real estate. There is no overall planning or zoning that directs or constrains house building, except for some restrictions on NATURA 2000 sites that meet bitter local resentment. Even agricultural land can be transformed into housing development, as Greek legislation allows development of parcels of cultivated land of at least 0.4ha. Larger fields are divided and sold, bringing large earnings to ex-farmers. This creates further demand to expand public infrastructure to service scattered development. The national land use plan does not consider putting restrictions on building; on the contrary, it promotes huge condo hotels and golf resorts (Ministry of Environment, Spatial Planning and Public Works, 2008)

At the same time, there are no restrictions on the type of houses that are built, except for some apparent regulations in settlements that are characterized as 'traditional'. Therefore, many of the buildings are far from sustainable: swimming pools are allowed with no restrictions even on islands with water scarcity instead of imposing the construction of cisterns for rainwater collection; there are no strict rules for energy use and new buildings tend to consume more energy than older ones for heating and especially cooling.

Water policy is another example. The overall state of the water resources is not audited or monitored. Even in cases where its quantity is not good, the response is not to attempt to reduce the demand, but to increase its supply. As surface water is lacking, drills reach deeper and deeper aquifers of decreasing quality. The construction of new dams, reservoirs and desali-

nation plants are proposed as the only solution in order to deal with supply limitations. There are no concrete measures to reduce water consumption or to reduce the pressure on the carrying capacity of the islands.

This development approach increases vulnerability of islands in two ways. Firstly, it increases pressure on natural resources through the combination of unsustainable consumption patterns and climate change trends. This threatens, for example, irreversible impacts of water and biodiversity resources. Secondly, it places tourism – the most dynamic and competitive and often the only important activity for many islands – under threat by eroding a significant proportion of its assets (i.e. beaches, landscape, flora and fauna, a part of infrastructure built on shoreline). At the same time, the costs of the inputs that are necessary for tourism production (i.e. water, energy, food, transport, etc.) are rising. There is, therefore, a need for a more sustainable development path.

We maintain that this new sustainable development path has to be based upon two basic principles:

1 Qualitative versus quantitative and low cost production Islands do not possess the resources necessary to sustain low cost and large-scale production without placing the overall system under stress. The formulation of policies that focus on the expansion of mass tourism and residential houses are unsustainable both locally and globally.

2 Proactive versus reactive policies[5] *in order to minimize risks* The preparation of realistic and applicable adaptation and mitigation strategies requires visionary implementation plans based on public engagement (Chapter 23). As Tompkins et al (2005, p52) stress, 'clear trade-offs [...] have to be made between minimising the cost of adapting to climate change, minimising the risk of damages occurring, and ensuring that local voices are heard in the decision making process, so that local views and values can be taken into account'.

The components of a spatial planning framework for the Aegean Islands in terms of mitigation and adaptation measures, that can address these challenges, are likely to be based on the following key areas:

• Increasing the social responsibility of the population and of economic operators. In particular, local populations have to be persuaded to recognize the limitations of the current model and to invest in the opportunities offered by the alternatives.
• Adaptive land and coastal planning, with the use of participative procedures, to protect natural resources and avoid human exposure to high risks (e.g. extreme events).
• Increased environmental efficiency of households and of the public and private sector.
• Formalized marine reserves to protect marine fauna and flora from pollution and over-use from human induced activities (e.g. aquaculture, fishing, maritime transport, yachting, etc.).
• Valorizing local natural and cultural resources to create high value-added tourism. As mass 3S (Sea, Sun and Sand) tourism efficiency is diminishing, it has to be substituted by other products incorporating sustainability principles (ESPON, 2006).
• The planning, implementation and institutionalization of monitoring schemes that will be used for evaluating current policies and planning for future ones.

Conclusion

Islands represent a particularly vulnerable type of territory. This chapter demonstrates that vulnerability is as much a function of socio-economic and institutional characteristics as it is of physical features. Place-centred solutions must recognize this vulnerability and seek to reduce it in tandem with mitigation actions. These solutions have to recognize that mitigation will play a relatively small role in the level of adaptation required relative to the size, population and extreme vulnerability of island territories. This is despite the fact that there are important opportunities for island communities to develop low carbon

systems as the basis of robust local economies. The Aegean Islands exhibit many of the shared challenges facing insular territories in relation to climate change. The major role that tourism and second homes play in the economy of the islands is a common feature of many islands worldwide. The chapter also establishes that this economic sector itself is very vulnerable to climate change as well as being a major driver of increasing vulnerability. There remain major conflicts between the current development path and the one that would reduce the islands' vulnerability. National leadership in policy and legislation needs to be instrumental in enabling the Aegean Islands to embrace a more sustainable development path. At the same time, the Aegean case also suggests the importance of securing support from local stakeholders in such a path.

Notes

1 8.9 per cent and 2.8 per cent of the GDP of the North and South Aegean Regions respectively came from agriculture in 2006, compared to 3.7 per cent for the country; 22 per cent and 8.7 per cent of the active population were employed in agriculture in 2001, compared to 14 per cent for the country.
2 The Statistical Office of the European Union (EUROSTAT) has developed a definition, for regional policy use: islands are all areas of size $1km^2$ at least, permanently populated, with at least 50 inhabitants, separated from the continent by a water channel of at least 1km, not connected with the continent by permanent structures (tunnels, bridges) and where no state capitals are located.
3 As Aegean Islands are mountainous (even the smaller of them), they do not risk 'disappearing', unlike a lot of small ocean islands.
4 Even if studies for Greek islands are not available, this scenario seems to be confirmed in other Mediterranean islands (World Trade Organization (WTO), 2003, pp45–47).
5 An integrated policy for islands has to be adopted for European islands, with European, national and local authority involvement, following the subsidiarity principle.

References

Allen, H. D. (2001) *Mediterranean Ecogeography*, Prentice Hall, London

Baldacchino, G. (2004) 'The coming of age of island studies', *Tijdschrift voor Economische en Sociale Geografie*, vol 95, no 3, pp272–283

Benoit, G. and Comeau, A. (eds) (2005) *Méditerranée: Les Perspectives du Plan Bleu sur l'Environnement et le Développement*, Éditions de l'Aube, Paris

Gilman, E., Van Lavieren, H., Ellison, J., Jungblut, V., Wilson, L., Areki, F., Brighouse, G., Bungitak, J., Dus, E., Henry, M., Sauni Jr., I., Kilman, M., Matthews, E., Teariki-Ruatu, N., Tukia, S. and Yuknavage, K. (2006) 'Pacific island mangroves in a changing climate and rising sea', *UNEP Regional Seas Reports and Studies No. 179*, United Nations Environment Programme, Regional Seas Programme, Nairobi, Kenya

Greek National Statistic Service (1951; 1991; 2001) *Population Census*, GNSS, Athens

Grove, A. T. and Rackham, O. (2002) *The Nature of Mediterranean Europe: An Ecological History*, Yale University Press, New Haven

ESPON (2006) 'Territory matters for competitiveness and cohesion. Facets of regional diversity and potentials in Europe', ESPON Synthesis Report III, European Commission, Luxembourg

EURISLES (2002) 'Off the coast of Europe', CPMR's Islands Commission, Rennes, France

Horden, P. and Purcell, N. (2000) *The Corrupting Sea: A Study of Mediterranean History*, Blackwell, London

IPCC (2000) 'Special Report on emissions scenario', Intergovernmental Panel on Climate Change, Geneva

IPCC (2007) *Climate Change 2007: Impacts, Adaptation and Vulnerability*, contribution of Working Group II to the Fourth Assessment Report of the Intergovernmental Panel on Climate Change, Cambridge University Press, Cambridge, UK, pp7–22

Kizos, T., Spilanis, I. and Koulouri, M. (2007) 'The Aegean Islands: A paradise lost? Tourism as a driver for changing landscapes', in B. Pedroli, A. Van Doorn, G. De Blust, M. L. Paracchini, D. Wascher and F. Bunce (eds) *Europe's Living Landscapes: Essays Exploring our Identity in the Countryside*, Landscape Europe, Wageningen/KNNV Publishing, Zeist, pp333–348

Mediterranean Action Plan (2005), *Mediterranean Strategy for Sustainable Development*, INFO/RAC MAP

Mimura, N., Nurse, L., McLean, R. F., Agard, J., Briguglio, L., Lefale, P., Payet, R. and Sem, G. (2007) 'Small islands', in M. L. Parry, O. F. Canziani, J. P. Palutikof, P. J. van der Linden and C. E. Hanson (eds) *Climate Change 2007: Impacts, Adaptation and Vulnerability*, contribution of Working Group II to the Fourth Assessment Report of the Intergovernmental Panel on Climate Change, Cambridge University Press, Cambridge, pp687–716

Ministry of Environment, Spatial Planning and Public Works (2008) www.minenv.gr/download/2008/kya.tourismos.e sxaa.pdf, accessed 1 August 2008

Royle, S. (2001) *Geography of Islands: Small Island Insularity*, Routledge, London

Spilanis, I. and Vayanni, H. (2004) 'Sustainable tourism: Utopia or necessity? The role of new forms of tourism in the Aegean Islands', in B. Bramwell (ed) *Coastal Mass Tourism: Diversification and Sustainable Development in Southern Europe*, Channel View Publications, Clevedon, UK, pp269–291

Tompkins, E. L., Nicholson-Cole, S. A., Hurlston, L.-A., Boyd, E., Hodge, G. B., Clarke, J., Gray, G., Trotz, N. and Varlack, L. (2005) *Surviving Climate Change in Small Islands: A Guidebook*, Tyndall Centre for Climate Change Research, Norwich

UN (1992) Report of the United Nations Conference on Environment and Development, www.un.org/Depts/los/consultative_process/doc uments/A21-Ch17.htm, accessed 1 August 2008

UNEP (2004) Islands web page, http://islands.unep.ch, accessed 1 August 2008

UNFCCC (2007) 'Vulnerability and adaptation to climate change in Small Island Development States', background paper for the expert meeting on adaptation for Small Island Developing States, United Nations Framework Convention on Climate Change

WTO (2003) 'Climate change and tourism', Proceedings of the 1st International Conference on Climate Change and Tourism, Djerba, Tunisia, 9–11 April

Part 2

Strategic Planning Responses

Introduction to Part 2

Jenny Crawford

Part 2 explores how strategic frameworks and planning processes have been responding to the climate change challenge in terms of both mitigation and adaptation. It sets out examples from countries leading the per capita emissions league, in which the highest levels of policy development, coordination and investment might be expected. Recent developments in strategic policy, at the transnational, state and city levels, in Europe, the US, Australia, Canada, the UK and the Netherlands, highlight the tensions between climate change policies and planning for other strategic objectives. Key themes are:

- the development of new paradigms for spatial planning;
- the challenges of policy integration and diversity.

New paradigms for spatial planning

The European Spatial Development Perspective represented a significant breakthrough in the development of spatial planning language, analysis and policy formulation across Europe. In Chapter 9, Sykes and Fisher describe the emergence of territorial cohesion as an important spatial planning concept for European cooperation and examine the extent to which social and economic objectives are being integrated with climate change issues through intergovernmental agreements around a 'territorial agenda'. They suggest there is evidence that such integration represents an important evolution in spatial planning. However, there remain clearly contested arenas. The discussion highlights the issue of sustainable consumption with respect to economic growth. Can the strengthening of urban areas and the promotion of increasing mobility based on large-scale road and rail infrastructure, and regional airports, for instance, be sustainable? They certainly contrast with policies aimed at reducing the need to travel at the sub-regional or local level and the concept of zero-emission settlements. Sykes and Fisher argue that clear thresholds for environmental and climate-change trade-offs need to be identified at the highest strategic levels.

Similar conflicts are revealed by Byrne et al (Chapter 13) in their analysis of the Australian experience in terms of the concept of ecological modernization. They point out that while this ranges from a weak 'techno-corporatist' approach to a strong 'ecological democracy' approach, Australian policy has focused on the former: 'subordinating ecological concerns to the imperatives of economic growth'. It can be argued that the fundamental relationship between spatial planning, the attributes of place and governance means that a spatial approach must be expected to move towards the 'ecological democracy' approach. Certainly, planning for climate change throws democratic structures and relationships into sharp relief with respect to environmental outcomes.

As Wheeler (Chapter 10) points out, mitigation policy lays particular emphasis on the control of environmental outcomes – specifically those of greenhouse gas (GHG) emissions. This implies a 'backcasting' approach to policy development: designing policy backwards from outcome targets as opposed to the visioning, goal-setting and policy evolution most familiar in development planning processes. California is a world-leader in many aspects of the development of climate change policy, not least in terms of investment, research and innovation. In 2006, it legislated for emissions targets that are only now being adopted in other nations and states and which went far beyond the Kyoto Protocol that the Federal Government had rejected. Wheeler cautions, however, that the absence of a wider US climate strategy, the weakness of stateside land use planning, and the failure to engage with key questions of demographic and economic growth, threaten to undermine the effectiveness of California's climate change policy, despite its highly advanced climate change legislation and institutional and regulatory capacity. Related environmental and resource thresholds (such as oil and water) have still to be fully recognized.

Adaptation policy, on the other hand, requires a focus on uncertainty and risk assessment. The impact of climate change on 'spatializing water management' in the Netherlands, as described in Chapter 15 by De Vries and Wolsink, has involved nothing less than the shifting of the land system paradigm. One of the most immediate results is the recognition of water storage and management as a major land use. This requires that by 2030 over twice as much land should be allocated for new 'water space' as for housing and economic development. Included in this shift of references for spatial planning is a focus on policies that maximize creative solutions, flexibility and resilience and integrate the implications of a risk-based approach to decision making.

A risk-based approach is described in technical detail by Coleman (Chapter 16) who plots the recent development of UK policy for dealing with flood risk in development decision making. The chapter highlights the uncertainty surrounding the impact of risk assessment processes on development strategies and plans. The development of 'Strategic Flood Risk Appraisals', for instance, are still at an early stage in terms of their influence on decision making. As De Vries and Wolsink make clear, perceptions and values are of major importance in the assessment and management of risk. The playing out of tensions between technical and value-based approaches will undoubtedly shape the future of spatial planning.

Policy integration and diversity

The global interconnectedness of the climate system has highlighted the need for policy integration between and within all spatial scales. At the same time, recognition of the regional and local diversity of climate change issues is an important, complementary principle. Spatial planning implies the integration of the concept of 'place' into development discourses, with a focus on the assets and potentials of areas as the basis of economic growth and development decision making. Chapters 12 and 13 underline the contrast between the integrative, place-based approach of spatial planning and predominantly sectoral approaches.

Robinson (Chapter 12) illustrates tensions between the relative roles of Canadian national, provincial and city governments in responding to climate change. City governments in Canada, as in many other parts of the world, have taken a leading role in responding to climate change. However, the focus has been on technological innovation and regulation which Robinson describes as a 'first generation' approach to the reduction of emissions. She argues that 'second generation' approaches involve effective growth management. Her description of the recent introduction of growth management policies at both provincial and city levels, in Ontario and Vancouver respectively, demonstrates, however, the political complexities involved. She highlights that spatial planners must engage with these complexities in order to achieve change.

Jay (Chapter 13) gives a comparative overview of the development of spatial policy frameworks for marine renewable energy developments. His examination of the drivers behind the adoption of marine spatial planning frameworks reveals fascinating processes of convergence from very different institutional arrangements and histories, as different national jurisdictions not only attempt to maximize the mitigation of emissions but also to cooperate in the management of limited spatial resources.

In Chapter 14, Rydin places discussion of spatial planning in the context of the integration of policies for low and zero carbon buildings in the UK. In particular, she explores the relationship between 'stakeholder-led strategic planning' and regulation. In doing so, she stresses the importance of the spatial planning processes of involvement, monitoring and enforcement and integration with building regulations and standards. In this context, she argues that professional and institutional learning must be improved in order to secure the delivery of sustainable development.

9

The Territorial Agenda of the European Union:
Progress for Climate Change Mitigation and Adaptation?

Olivier Sykes and Thomas Fischer

Introduction

During the 1990s, intergovernmental efforts by EU member states with the support of the European Commission contributed to a growing awareness of the spatial impacts and dynamics of the European project. This culminated in 1999 in the agreement of an indicative statement of principles to guide the balanced and sustainable development of the EU's territory – the 'European Spatial Development Perspective' (ESDP). This document was to have a varying impact across Europe in the following years, with different territorial contexts at different state and sub-state scales playing a significant role in conditioning the attention it was accorded, and the ESDP policy principles which were seen as being pertinent in different places. In the mid-2000s, after something of a lull, the momentum for intergovernmental working within the EU on spatial issues returned, with the new debates being increasingly framed in the language of territorial development and the stated objective

of achieving 'territorial cohesion'. For its part, the European Commission increasingly emphasized and sought to give definition to the concept of territorial cohesion in documents such as the Third Report on Economic and Social Cohesion of 2004.

In 2007, EU member states meeting at Leipzig agreed the 'Territorial Agenda of the European Union: Towards a More Competitive and Sustainable Europe of Diverse Regions' (Territorial Agenda). This is a non-binding statement of principles which intends to inform sustainable territorial development across the EU. In many respects, it can be seen as a successor to the earlier ESDP document adopted in 1999 (Commission of the European Communities (CEC), 1999; Faludi, 2007a). The Territorial Agenda aims at 'strengthening territorial cohesion' in Europe, as well as supporting the 'growth and jobs', cultural, social, environmental and economic, sustainable development objectives, of the EU's Lisbon and Gothenburg strategies. In this context, it aims at contributing to 'sustainable economic growth and job creation, reconciling it with social and ecological development in all EU regions' (EU Ministers for Spatial Planning and Development, 2007a, p23). Since the 1990s and the debates that culminated in the agreement of the ESDP, many of the challenges and dynamics affecting the European territory have become amplified and more clearly defined (for example, through the more systematic investigation of these issues through the European Spatial Planning Observation Network (ESPON). The growing cognisance of the implications of climate change for Europe's diverse territories is particularly significant in this respect. The Territorial Agenda acknowledges that climate change is one of the 'major new territorial challenges today' and that this includes 'the regionally diverse impacts of climate change on the EU territory and its neighbours...' (EU Ministers for Spatial Planning and Development, 2007a, p25). Furthermore, the background document to the Territorial Agenda – the 'Territorial States and Perspectives of the European Union' (TSPEU), states that 'climate change is identified as the first challenge to the "regions and cities of Europe"' and that 'the impacts of climate change are of increasing importance for European regional economies and their need to adapt' (EU Ministers for Spatial Planning and Development, 2007b, p3 and pp21–22).

In adopting the Territorial Agenda, the EU member states called on the European Commission to prepare a report on territorial cohesion by 2008. In autumn 2008, the Commission duly published a 'Green Paper on Territorial Cohesion – Turning Territorial Diversity into Strength' (the Green Paper) (CEC, 2008a). In this, the problems associated with climate change are mentioned as an issue requiring cooperation. Climate change was also one of the key themes of the debate on territorial cohesion and the future of cohesion policy, which took place under the French EU Presidency during the latter part of 2008.

Informed by the context outlined above, this chapter discusses whether the Territorial Agenda and the territorial cohesion objective and guiding principle can be considered to represent progress for climate change mitigation and adaptation when compared with the earlier ESDP and the EU's wider evolving policy response to climate change. In order to provide a backdrop for this discussion, the following section outlines the current climate change context and prospects for mitigation and adaption measures in Europe.

Climate Change: The evidence and context for mitigation and adaptation actions in Europe

One way of interpreting the current rise of climate change as a topic and theme of policy maker and academic attention, is by drawing on Anthony Downs' (1972) five stage 'issue-attention cycle' of public and policy makers' interest. It is arguable that current societal perception of climate change in Europe is now clearly past Stage 1 of the cycle – the 'pre-problem stage' – and is within Stage 2 – 'alarmed discovery and euphoric enthusiasm' – which occurs when a dramatic series of events means that the public suddenly becomes 'alarmed about the evils of a particular problem' (Downs, 1972, p39). For example, a series of extreme weather events and patterns with tangible consequences such as heat waves, droughts, flooding and forest fires, which affected different parts of Europe in the mid-2000s, has contributed to raising the profile of climate change in the media and political agendas (even though the relationship between such events and wider climatic change is complex and sometimes contested). We thus appear to be at a particularly crucial point in time in relation to societal awareness of, and potential response to, climate change that might constitute what Kingdon (1995) describes as a 'window of opportunity' for action. What is undeniable is that the currently available scientific evidence is quite compelling and overwhelming in identifying climate change as a serious global threat that demands an urgent and comprehensive global response. Attention has also been increasingly focused on seeking to assess the costs associated with pursuing different responses to the challenges posed by climate change (e.g. European Commission Joint Research Centre's PESETA study (CEC, 2007a))

As the relative costs of expected damage and those associated with enacting adequate mitigation and adaptation measures become more apparent, it might appear plausible to assume that policy makers and populations would become more prepared to act now to avert having to bear substantially higher costs in future. However, the possibility also exists that – as has been witnessed with numerous other social and environmental problems – the realization of the costs of significant progress in tackling the problem of climate change, even if much smaller than potential damage, may lead to a gradual decline of intense public, and subsequently, policy makers' interest. Climate change as a political issue may thus begin to move into Stage 3 of Downs' cycle – 'realizing the cost of significant progress'. This occurs when there is a 'gradually spreading realization that the cost of "solving" the problem is very high indeed' and that 'really doing so would not only take a great deal of money but would also require major sacrifices by large groups in the population' (Downs, 1972, pp39–40).

Given the mixed evidence that is currently available on the 'sustainability' of societal interest in, and commitment to addressing climate change (Brown, 2008), and Downs' propositions regarding the trajectory of environmental issues as matters of public and policy makers attention, a persuasive case might be mounted that the current period does indeed present a precious Kingdonian 'window of opportunity' for concerted societal and political action to address the challenges posed by anthropogenic climate change. In order to provide a baseline against which to gauge whether climate change is becoming increasingly inscribed as an issue requiring concerted action within the new European territorial development policy agenda, the following section considers the treatment of climate change in the 'predecessor' document to the Territorial Agenda: the ESDP document of 1999.

The European Spatial Development Perspective

Following the publication of the ESDP in 1999 (CEC, 1999), different views emerged on its environmental credentials. Whilst the document was criticized in some quarters for focusing too heavily on economic competitiveness and GDP (gross domestic product) growth, it should be noted that this was in keeping with a context framed by the aims of the 1999 Amsterdam Treaty on the European Union. In addition, social justice was also a major concern (Roberts, 2003). The ESDP explicitly introduced a spatial model of polycentricity for Europe, which, though subsequently subject to a range of different interpretations, has proved to be one of its most enduring influences on spatial planning thinking and practice (Baudelle and Castagnède, 2002; Davoudi, 2003, 2005a; Shaw and Sykes, 2004; Meijers et al, 2007; Waterhout, 2007). Its position on environmental issues (including climate change) was interpreted by some as being grounded in an 'ecological modernization' approach (see also Davoudi et al, Chapter 1, and Byrne et al, Chapter 11), which looked mainly towards technological innovation and institutional reform to address environmental problems, in a manner similar to the thinking articulated by Joseph Huber's works in the early 1980s (see Mol and Sonnenfeld, 2000, for a summary).

At various points, the ESDP referred to the greenhouse effect, stating that 'spatial development policy can make an important contribution to climate protection through energy-saving from traffic-reducing settlement structures and locations' (CEC, 1999, p31). Furthermore, CO_2-neutral energy sources and sustainable forest management were mentioned. An overarching sustainability 'magic triangle' (CEC, 1999, p10) was presented in an attempt to reconcile the spatial policy objectives of economic (GDP) growth, social equity and environmental protection. Several commentators, however, subsequently criticized this approach, suggesting that 'the central objectives of spatial policy, relating to growth, equity and the environment' are 'riven by internal contradictions' (Jensen and Richardson, 2004, p226).

One of the most important criticisms of the ESDP which was advanced by those critical of its treatment of environmental issues, was that it promoted a vision of European territorial development that would result in increased 'hyper-mobility' and associated substantial negative environmental consequences, including energy consumption and CO_2 emissions. In particular, the document's acceptance of the rationale and specific infrastructure project suggestions of the trans-European Transport Networks (or TEN-Ts), was subject to some fierce criticism. For some, the TEN-Ts seemed incompatible with the ESDP's promotion of energy-saving from traffic-reducing settlement structures and locations. The ESDP, it was argued, seemed to be grounded in a rationality of 'frictionless mobility' which did not appreciate, or pay heed to, the physical, environmental and cultural/identity consequences of conceptually and literally (e.g. through the realization of large-scale infrastructure projects) constructing European space as a 'space of flows' rather than a 'space of places' (Jensen and Richardson, 2004).

Other less critical voices, however, observed that a sustainability discourse had actually 'successfully penetrated' the ESDP (Waterhout, 2007, pp37–59). Furthermore, there were suggestions that the document had demonstrated 'great concern about ecologically sensitive areas, which in the densely populated EU are often being threatened by urban development' (Waterhout, 2007, pp37–59). The ESDP was also credited with contributing to a revival in strategic spatial planning in many European countries and regions (see Sykes and Shaw, 2005; Town and Country Planning Association, 2006). Importantly, it was argued that, although the ESDP did contain manifest contradictions between the different dimensions of sustainability, which it sought to promote, it did not originate these. In this context, the ESDP merely served as a reminder of the great problems and the value choices that arise when attempting to balance social, environmental and economic dimensions of sustainability, taking into account the territorial impacts of different policy and development choices and assessing the spatial impacts of large-scale projects and investments. Furthermore, the ESDP did seek to promote the use of the tool of Territorial Impact Assessment (TIA), though to date this technique has mainly been developed within the context of ESPON evaluation studies.[1] The following section now moves on to consider how the Territorial Agenda of the EU addresses the issues of climate change mitigation and adaptation.

A new 'Territorial Agenda' for the European Union

An initial observation which can be made about the Territorial Agenda is that it is a much thinner document than the ESDP, consisting of roughly 11 pages compared to the earlier document which was over 80 pages in length.[2] However, it should be noted that the Territorial Agenda is based on an earlier more voluminous ESPON-informed background document, the 'Territorial State and Perspectives of the European Union' (EU Ministers for Spatial Planning and Development, 2007b). This sought to apply an 'evidence based' approach to the analysis of territorial trends in Europe and the formulation of appropriate policy orientations in response to these (Faludi and Waterhout, 2006). This explains in part why the Territorial Agenda is a much thinner document than the ESDP, which was composed of an analytical Part B as well as a policy-focused Part A that contained its statement of spatial development policy principles. In developing the Territorial Agenda, background maps and information used included those on changes in temperatures and precipitation across Europe by the end of the 21st century (EU Ministers for Spatial Planning and Development, 2007d, pp30–31). Other maps produced to support the dialogue surrounding the agreement and implementation of the Territorial Agenda also sought to give an impression of the issues that adaptation measures will need to address (EU Ministers for Spatial Planning and Development, 2007d, p23). These included 'natural hazards' which can be associated with global climate change such as: 'high probability of winter storms'; 'high or very high forest fire potential'; 'risk of avalanches'; and 'flood endangered settlement areas' (considering flood potential and share of artificial areas).

The final Territorial Agenda document consists of four main parts. Part I introduces the 'Future Task' of strengthening territorial cohesion in Europe. Here, and in keeping with the objectives set out in the EU 'Lisbon' strategy for jobs and growth (CEC, 2000), and the 'Gothenburg' EU Sustainable Development Strategy (CEC, 2001), the focus is on contributing to a Europe that is culturally, socially, environmentally and economically sustainable and on promoting 'territorial solidarity'.

Part II identifies six 'major new territorial challenges' for the EU, focusing on 'strengthening regional identities', and 'making better use of territorial diversity'. The first challenge is identified as being the 'regionally diverse impacts of climate change on the EU territory and its neighbours, particularly with regard to sustainable development', with the second being 'rising energy prices, energy inefficiency and different territorial opportunities for new forms of energy supply' (EU Ministers for Spatial Planning and Development, 2007a, p25). Here, it is highlighted that 'every region and city may, through their engagement, contribute to saving energy and to its decentralized supply and to mitigating climate change, for example, by supporting the development of low or zero-emission settlements, developing potential new renewable sources of energy supply and promoting energy efficiency particularly of the building stock', and argued that 'our cities and regions need to become more resilient in the context of climate change' (p26). The Leipzig Charter of Sustainable European Cities, adopted at the same meeting as the Territorial Agenda, also states that a 'well designed and planned urban development can provide a low carbon way of accommodating growth, improve environmental quality and reduce carbon emissions' (p13).

In Part III of the Territorial Agenda six 'Territorial Priorities' are identified, namely the:

1 strengthening of polycentric development and innovation through networking of city regions and cities;
2 promotion of new forms of partnership and territorial governance between rural and urban areas;
3 promotion of regional clusters of competition and innovation in Europe;
4 strengthening and extension of trans-European networks;
5 promotion of trans-European risk management, encompassing the impacts of climate change;
6 strengthening of ecological structures and cultural resources.

Section IV of the Territorial Agenda document deals with the implementation of these priorities, asking European institutions for support, expressing a commitment by the 27 ministers to take into account the priorities of the Territorial Agenda and outlining joint activities that European ministers will pursue to promote its implementation. Here, the ministers state that as a first step for promoting the implementation of the Territorial Agenda, they commit themselves to contributing to a 'sustainable and integrated climate and energy policy in the EU' (EU Ministers for Spatial Planning and Development, 2007a, p10). Later in 2007 during the Portuguese EU Presidency, the EU ministers also agreed a joint statement as a contribution to the discussions surrounding the European Commission's 'Adapting to Climate Change in Europe' Green Paper (CEC, 2007b) at their meeting in the Azores in November 2007 (EU Ministers for Spatial Planning and Development, 2007e). This re-emphasized the Territorial Agenda's message that cities and regions need to become more resilient in the context of climate change. Of significance here, the role of spatial planning and development in contributing to climate change mitigation and adaptation measures is also stressed, and the ministers commit themselves to highlighting the importance of the territorial dimensions of climate change at the national and European levels.

It is interesting to note that climate change as a term is referred to 26 times in the main background document to the Territorial Agenda – the 'Territorial State and Perspectives of the European Union'. In contrast, the ESDP only mentioned 'climate change' once, with the 'greenhouse effect' being mentioned five times. Clearly, there is a need to be cautious in imputing too much significance to this simple quantitative indicator of the content of the two documents and using it to infer the relative attention they accord to climate change. However, the policy rhetoric being articulated in the more recent Territorial Agenda does seem to indicate an increase in the attention that is being paid to the issue of climate change compared to the treatment of the issue in the earlier ESDP document. It is also true, as noted above, that the first of the six 'major new territorial challenges' identified by the document is seen as being the 'regionally diverse impacts of climate change on the EU territory and its neighbours, particularly with regard to sustainable development'. The second challenge is seen as being 'rising energy prices, energy inefficiency and different territorial opportunities for new forms of energy supply' (EU Ministers for Spatial Planning and Development, 2007a, p25). As always it is important when seeking to interpret what a policy text establishes as being significant, to think about the wider context in which policy statements are elaborated. If one thinks about the Territorial Agenda in terms of a possible policy 'window of opportunity' for concerted policy responses, it is interesting, for example, to note Faludi's assessment that during the Territorial Agenda process climate change was a topic which received 'ever-increasing levels of emphasis, due among other things to the volume of media attention given to the topic, with the Al Gore film *An Inconvenient Truth* discussed during at least one of the meetings'(Faludi, 2007a, p10). Having considered the treatment of climate change in the Territorial Agenda and its policy orientations the section below reflects on the possible climate change implications of the document and the territorial cohesion model that underpins it.

Reflecting on the climate change implications of the Territorial Agenda and territorial cohesion

As noted above, the Territorial Agenda's policy goals are firmly rooted in the concept of 'territorial cohesion'. The 'EU Reform' or Lisbon Treaty, signed by member states in 2007 and currently in the process of being ratified, makes territorial cohesion into an objective of the Union alongside economic and social cohesion as a shared competence of the EU and its member states. Despite the uncertainty introduced into the process of ratifying the Treaty following its rejection by Irish voters in June 2008, the European Commission published its Green Paper on territorial cohesion 'Turning Territorial Diversity into Strength' in October 2008 (CEC, 2008a). This seeks to launch a debate which can work towards a (more) common interpretation of the concept (Hübner, 2008). The paragraphs below reflect on the possible climate change implications of the Territorial Agenda document and the territorial cohesion model that underpins it.

The concept of territorial cohesion

Territorial cohesion has been defined as 'a goal of spatial equity that tends to favour development-in-place over selective migration to locations of greater opportunity' (Carbonell, 2007, pvii). Commentators, such as Davoudi (2005b) and Faludi (2007a), have suggested that this model can be compared and contrasted with the American model of economic and social development, where 'selective migration to locations of greater opportunity' plays a greater role. An important assumption, which underpins the notion of territorial cohesion, is articulated in the third EU Report on Economic and Social Cohesion which argues that 'people should not be disadvantaged by wherever they happen to live or work in the Union' (CEC, 2004, p27).

An increasingly important goal of EU regional policy is also to seek to ensure that the diverse potential, or 'territorial capital', of different territories in Europe is mobilized in the interests of enhancing the competitiveness of Europe's regions and nations and the EU as a whole. As Tatzberger (2003, p18) notes, as well, focusing on the territorial dimensions of policies and better tailoring these to the specific needs of different territories, territorial cohesion also represents an evolution of cohesion policy away from a purely redistributive logic of making transfer payments to poorer areas to reduce regional economic development disparities, to an emphasis on the 'optimal use of potentials of territorial units throughout Europe'. It therefore reflects the idea of the 'European model of society understood to foster competitiveness whilst keeping in mind concerns for social welfare, good governance and sustainability' (Tatzberger, 2008, p106).

The Green Paper on territorial cohesion – 'Turning Territorial Diversity into Strength' – was published in autumn 2008 and a debate on the definition of the concept was launched (CEC, 2008a). The Green Paper argues that 'territorial cohesion is about ensuring the harmonious development of all places and about making sure that their citizens are able to make the most of inherent features of these territories' (CEC, 2008a, p3). Territorial cohesion is presented as 'a means of transforming diversity into an asset that contributes to sustainable development of the entire EU' (CEC, 2008a, p3). The 'endogenous' view of growth which Tatzberger (2003, 2008) identifies as being one feature of the concept is thus articulated once more. It is stressed that 'increasingly, competitiveness and prosperity depend on the capacity of the people and businesses' located in particular places 'to make the best use of all territorial assets' (CEC, 2008a, p3). The environmental problems associated with climate change are mentioned as an issue requiring cooperation, and it is asserted that the concept of territorial cohesion 'builds bridges between economic effectiveness, social cohesion and ecological balance, putting sustainable development at the heart of policy design' (CEC, 2008a, p3). The Green Paper thus reiterates and reinforces many of the core ideas and principles which have characterized debates on territorial development in Europe since the 1990s, including those articulated in the ESDP (1999) and Territorial Agenda (2007). The strengthened emphasis on sustainability and environmental issues, and the integration of these among the core concerns of territorial cohesion policy, is one area where there appears to have been an evolution in thinking. The goal of securing greater territorial coherence and coordination in EU sectoral policies beyond the structural and cohesion funds is a long-standing aspiration of intergovernmental and (parts of) Commission thinking.

The different dimensions, interpretations and interests that are subsumed within the concept and discourse are stressed by Waterhout (2007) who identifies four main territorial cohesion 'storylines': 'Europe in Balance'; 'Coherent European policy'; 'Competitive Europe'; and, 'Green and Clean Europe'. For Waterhout, writing in 2007 (p50), the last of these – 'Green and Clean Europe' – can be seen as looming 'in the background' and currently the 'Europe in balance' storyline – with its focus on addressing regional development disparities and fostering spatial equality of opportunity – is the most influential. He also alludes to the possibility that the 'Competitive Europe' storyline, with its emphasis on territories making the most of their territorial capital to 'autodevelop' themselves, might in time come to challenge this dominance. Overall, it is clear therefore that the territorial cohesion concept is subject to different interpretations and might be invoked by different interests in support of different values (Hübner, 2008, p7).

The emphasis on different dimensions of territorial cohesion and development in Europe can be traced across the different spheres of the territorial policy field. One example is in the generation of evidence to substantiate the policy prescriptions of documents such as the Territorial Agenda. In light of this, it is interesting to note, given the concerns of this chapter, that the first ESPON programme which ran from 2002 to 2006 and informed the Territorial Agenda and Territorial States and Perspectives of the European Union documents, seems to have been less focused on environmental issues than on economic and social issues. The growing emphasis on the territorial dimensions of climate change is, however, reflected by the new ESPON 2013 programme where a specific call for work on 'Climate Change and Territorial Effects on Regions and Local Economies' was launched in summer 2008. Furthermore, in 2006, the 'ESPON Atlas' (Federal Office for Building and Regional Planning et al, 2006) which provides a 'synoptic and comprehensive overview of the findings of the ESPON projects', already featured maps showing potential future CO_2 emissions for inter-urban traffic in 2030 (based on qualitative predictions) for three European development scenarios – including business as usual, cohesion and competitiveness. However, an assumption in this exercise was that in the cohesion scenario, policies would not focus on global competitiveness, something that, at least based on current EU policy intentions (e.g. the overarching 'Lisbon Strategy' which aims to make the EU into the world's most competitive knowledge-based economy), is very hard to imagine. In any case, overall, both cohesion and competitiveness scenarios show much higher CO_2 emissions than a 'business-as-usual' scenario. It is clear therefore that the interpretation and emphasis that is given, through territorial policy making and investment decisions, to territorial cohesion and its different constituent elements is likely to be a significant factor that influences how far the territorial policy agenda contributes to the EU's response to the challenges posed by climate change.

It is also necessary to think more specifically about the climate change implications of the spatial development model promoted by the Territorial Agenda and the territorial cohesion concept, which underpins it. Some potential substantive implications of the spatial dimensions of these for climate change mitigation, focusing mainly on mobility and transport, are therefore considered below.

Potential climate change implications of the Territorial Cohesion Agenda and Spatial Model

In scalar terms, territorial cohesion is commonly held to have three main dimensions: a European dimension, a national / transnational dimension and a regional dimension. Framing of these dimensions is important, as it allows for a first reflection and discussion on the potential substantive climate change mitigation 'credentials' of the EU's territorial policy agenda.

The European dimension of the, currently dominant, 'Europe in balance' (Waterhout, 2007) interpretation of territorial cohesion is concerned with reducing disparities between different parts of the European territory and the promotion of more balanced development. In the Territorial Agenda this is reflected in the promotion of the enlargement of economic growth zones or European Integration Zones 'beyond the economic core area of the EU' territory (EU Ministers for Spatial Planning and Development, 2007a, p29). This area was defined in the ESDP as the area between the metropolises of London, Paris, Milan, Munich and Hamburg and referred to as the 'pentagon' (other definitions and appellations of this area include the 'blue banana' (Brunet et al, 1989) or Central Nucleus). The 'pentagon' covers the south east of England, the Netherlands, Belgium, northeastern France, Luxembourg, northern Italy, western Austria, Switzerland and western Germany. There have been a number of attempts and initiatives to try and identify other potential European Integration Zones 'beyond' this area. Though it may be argued that such a spatial outworking of the territorial cohesion concept and agenda responds to the equity focused 'Europe in balance' and economically orientated 'Competitive Europe' dimensions of territorial cohesion, its relationship with the environmentally inspired 'Green and Clean Europe' storyline (Waterhout, 2007) appears more equivocal. In particular, regarding transport efficiency (translated into a minimization of climate change relevant energy consumption and associated CO_2 emissions), if no other concrete transport policies are introduced in parallel, the climate change mitigation credentials of a spatialization of the European dimension of territorial cohesion which results in the creation of additional growth zones across Europe can only be said to be more than doubtful.

Introducing adequate additional transport policy to accompany the Territorial Agenda's aim of promoting the development of growth zones outside the EU core will therefore be crucially important; otherwise, climate change mitigation will be very unlikely. This will involve establishing links between policy sectors such as transport and regional policy to identify how the impacts of the increased mobility and accessibility likely to result from this aspect of the Territorial Agenda's policy orientations can be mitigated. It should be noted here though that the horizontal coordination of EU policy sectors has proved to be something of a struggle in the past.

There are other potential implications of the European dimension that may – directly or indirectly – have an impact on climate change mitigation, including, for example, the increased urban sprawl that is frequently observed in core economic regions. Urban sprawl is mentioned once in the Territorial Agenda and once also in the Leipzig Charter on Sustainable European Cities, which argues for compact settlement structures achieved by spatial and urban planning to aid efficient use of resources and prevent urban sprawl 'by strong control of land supply and of speculative development' (EU Ministers for Spatial Planning and Development, 2007c, p13). In the Territorial Agenda there appears, however, to be little 'working through' of the potential impacts that the pursuit of some of its identified 'priorities for territorial development' may have on the phenomenon of urban sprawl.

In summary, regarding the European dimension of territorial cohesion, the Territorial Agenda seems to be targeting mainly the economic and social dimensions of development, and is (currently at least) neglecting to fully consider in an integrated manner the possible environmental and climate change implications of some of the policy orientations it promotes.

The national/transnational dimension of territorial cohesion is reflected in the Territorial Agenda's aim of supporting the development of networks of competitive regions composed of networks of core cities, city regions and their surrounding towns and rural areas, and interdependent regional centres and medium-sized towns in more remote rural areas. Categorization of European cities into 'Global Nodes', 'European Engines', 'Metropolitan European Growth Areas' (MEGAs), 'potential' and 'weak MEGAs', and smaller Functional Urban Areas (FUAs), developed through the ESPON programme, might provide an indication of the pattern of such a network of cities and city regions (ESPON 1.1.1, 2004; ESPON, 2006, p29). Following the rationale of the first three 'priorities for territorial development' articulated by the Territorial Agenda it would seem logical to seek to strengthen those urban areas characterized as being 'potential' and 'weak' MEGAs. It is difficult to comment on the implications for climate change mitigation of pursuing and achieving such an objective. Overall though, it seems plausible to surmise that the impacts would mirror those associated with the European/transnational dimension discussed earlier, and that positive outcomes would be highly unlikely, as additional growth areas would imply additional carbon intensive economic and transport activities.

At the regional scale, the Territorial Agenda continues the European spatial planning 'tradition' of encouraging polycentric development at the intra-regional and city-regional/intra-urban level, although the document does not provide a concrete model of what a polycentric region should or may look like. There has been a substantial body of work in recent years on polycentric regional development and Polycentric Urban Regions (PURs) (Kloostermann and Musterd, 2001; Meijers and Romein, 2003; Davoudi 2003, 2005a; ESPON 1.1.1, 2004). In Europe it appears that a polycentric region could be equated with what some authors have termed 'a transport efficient' or 'sustainable' spatial structure (see Rothengatter and Sieber, 1993; Newman and Kenworthy, 1999; Fischer, 2001).

Polycentric settlement structure, however, is only one of many factors influencing commuting distance and modal choice. Issues such as workplace/residential balance, the distribution of retail structures, and the availability of alternatives to private motorized transport are also influential. Dielemen et al (2002, p209) have also pointed to the influence of personal, socio-economic and lifestyle attributes, stating that:

Apart from urban form and design, personal attributes and circumstances have an impact on modal choice and distances travelled. People with higher incomes are more likely to own and use a private car than low-income households. Families with children use cars more often than one-person households. The purpose of a trip – work, shopping and leisure – also influences travel mode and distance.

Therefore gauging the climate change mitigation credentials of the polycentric region as a spatial model (in terms of transport efficiency and potential for minimization of climate change relevant transport-related energy consumption), is a complex matter. It is clear that key contextual aspects need to be considered, including transport choice, the extent and mix of transport networks and modes, the economic composition and occupational structures of cities and regions, and societal and cultural factors such as lifestyles.

The points raised above hint at a final dimension which needs to be considered in reflecting on the probable climate change mitigation impacts of the Territorial Agenda: the infrastructural logic which it appears to subscribe to. In this respect, it is important to note that the Territorial Agenda seems to accept that the achievement of territorial cohesion requires the facilitation of mobility and accessibility in all EU regions, particularly in the more remote areas of the EU (EU Ministers for Spatial Planning and Development, 2007a, pp27–29).

The first territorial priority for the development of the EU introduced earlier can be seen as being highly controversial, at least from a climate change mitigation viewpoint, in recommending that 'infrastructure networks within and between regions in Europe ... need to be extended and updated on a *continuous basis*' (added emphasis) (EU Ministers for Spatial Planning and Development, 2007a, p28). Here, there seems to be an acceptance of the assumption that 'mobility and accessibility are key prerequisites for economic development in all regions of the EU' (EU Ministers for Spatial Planning and Development, 2007a, p29). Yet the work of Hart (1993) on transport investment and disadvantaged regions and more recent evidence provided by ESPON (2006, p39) suggests that such causal assumptions are debatable.

The apparent importance attached to 'networks of viable regional' airports under Priority 4 (Strengthening and Extension of Trans-European Networks), also needs to be set against the contribution of aviation to GHG emissions and mixed evidence regarding the impacts that airport expansion and increased accessibility by air can have on national and regional economies. The UK Royal Commission on Environmental Pollution (RCEP, 2002, p37), for example, has examined the environmental impacts of civil aircraft in flight, and concluded that 'short-haul passenger flights, such as UK domestic and European journeys, make a disproportionately large contribution to the global environmental impacts of air transport' and that 'these impacts are very much larger than those from rail transport over the same point-to-point journey'. Other research has investigated the impacts that proposed airport expansion in the UK will have on regional economies, and concluded that, other than for London, increased air travel will result in negative impacts on regional economies (Friends of the Earth, 2005).

Overall, what emerges from the discussions above is that, as might have been expected, different priorities and interests and dimensions, or 'storylines' (Waterhout, 2007), of territorial cohesion are represented in the Territorial Agenda, and that the relationships and potential contradictions between these do not yet appear to be fully resolved. In particular, and in this the Territorial Agenda is by no means remarkable amongst spatial development strategies, there appears to be some degree of disjunction between the competitiveness, growth and mobility oriented Territorial Priorities of the strategy and its ostensible promotion of trans-European risk management, mitigation of GHG emissions and stewardship of irreplaceable ecological structures.

Conclusion

In terms of the language and rhetoric used and the aims and objectives formulated, the Territorial Agenda of the European Union document can be seen as indicating some progress for climate change mitigation and adaptation if it is compared with the ESDP which was adopted eight years earlier. At various stages in the document the text alludes to the need to address both climate change mitigation and adaptation. However, there remain potential incompatibilities between its economic and accessibility priorities on the one hand (particularly promotion of growth in GDP and the continuous extension of the trans-European networks) and its environmental objectives. Priorities for the strengthening and 'continuous' extension of the trans-European transport networks and on strengthening ecological structures can also be expected to prove more difficult to reconcile 'on the ground' than on paper.

At a more general level, the territorial cohesion model that underpins the Territorial Agenda is currently seen as a 'prerequisite for achieving sustainable economic growth and implementing social and economic cohesion – a European social model' (EU Ministers for Spatial Planning and Development, 2007e, p25). Despite the reference to sustainable growth here, there are substantial environmental issues to be resolved surrounding the potential spatial 'outworking' of the territorial cohesion agenda, particularly regarding the encouragement of the creation of several powerful economic zones of international importance outside the current 'pentagon' zone.

Such questions are important in light of current attempts to work towards a (more) common interpretation of territorial cohesion, for example, through the debate which was launched with the publication of the EU Green Paper on Territorial Cohesion in 2008, especially given that the concept is frequently considered to be grounded in a European model of society that has both environmental protection and social equity as two of its defining characteristics. Indeed, if it is accepted that these goals are viewed as fundamental shared European values (EU Member States, 2007), then it follows that their articulation through territorial cohesion policy will require careful monitoring and balancing of competing claims on resources. Spatial planning's role as a platform for mediating controversy and ensuring transparency means that planners need to make clear what development paths are being adopted, where the trade-offs lie and what action is required. It is important that such trade-offs are not entirely left to be resolved at subsequent levels of decision making through the operation of impact assessment instruments such as TIA and SEA (Strategic Environmental Assessment), as these work most effectively in the presence of clear and compatible objectives. Considering the current scientific evidence from the IPCC and the Stern Review, and a growing evidence base on the territorial impacts of climate change from bodies such as the European Environment Agency (EEA), there is a strong case to be made for some additional policies and clear thresholds (or 'red-lines') for environmental and climate change trade-offs to be identified.

In making such a case the wider context of EU climate protection and adaptation policy within which the territorial debate sits is of crucial importance. For example, during the latter part of 2008 the French EU Presidency prioritized securing an agreement on the EU 'Energy-Climate' package. In December 2008 this was finally reached between EU member states with binding targets to reduce GHG emissions by 20 per cent, achieve a 20 per cent share of energy use from renewable energies and to improve energy efficiency by 20 per cent by 2020. Climate change was also one of the key themes of the debate on territorial cohesion and the future of cohesion policy which took place under the French EU Presidency during the latter part of 2008. A major conference in Paris in October 2008 highlighted the territorial dimension of the 'struggle against climate change' as one of four key areas of discussion on territorial cohesion, alongside the Common Agricultural Policy and rural development, the Lisbon Strategy and integrated territorial development and governance (Gizard, 2008). The Marseille meeting of EU Ministers for Housing, Urban Development, Spatial Planning, and Cohesion Policy also noted that 'mitigation of greenhouse gases and adaptation policies to climate change should not be disassociated and should be dealt with together as two complementary dimensions within the framework of integrated territorial sustainable development strategies' (Présidence Française de l'Union Européenne, 2008, p16). One of the goals of the 'EU Economic Recovery Plan', adopted in December 2008 in response to the economic downturn, is to 'speed up the shift to a low carbon economy' (CEC, 2008b). It is also noteworthy that the 'vulnerability index' on challenges facing European regions produced by the Commission in its 'Regions 2020' report of December 2008 includes the 'far-reaching impacts of climate change' as one of four key issues. It notes that 'most European regions are anticipated to be negatively affected by future impacts of climate change', stating that 'the impact of climate change on Europe's environment and its society has become central to the European agenda, challenging policymakers to reflect on how best to respond with the policy instruments at the EU's disposal' and that 'this applies both to efforts to mitigate climate change by tackling the growth in greenhouse gas emissions and the need for measures to adapt to the consequences of climate change' (CEC, 2008c, pp5, 11).

In such a context, it does seem important to note the apparent attention which is being given to measures to respond to climate change at the European level, and to capitalize on the potential 'policy window' that this provides to sustain debates on, and address some of the issues identified above in relation to, the climate change dimensions of Europe's putative territorial cohesion policy agenda.

Notes

1   The first ESPON programme started in 2002 with the successor 2007–2013 programme currently underway. See www.espon.eu
2   The consolidated version of the Leipzig Charter and the Territorial Agenda documents (German Presidency of the EU, 2007) adopts a different format in which the Territorial Agenda runs to 14 pages in total.

References

Baudelle, G. and Castagnède, B. (2002) *Le Polycentrisme en Europe*, Editions l'Aube & DATAR, La Tour d'Aigues

Brown, G. (2008) 'No excuse for inaction when times get tough', *The Guardian*, 16 July, p3, see also www.guardian.co.uk/climatesummit

Brunet, R. (1989) *Les Villes Européennes – Rapport pour la DATAR*, under the supervision of Roger Brunet, with the collaboration of Jean-Claude Boyer, RECLUS et la Documentation Française, Paris

Carbonell, A. (2007) 'Foreword' in A. Faludi (ed) *Territorial Cohesion and the European Model of Society*, Lincoln Institute of Land Policy, Cambridge, MA, ppvii–viii

Commission of the European Communities (CEC) (1999) *European Spatial Development Perspective – ESDP*, Potsdam, http://ec.europa.eu/regional_policy/sources/docoffic/official/reports/som_en.htm, accessed 14 January 2009

CEC (2000) 'The Lisbon European Council – An agenda of economic and social renewal for Europe', www.euractiv.com/en/future-eu/lisbon-agenda/article-117510#links, accessed 14 January 2009

CEC (2001) 'A sustainable Europe for a better world – A European strategy for sustainable development', http://ec.europa.eu/sustainable/sds2001/index_en.htm, accessed 14 January 2009

CEC (2004) 'A new partnership for cohesion – convergence, competitiveness, cooperation: Third Report on Economic and Social Cohesion', Office for Official Publications of the European Communities, Luxembourg

CEC (2007a) 'The Peseta Project: Impacts of climate change in Europe', http://peseta.jrc.es/docs/ClimateModel.html, accessed 14 January 2009

CEC (2007b) 'Adapting to climate change in Europe – Options for EU action', Green Paper, SEC

(2007) 849final, Brussels, Commission of the European Communities, http://eur-lex.europa.eu/LexUriServ/LexUriServ.do?uri=CELEX:52007DC0354:EN:NOT, accessed 14 January 2009

CEC (2007c) 'Trans-European networks', http://ec.europa.eu/ten/, accessed 14 January 2009

CEC (2008a) 'Green communication from the Commission to the Council, the European Parliament, the Committee of the Regions and the European Economic and Social Committee – Green Paper on territorial cohesion : Turning Diversity into Strength', {SEC(2008)2550}, Brussels, 6 October 2008, COM(2008) 616 final, http://ec.europa.eu/regional_policy/consultation/terco/paper_terco_en.pdf, accessed 14 January 2009

CEC (2008c) 'Communication from the Commission to the European Council – A European economic recovery plan', COM(2008) 800 final, 26 November 2008, Brussels

CEC (2008c) 'Commission staff working document: Regions 2020 – An assessment of future challenges for EU regions', SEC(2008) 2868 final, 14 November 2008, Brussels.

Davoudi, S. (2003) 'Polycentricity in European spatial planning: From an analytical tool to a normative agenda', *European Planning Studies*, vol 11, no 8, pp979–999

Davoudi, S. (2005a) 'The northern way: A polycentric megalopolis', *The Yorkshire and Humber Regional Review*, Spring 2005, pp2–4

Davoudi, S. (2005b) 'Understanding territorial cohesion', *Planning Practice and Research*, vol 20, no 4, pp433–441

Dieleman, F. M., Dijst, M. and Burghouwt, G. (2002) 'Urban form and travel behaviour: Micro-level household attributes and residential context', *Urban Studies*, vol 39, no 3, pp507–527

Downs, A. (1972) 'Up and down with ecology: "The issue attention cycle"', www.anthonydowns.com/upanddown.htm, accessed 14 January 2009

EU Member States (2007) 'Declaration on the fiftieth anniversary of the signature of the Treaties of Rome', www.eu2007.de/de/News/download_docs/Maerz/0324-RAA/English.pdf, accessed 14 January 2009

EU Ministers for Spatial Planning and Development (2007a) 'Territorial agenda of the European Union: Towards a more competitive and sustainable Europe of diverse regions', www.bmvbs.de/Anlage/original_1005295/Territorial-Agenda-of-the-European-Union-Agreed-on-25-May-2007-accessible.pdf, accessed 23 January 2009

EU Ministers for Spatial Planning and Development (2007b) 'The territorial state and perspectives of the European Union: A background document for the Territorial Agenda for the European Union' www.bmvbs.de/Anlage/original_1005296/The-Territorial-State-and-Perspectives-of-the-European-Union.pdf, accessed 23 January 2009

EU Ministers for Spatial Planning and Development (2007c) 'Leipzig Charter on sustainable European cities', www.eu2007.de/en/News/download_docs/Mai/0524AN/075DokumentLeipzig Charta.pdf, accessed 14 January 2009

EU Ministers for Spatial Planning and Development (2007d) 'Maps on European territorial development', www.bmvbs.de/Anlage/original_1002205/Maps-on-European-territorial-development.pdf, accessed 14 January 2009

EU Ministers for Spatial Planning and Development (2007e) 'Contribution by the ministers responsible for spatial planning and development to the on-going public discussion on the Green Paper "Adapting to Climate Change in Europe – Options for EU actions"', www.mrr.gov.pl/NR/rdonlyres/F73098BF-42D2-409A-B41A-B11FDF5B3D98/41195/Contribution_Green Paper_16Nov_vFf.pdf, accessed 14 January 2009

European Spatial Planning Observation Network (ESPON), www.espon.eu/, accessed 14 January 2009

ESPON 1.1.1 (2004) 'Potentials for polycentric development in Europe', www.espon.eu/mmp/online/website/content/projects/259/648/file_1174/fr-1.1.1_revised-full.pdf, accessed 14 January 2009

ESPON (2006) 'ESPON Atlas – Mapping the structure of the European territory (ESPON Project 3.1)', www.espon.eu/mmp/online/website/content/publications/98/1235/index_EN.html, accessed 14 January 2009

Faludi, A. (2007a) 'Making sense of the Territorial Agenda of the European Union', *European Journal of Spatial Development*, www.nordregio.se/EJSD/refereed25.pdf, accessed 14 January 2009

Faludi, A. and Waterhout, B. (2006) 'Introducing evidence-based planning', *disP*, vol 165, no 2, pp4–13

Federal Office for Building and Regional Planning, ESPON Monitoring Committee (2006) 'ESPON Atlas – Mapping the structure of the European territory', October, Bonn

Fischer, T. B. (2001) 'Towards a better consideration of climate change and greenhouse gas emission targets in transport and spatial/land use policies, plans and programmes', paper prepared for the Open Meeting of the Global Environmental Change Research Community, Rio de Janeiro, 6–8 October

Friends of the Earth (2005) 'Why airport expansion is bad for regional economies', www.foe.co.uk/resource/briefings/regional_tourism_deficit.pdf, accessed 14 January 2009

German Presidency of the EU (2007) *Essentials for European Territorial and Urban Development Policies: Leipzig Charter on Sustainable European Cities – The Territorial Agenda of the European Union*, Federal Ministry of Transport, Building and Urban Affairs, Berlin www.bmvbs.de/Anlage/original_1010200/German-EU-Presidency-Essentials-for-European-Territorial-and-Urban-Development-Policies.pdf, accessed 14 January 2009

Gizard, X. (2008) 'Synthesis of the workshops – Conclusions', presented at Territorial Cohesion and the Future of Cohesion Policy, Paris, 30–31 October 2008. see www.conference-cohesionue2008.fr/

Hart, T. (1993) 'Transport investment and disadvantaged regions: UK and European policies since the 1950s', *Urban Studies*, vol 30, no 2, pp417–435

Hübner, D. (2008) 'The future of regional policy', *Regions and Cities of Europe: Newsletter of the Committee of the Regions*, vol 60, no 7, May–June, p7

Jensen, O. B. and Richardson, T. (2004) *Making European Space: Mobility, Power and Territorial Identity*, Routledge, London

Kingdon, J. W. (1995) *Agendas, Alternatives and Public Policies*, 2nd Edition, Longman, New York

Kloostermann, R. C. and Musterd, S. (2001) 'The polycentric urban region: Towards a research agenda', *Urban Studies*, vol 38, no 4, pp623–633

Meijers, E. J. and Romein, A. (2003) 'Realising potential: Building regional organising capacity in polycentric urban regions', *European Urban and Regional Studies*, vol 10, no 2, pp173–186

Meijers, E. J., Waterhout, B. and Zonneveld, W. A. M. (2007) 'Closing the GAP: Territorial cohesion through polycentric development', *European Journal of Spatial Development*, October, no 24, www.nordregio.se/EJSD/refereed24.pdf, accessed 14 January 2009

Mol, A. P. J. and Sonnenfeld, D. A. (2000) 'Ecological modernization around the world: An introduction', *Environmental Politics*, vol 9, no 1, pp3–16

Newman, P. and Kenworthy, J. (1999) *Sustainability and Cities: Overcoming Automobile Dependence*, Island Press, Washington DC

Présidence Française de L'Union Européenne (2008a) 'Cohesion serving the regions', Informal Meeting of Ministers for Spatial Planning and Cohesion Policy, www.ue2008.fr/webdav/site/PFUE/shared/import/1126_reunion_cohesion_territoriale/1126_reunion_cohesion_territoriale_Press_Kit_EN.pdf, accessed 14 January 2009

Roberts, P. (2003) 'Sustainable development and social justice: Spatial priorities and mechanisms for delivery', *Sociological Inquiry*, vol 73, no 2, pp228–244

Rothengatter, W. and Sieber, N. (1993) *Verkehrspolitisches Handlungskonzept für den raumordnungspolitischen Orientierungsrahmen – Heft 60*, Bundesforschungsanstalt für Landeskunde und Raumordnung, Bonn

Royal Commission on Environmental Pollution (RCEP) (2002) *The Environmental Effects of Civil Aircraft in Flight*, Royal Commission on Environmental Pollution, London

Shaw, D. and Sykes, O. (2004) 'The concept of polycentricity in European spatial planning: Reflections on its interpretation and application in the practice of spatial planning', *International Planning Studies*, vol 9, no 4, pp283–306

Sykes, O. and Shaw, D. (2005) 'Tracing the influence of the ESDP on planning in the UK', *Town and Country Planning*, March 2005, pp108–110

Tatzberger, G. (2003), 'Territorial cohesion in Europe: Its genesis and interpretations', paper presented at the Third Joint Congress of the Association of European Schools of Planning and the Association of Collegiate Schools of Planning, Leuven, Belgium, 8–12 July

Tatzberger, G. (2008) *A Global Economic Integration Zone in Central Europe? Vienna-Bratislava-GyŒr as a Laboratory for EU Territorial Cohesion Policy*, Austrian Institute for Regional Studies and Spatial Planning (ÖIR), Vienna

Town and Country Planning Association (2006) 'Connecting England: A framework for regional development', www.tcpa.org.uk/press_files/pressreleases_2006/CONNECTING_ENGLAND.pdf, accessed 14 January 2009

Waterhout, B. (2007) 'Territorial cohesion: The underlying discourses', in A. Faludi (ed) *Territorial Cohesion and the European Model of Society*, Lincoln Institute of Land Policy, Cambridge, MA, pp37–59

10

California's Climate Change Planning:
Policy Innovation and Structural Hurdles

Stephen Wheeler

Introduction

In the absence of national-level action in the United States during most of the 2000s to plan for climate change, states, regions and local governments have taken the lead. In the US system of federalism, states can adopt a wide range of policies that go well beyond those of the national government. The State of California, for example, has air quality regulation substantially stronger than the US as a whole, and many states have passed environmental quality acts and energy policies tougher than those approved in Washington. Local governments for their part have great authority over land use planning, building codes, transportation systems, recycling, water systems and other areas of activity important to reducing greenhouse gas (GHG) emissions and adapting to climate change.

As of 2008, 29 states had prepared some sort of climate change plan, and more than 170 local governments had joined the Cities for Climate Protection (CCP) campaign which requires that a plan be developed. However, most of these plans are only a first step towards addressing the problem (Wheeler, 2008). They typically establish policy to green the public sector by requiring public buildings to be certified under the Leadership in Energy and Environmental Design (LEED) standards, public fleets to be energy efficient or to use alternative technologies or fuels, and public agencies to use energy audits to improve the efficiency of their facilities. Most states and a few cities have also adopted renewable portfolio standards for utilities, requiring that a certain percentage of electricity sold in their jurisdiction be generated from renewable sources.

However, neither states nor cities have developed or implemented the full range of programmes needed to reduce GHG emissions. Few have adopted regulation for private sector activities or allocated substantial resources towards climate change programmes. Additional legislative approval is needed for many proposed actions, and this will be politically difficult to obtain. Almost no jurisdictions have adopted programmes for adapting to a changed climate. Existing US state and local climate change plans are in short largely aspirational, setting out ambitious goals and developing initial inventories of GHG emissions, but without the regulatory changes, funding or political backing needed to begin actually reducing emissions.

Still, some jurisdictions are beginning to put together the sort of sustained, ongoing planning effort needed to address the global warming problem in the long run. California is among these leaders at the state level. Many other states are following its lead, for example, by adopting California requirements for reduced motor vehicle emissions. Given that state programmes have often been a laboratory for future national efforts (Rabe, 2002), California's climate change initiatives may well influence future federal actions. California's programmes are still in the early stages of implementation, with most relevant policy and legislation having been established only in 2005 and 2006. But, as with efforts to address local air pollution in recent decades, it looks likely that the state will become a trendsetter on climate change planning.

This chapter analyses California's climate change planning framework, seeking to identify elements that are particularly innovative or promising as well as obstacles to implementation and achievement of the state's GHG reduction goals. This analysis is based on review of planning documents, research reports, staff presentations and third-party comments, as well as interviews with state and local officials and a review of the broader literature on climate change policy nationally and internationally. It builds upon a previous project in which the author reviewed climate change plans of 29 states and more than 50 municipalities around the US, seeking to identify characteristics and limitations of American climate change planning through 2008 (Wheeler, 2008).

The California context

It should be said at the outset that California is different from other US states in ways that affect its ability to plan for climate change. In terms of climate, most parts of the state have milder winters than most of the rest of the country, reducing structural heating needs. The state also tends to have abundant sunshine, making solar

energy a more viable option, and has benefited from extensive hydropower, geothermal and wind resources. These factors lower per capita energy consumption and increase the renewable portion of the state's electric generation portfolio relative to other states. Since adoption of its Title 24 energy efficiency codes in 1978, California has also had the nation's strictest regulations for building construction, further lowering GHG emissions per capita.

However, California is also known for sprawling, automobile-oriented urban form, epitomized by Los Angeles, and for high levels of motor vehicle use. Some 40 per cent of GHG emissions in the state come from transportation (California Energy Commission (CEC), 2006), compared with 28 per cent in the rest of the country (United States Environmental Protection Agency (USEPA), 2007). In some jurisdictions, more than 50 per cent of GHGs arise from transportation. The state's development has also depended on pumping enormous volumes of water long distances, a very energy intensive activity with corresponding GHG emissions. Such factors reduce the energy savings gained from the mild climate and tough building regulation, and skew the source distribution relative to other states. Overall, Californians produce fewer GHG per capita than other Americans, an average of about 14 tons of CO_2 equivalent gases each year, only 59 per cent of the national average (23.4 tons).

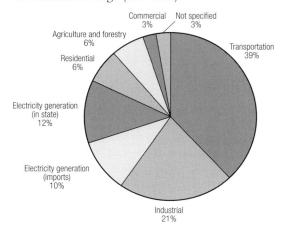

Source: California Air Resources Board (2009)

Figure 10.1 GHG emissions sources in California

Politically the state's electorate and legislature have been Democratic in recent years. However, since the early 1980s governors have been Republican, with the brief exception of former Governor Gray Davis in the early 2000s, a Democrat who was recalled in a special election in 2003 and replaced by Republican Arnold Schwarzenegger. California is known for a history of proactive environmental policy, especially regarding air quality, and is widely considered a trendsetter on political, cultural and economic issues. This unique culture has made the state a fertile ground for the development of climate change policy.

California's climate change planning to date

Building on the state's tradition of cutting-edge environmentalism, California policy makers have expressed concern about global warming for the best part of three decades (Franco et al, 2008). In 1988 a pioneering state law, Assembly Bill (AB) 4420 (Sher), led to a study of global warming risks and early efforts to develop a GHG inventory. The resulting 1991 report by the California Energy Commission helped move the climate change issue into public discussion. During the 1990s individual cities such as San Francisco, San Jose and Santa Monica initiated sustainable city programmes, and the state was home to the CCP campaign initiated by the International Council on Local Environmental Initiatives (ICLEI; recently renamed ICLEI – Local Governments for Sustainability). This campaign assisted local governments across the nation and internationally in developing climate change policy. Study of policy options for GHG mitigation was conducted at the staff level within state government (CEC, 1998), although until Schwarzenegger leadership did not exist in the governor's office or legislature to take action on such analysis.

The state's climate change planning efforts moved to another level in the early 2000s. In 2000, Senate Bill (SB) 1771, authored by long-time environmental legislator Byron Sher, established the California Climate Action Registry. This non-profit agency enables public and private entities throughout the state to voluntarily record their emissions and has played a key role in standardizing emissions reporting protocols. Such standardization is essential to future implementation of any market-based emissions trading framework. In 2002, AB 1493 (Pavley) set forth lower standards for CO_2 emissions from motor vehicles sold in the state, a step that was widely seen as a way around the federal government's long-time refusal to raise mileage standards for cars and light trucks. Sixteen other states then announced that they would implement the California standard. To enter into effect, this regulatory measure required a waiver from the US Environmental Protection Agency (EPA). The Bush Administration stalled this request and eventually declined it in late 2007. However, the state litigated the Bush decision, and standards similar to California's were eventually endorsed by the Obama administration.

In 2005 Governor Schwarzenegger's Executive Order S-3-05 set emissions reduction targets of 2000 levels by 2010, 1990 levels by 2020 (approximately 30 per cent below 2020 business-as-usual levels and 15 per cent below 2008 levels), and 80 per cent below 1990 levels by 2050. This trajectory of reductions went far beyond the Kyoto goal (for the US, 7 per cent below 1990 levels by 2008–2012) that had been widely promoted in US public discourse, for example, through the Mayor's Agreement on Climate Protection initiated by Seattle Mayor Greg Nickels in 2005. The California targets can be seen as heralding a new generation of climate change planning in the US, stemming from international acknowledgement during the mid-2000s that far greater GHG reductions are necessary in order to avoid dangerous climate change. The 2005 Executive Order also directed the Secretary of the California Environmental Protection Agency (Cal EPA) to convene meetings with seven other agencies to coordinate actions on this topic, and to issue biannual reports on progress towards reducing the state's emissions as well as the impacts of global warming on the state. The resulting Climate Action Team (CAT) now includes representatives from 19 agencies, and has been the central coordinating body for the state's climate change planning.

In 2006 two pieces of legislation addressed electricity generation in the state, in particular SB 107 (Simitian) that established a renewable portfolio standard of 20 per cent by 2010 for the state's investor-owned utilities. Most public utilities have announced that they will meet or surpass this target. But the biggest breakthrough was AB32 (Nunez), the California Global Warming Solutions Act of 2006, which set the governor's 2020 reduction target into law and directed the state's Air Resources Board (ARB) to begin implementation of measures to meet this goal. Among other steps, AB32 required ARB to set a statewide emissions limit equal to 1990 levels, to require mandatory emissions reporting for large GHG emitters, to identify early actions that could begin reducing emissions, and to consider environmental justice implications of climate change policies. The Act also required the ARB to prepare a Scoping Plan

of proposed actions, and set a series of deadlines for this and other activities, culminating in regulations becoming operational by 1 January 2012.

Since 2006 AB32 has thus been the driving force behind California's climate protection planning, activating powerful regulatory institutions in the state such as the ARB that had previously been developed to combat southern California's notoriously bad local air pollution. Unlike plans in most other states, which were developed as advisory documents by state agencies or governors' offices, AB32 is a law passed by the legislature and signed by the governor directing state agencies to take the necessary actions necessary to reach GHG reduction goals. These agencies now have substantial authority to develop a wide range of implementing regulations and programmes themselves.

In September 2007 the ARB proposed 44 early action measures for the 2007–2011 time frame, including the following steps (CAT, 2007):

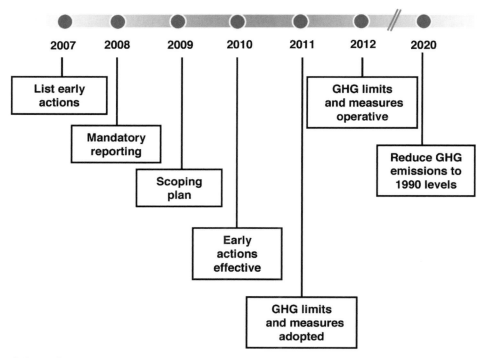

Figure 10.2 Timetable for implementation of AB32 in California

- establish low-carbon fuel standard (reduce carbon intensity of CA fuel 10 per cent by 2020);
- restrictions on high global warming potential (GWP) refrigerants;
- landfill methane capture;
- regulate off-road diesel emissions;
- ban SF6 in many applications;
- restrict high GWP applications in consumer products;
- 'SmartWay Transport': improvements in truck efficiency;
- reduction of perfluorocarbons from the semiconductor industry;
- 'Green Ports': provide electricity to ships in port;
- refrigerant tracking, reporting, and recovery;
- low carbon-fuel-based production of cement;
- anti-idling enforcement for trucks.

A few of these actions, such as providing electricity to ships in ports, were quickly implemented by the ARB. Others will require more lengthy implementation processes or action by local governments.

In June 2008 the ARB released its draft Scoping Plan (ARB, 2008). This document, based on the work of 3 advisory groups and 12 CAT sub-groups in different issue areas, outlined a wide range of more substantial longer-term actions in 18 different areas calculated to help the state reach the 2020 target. Main proposed policies included adoption of a market-based cap-and-trade system linked with other states in the western US, an increase in the renewable portfolio standard for utilities to 33 per cent, strengthening of building and appliance efficiency standards, and implementation of the Pavley standards for motor vehicle efficiency and other transportation measures. The Scoping Plan quantified prospective emissions reductions from each strategy. In essence, this plan and other state plans like it represent a real-world version of the 'wedges' approach to GHG reduction popularized by Steven Pacala and Robert Socolow in 2004, which sought to identify a handful of particularly promising strategies to reduce global GHG emissions by

the necessary amount (Pacala and Socolow, 2004). In this case, strategies have been screened for political and financial feasibility in the State of California, and no feasible strategy has been omitted no matter how small the potential reductions.

The draft Scoping Plan is notable for what it leaves out as well as what it includes. Land use is barely mentioned, although factors such as the compactness of communities and balance of land uses are widely viewed as affecting levels of driving and emissions (Ewing et al, 2007). Broad-based pricing initiatives, such as carbon taxes, gas taxes, road tolls, congestion pricing and feebates for purchase of efficient/inefficient vehicles, are also absent. Agricultural measures are not mentioned beyond recommended use of manure digester systems. The politically touchy issue of population is not discussed, although the state is projected to grow from 38.1 million residents currently to 59.5 million by 2050 (California Department of Finance, 2008), and it can be argued that such population growth will make efforts to cut overall emissions extremely difficult. Perhaps most surprisingly, the Scoping Plan does not aim to reduce motor vehicle use, even though a previous report from the ARB's Environmental Technical Assistance Advisory Committee (ETAAC) had stated that 'it is time to rethink current methods of mobility for both freight and people ... decreasing Vehicle Miles Traveled (VMT) is critical to meeting AB32 GHG emission reduction goals' (ETAAC, 2008, pp1–9).

The Scoping Plan does indicate that a number of these missing policy areas are under study for potential future adoption. Still, this document shows the ARB to be taking a highly pragmatic approach in which some of the most controversial potential strategies (changing land use, reducing motor vehicle use, instituting fees, promoting family planning to reduce population growth) are downplayed or omitted. Lack of land use strategies also reflects the ARB's predominant focus on 2020 (a 12-year time frame in which land use changes will have relatively small effect) rather than 2050 (by which these changes will presumably produce far larger results). Legislation approved by the legislature and

Table 10.1 Proposed California strategies for meeting 2020 goal of 1990 levels

Recommended strategy	Millions of metric ton CO2-eq. reduction by 2020
GHG emissions standards for vehicles	31.7
Increased efficiency for new appliances and buildings	26.4
Require utilities to provide 33% of electricity from renewables	21.6
Reformulated motor vehicle fuels	16.5
Reducing refrigerants and other non-CO_2 greenhouse gases	16.2
Forest management/forest fire prevention	5.0
Efficiency measures for existing vehicles, such as improved tyre maintenance	4.8
Increased water-related energy efficiency	4.8
Requiring more energy-efficient transportation of goods, such as electrification of ships in port	3.7
Increased efficiency standards for medium/heavy duty vehicles	2.5
California solar programme	2.1
Encourage local governments to build more walkable communities/reduce commuting	2.0
Reduction in state government carbon footprint	1.0–2.0
High speed rail	1.0
Landfill methane capture standards	1.0
Voluntary dairy methane capture	1.0
Unspecified cuts through cap-and-trade programme	35.2
Energy audits for large industrial emitters	unknown
TOTAL	At least 176.1
GOAL	169

Source: California Air Resources Board/Sacramento Bee (2008).

signed by Schwarzenegger in late 2008 will require development of regional 'blueprint' plans, coordinating land use, housing and transportation, and will give ARB the authority to require development of alternative plans if those submitted do not appear likely to reduce GHG emissions. But such plans will be advisory, and there is much political resistance to a stronger state role in regulating land use.

Policy innovation

Through such efforts to date, the State of California has been able to gear up a remarkably strong planning effort related to climate change. Many state agencies have been involved, coordinating their work with one another, meeting deadlines and producing high quality reports and planning documents. Separate pieces of legislation and executive action have built on one another, as, for example, AB32 built on Executive Order S-3-05 and both built on AB1493 and SB1771. Relatively strong targets have been set,[1] and with the release of the early action items and the draft Scoping Plan systematic action to meet them has been initiated. Media reaction to these initiatives has been primarily positive; Schwarzenegger, in particular, has received favourable press coverage around the world. The most serious crisis in the state's process arose in 2007, when Schwarzenegger fired ARB chairman Robert Sawyer and the agency's Executive Director Catherine Witherspoon resigned in protest. Both cited political pressure from the governor's office against strong climate change action (Wilson, 2007). However, Schwarzenegger avoided a major derailment of climate change planning by appointing Mary Nichols to the Executive Director position. An experienced Sacramento veteran, Nichols is highly regarded by environmentalists and has helped keep the agency on track towards its 2012 implementation deadline.

Although a number of other states have also been able to develop broad-based climate change strategies, the scope and depth of climate change planning in California as well as the potential for implementation of a broad range of policies goes beyond virtually all of these. The process used by many states, facilitated by the non-profit Center for Climate Strategies, convenes stakeholder groups over many months to produce a plan with around 55 action items. Some elements of these plans can be implemented through executive order, but most require additional legislative action. Once developed, state plans often languish awaiting legislation or executive action. Changes in political leadership have often sidetracked state efforts as well.

Particular innovations in California's climate change planning fall into several categories. The first has to do with goals. California was among the first jurisdictions to move well beyond Kyoto by adopting a very deep long-term target for GHG emissions reductions (80 per cent below 1990 levels by 2050). Three years later, this 2050 goal is still stronger than the vast majority of state climate change plans, most of which do not aim beyond 2020 at all (Wheeler, 2008). Seen in a global context, setting such a long-range target supported by science can be seen as a major step towards a successful post-Kyoto framework of climate change policy.

Secondly, California's climate change planning represents perhaps the fullest expression to date of a 'backcasting' approach to planning in which necessary targets are set and policy makers work backwards from that point to determine the necessary steps required. Such an approach is radically different from many other planning processes in which general goals or visions are set forth as well as policies aiming in their direction, but without rigorous quantification of likely success or consistent follow-up and revision to ensure success. These usual 'muddling through' processes have been seen as a pragmatic response to political realities in a pluralistic society (Lindblom, 1959). However, they often never reach their desired goals, and are inadequate to the task of addressing environmental crises in

which the end state is determined by scientific reality rather than the more flexible needs of social systems and political acceptability. Backcasting approaches to planning have been employed before, for example, in air quality regulation and efforts to preserve habitat for endangered species. But never has this style of planning been attempted on a scale that will require change in virtually every aspect of economy and society. California's climate change planning may thus be seen as a significant step towards a new style of planning appropriate to sustainable development generally.

For such a backcasting approach to gain political traction, it must be thoroughly supported by science so as to be credible. Scientific research into climate change effects and technology options is a third main area in which California has been an innovator. Using funds from a surcharge on utility bills in the state, the California Energy Commission's Public Interest Energy Research (PIER) programme annually awards up to $62 million for energy research and has sponsored 146 technical studies related to climate change since 1998. The CEC has also sponsored summary documents such as the 2006 'Our Changing Climate' report that have helped galvanize public concern about climate change (California Climate Change Center, 2006a). It is currently sponsoring a Scenarios project looking at implications for the state of different IPCC scenarios for future emissions (Cayan et al, 2006). California universities, corporations and labs have undertaken much additional research related to climate change, with or without state support; the role of national labs such as Lawrence Berkeley Laboratories is particularly worth mentioning. Although more research into social marketing and other political, social or institutional aspects of implementing climate change policy is needed, this massive effort at directed research is an important element of the state's climate planning process.

A fourth main area of innovation, as mentioned previously, lies in inter-agency coordination and stakeholder involvement. Rabe (2002) concluded that bipartisan support and coalition-building are important elements of successful state climate change planning efforts, and California seems to be incorporating these elements. Groups such as the CAT, the ARB and the CEC represent high-level, public involvement by state officials responding to climate change. Whereas officials in many other states were brought together in one-time processes to create climate change plans, California's efforts represent ongoing coordination that will presumably continue until at least 2020 or even 2050. At a staff level, the CAT's 12 sub-groups have provided extensive opportunities for networking and information-sharing between agencies. Other forums such as the annual Climate Change Research Conferences in Sacramento sponsored by the CEC and Cal EPA develop processes of education, diffusion and coordination further.

State Attorney General Jerry Brown's creative application of the California Environmental Quality Act (CEQA) to local planning represents yet another policy innovation. A colourful character, who is a former governor, former mayor of Oakland, and former candidate for President of the United States, Brown has used his current office to force local governments in the state to consider the climate change impacts of their policies. CEQA requires environmental impact review of all public and private projects in California, and has been a centrepiece of environmental protection in the state since its inception in 1970. However, climate change impacts of projects have never been considered until recently. In 2007 Brown sued San Bernadino County for not addressing GHG emissions within the Environmental Impact Report (EIR) for its General Plan, and reached a settlement under which the county agreed to include such analysis. His office has since sent letters to dozens of other California jurisdictions requesting specific changes within EIRs to add climate change analysis, striking fear into the hearts of many local officials, many of whom are pre-emptively adding climate change analysis and policy alternatives in order to avoid litigation. Through his interpretation of CEQA Brown is in effect establishing policy for the entire state regarding environmental review of climate change impacts, a course of action with

very large implications for day-to-day local planning. Somewhat belatedly, the Governor's Office of Planning and Research is drafting new CEQA guidelines that will formalize such requirements.

California's nascent cap-and-trade emissions trading system represents a final area of innovation for the state. Such a market-based system was previously applied to sulphur dioxide emissions in southern California, but has not been otherwise used in the US. Indeed, the European Union represents the only large-scale GHG emissions trading system in the world at this point. Because of its size, California can unilaterally initiate such a system without waiting for multi-state regional systems to develop. California intends its system to be a main part of the Western States Initiative in the long term, but will initiate activities of its own if need be. Within the state, intensive thought is now going into the potential design of such a system, for example, how to apportion or auction the emissions permits to begin with (e.g. California Air Resources Board Market Advisory Committee (MAC), 2007).

These policy innovations have been made possible by a number of factors. Political support from Governors Davis and Schwarzenegger as well as a Democratic-controlled legislature has certainly been important. The large size and economic clout of the state are also factors. But a major element has been the state's history of air quality and energy initiatives, which have left it with strong agencies such as the ARB, CEC and Cal EPA that are used to taking aggressive regulatory action. AB32 specifically empowers the ARB to use its regulatory authority to reduce emissions, and to levy fees in order to fund these steps. Such authority gives California agencies unusual power in implementing climate change policy and raising funds to support oversight, compared with other states.

Structural obstacles

Although California's process bears considerable promise, the state faces large structural obstacles in reaching its 2020 goal let alone the 2050 target. These obstacles are typical of other US states, but in some cases are amplified by California's size, history and diversity. The state clearly has the technical ability and resources to mitigate emissions – there is evidence not only that it can do this but that many climate change mitigations may yield a net economic benefit (California Climate Change Center, 2006b). But whether California or any other state can take action institutionally and politically to reach such climate change goals is open to question.

One basic problem is that control over land use is almost entirely local: as in most other parts of the US California has no history of statewide land use planning, as do Oregon, Washington, Maryland, Vermont, Florida and a few other states. Furthermore, California regional agencies have little authority to mandate 'smart growth' or other land use planning strategies. The regional 'Blueprint' plans that are currently being touted as the state's approach to smart growth are entirely voluntary in nature, meaning that they can simply be ignored by local governments intent on growth. While they do have some impact through education of policy makers and peer pressure upon local governments, these plans are only likely to have teeth if state infrastructure funding is conditioned upon local compliance, a step that has not yet happened and that would be resisted politically. Since transportation is such a large share of California's GHG emissions, and since land use changes are likely to be necessary to bring about deep reductions in transportation emissions, the state has a problem.

A related challenge is the state's tax structure which, following the adoption of Proposition 13 in 1978 and subsequent tax-cutting measures, actively encourages motor vehicle-dependent sprawl and leaves state and local governments strapped for resources. 'Fiscalization of land use' is a major problem in the state (Fulton and Shigley, 2006): local governments zone far-flung parcels of land for malls, big box stores and motor vehicle dealerships in an attempt to gain sales tax revenue, while avoiding much-needed land uses such as multifamily housing that would require services without providing compensating funds. The result is to encourage growth in

motor vehicle use and related GHG emissions. Without sufficient resources and faced with requirements for two-thirds votes of the electorate to raise taxes, local governments are having difficulty in implementing programmes that could reduce emissions, ranging from energy and water conservation to transportation and recycling. Such local programmes are likely to depend on assistance from higher-level government (Bailey, 2007), but the state itself is perpetually broke, unable to adequately fund important services such as schools. Proposition 13 is deeply entrenched politically, and is unlikely to be revoked anytime soon.

A related challenge is political: there is a deep divide between the two main political parties in the state, one of which is largely responsible for creating the current structural problems. With the Republican Party and its business allies sceptical of climate change in general, against policy measures perceived as regulatory burdens, and fiercely opposed to any actions resembling tax increases, the challenge of developing climate change policy, funding initiatives and changing pricing structures to promote energy conservation and reduce emissions is very large. For example, lobbyists from the oil industry, auto dealers, the California Chamber of Commerce and the Howard Jarvis Taxpayers Association have successfully mobilized against bills such as AB 2558 in 2008, a measure which would have allowed California regions the ability to impose climate mitigation fees on gasoline or vehicle registrations, and to use the proceeds to fund programmes such as public transit.

The political problem is amplified at the state level by a constitutional requirement that budgets be passed with a two-thirds majority. Although the Legislature is strongly Democratic, this provision essentially gives the minority Republicans veto power over new spending. The ARB can finance some initiatives through fees that it adopts itself, but major new financing for efforts such as retrofitting state buildings or providing financial incentives for alternative energy would need financing through the Legislature. Other initiatives, such as for a proposed high speed rail line between Los Angeles and the Bay Area, would be referred to the voters through referenda subject to the same two-thirds requirement, and are likely to be vigorously opposed by conservative political forces.

A final, more general structural problem has to do with the extent to which resource consumptive lifestyles are entrenched within California society, like US society generally. The ARB's approach to the 2020 target has been to identify relatively feasible regulatory changes that can produce the needed emission reductions, such as improving motor vehicle efficiency, reformulating fuels and raising renewable requirements for electric utilities. This strategy may work for 2020. But the much deeper cuts in emissions that will be needed for 2050 essentially require an end to the fossil fuel-based economy so central to California lifestyles. People will almost certainly need to drive and fly far less, challenging the ethic of mobility central to the culture. Vacation homes, recreational vehicles, powerboats and large houses may become problematic due to their resource demands. Living in hot, desert areas such as much of southern California may also become problematic due to needs for air conditioning and imported water. Californians do not seem ready to question such basic elements of their lifestyles, and a fundamental reorientation of social and individual values would be required to do so. Many experts view changes in lifestyle and related pricing of resources as among the most difficult measures to implement (Shaheen, 2008).

Conclusion

California, then, illustrates some of the more creative planning efforts to address global warming, and some of the most deeply entrenched structural obstacles to doing so. Whether the state can maintain the rate of progress on this topic made during the 2000s and achieve its long-term goals remains to be seen. Prospects for the short term are good; longer-term change will be more difficult. Much may depend on exogenous factors such as whether the price of petroleum continues to rise, whether the US adopts tough climate

change policy at a national level, and whether American social norms and lifestyles change across the board. The stakes are high: if California can help lead the US as a whole towards a more sustainable way of life, chances are greater that developing nations such as China and India will be motivated to reduce their own emissions, and that the world overall can respond to the global warming challenge. If California and the US cannot change in this way, future prospects are not nearly as bright.

Note

1 It is still debatable whether these targets are strong enough. The goal of 80 per cent reduction from 1990 levels by 2050 is the strongest target actually being adopted by jurisdictions internationally as of 2008, but offers only a 50 per cent likelihood of holding climate change to two degrees Celsius (Luers et al, 2007). Monbiot (2007) argues that 90 per cent reductions by 2030 are needed to do this. Also, the target of reaching 1990 levels in 2020 represented only a 16 per cent reduction from 2005 levels, or about 1 per cent a year. This trajectory is not nearly strong enough to reach the 2050 target, which would require annual reductions of between 3 and 4 per cent.

References

Bailey, J. (2007) *Lessons From the Pioneers: Tackling Global Warming at the Local Level*, Institute for Local Self-Reliance, Washington, DC

California Air Resources Board (2007) 'Proposed early actions to mitigate climate change in California', California Air Resources Board, Sacramento, CA

California Air Resources Board (2008) 'Climate change draft scoping plan: A framework for change', California Air Resources Board, Sacramento, CA

California Air Resources Board (2009) 'Greenhouse gas inventory data – graphs', California Air Resources Board, Sacramento, CA, www.arb.ca.gov/cc/inventory/data/graph/graph.htm, accessed 1 June 2009

California Air Resources Board Market Advisory Committee (MAC) (2007) 'Recommendations for designing a greenhouse gas cap-and-trade system for California', Sacramento, www.climatechange.ca.gov, accessed 7 September 2008

California Climate Action Team (CAT) (2007) 'Proposed early actions to mitigate climate change in California', California Environmental Protection Agency, Sacramento

California Climate Change Center (2006a) 'Our changing climate: Assessing the risks to California', Berkeley, www.climatechange.ca.gov/publications/biennial_reports/index.html, accessed 7 September 2008

California Climate Change Center (2006b) 'Managing greenhouse gas emissions in California', Berkeley, http://calclimate.berkeley.edu, accessed 7 September 2008

California Department of Finance (2008) 'E-1 city/county population estimates' and 'P-3 population projections', www.dof.ca.gov/HTML/DEMOGRAP/ReportsPapers/ReportsPapers.php#projections, accessed 1 September 2008

California Energy Commission (CEC) (1998) 'Greenhouse gas emissions reductions strategies for California', Staff Report, Sacramento

CEC (2006) 'Inventory of California greenhouse gas emissions and sinks, 1990–2004', Staff Final Report, Sacramento

Cayan, D., Luers, A. L., Hanemann, M., Franco, G. and Croes, B. (2006) 'Scenarios of climate change in California: An overview', California Climate Change Center, Sacramento

Economic and Technological Advisory Committee (ETAAC) (2008) 'Technologies and policies to consider for reducing greenhouse gas emissions in California', Final Report to the California Air Resources Board, Sacramento

Ewing, R., Bartholomew, K., Winkelman, S., Walters, J. and Chen, D. (2007) *Growing Cooler: The Evidence on Urban Development and Climate Change*, Urban Land Institute, Washington, DC

Franco, G., Cayan, D., Luers, A. L., Hanemann, M. and Croes, B. (2008) 'Linking climate change science with policy in California', *Climate Change*, vol 87, no 1, pp7–20

Fulton, W. and Shigley, P. (2005) *Guide to California Planning*, Third Edition, Solano Press Books, Point Arena, CA

Lindblom, C. (1959) 'The science of muddling through', *Public Administration Review*, vol 19, pp79–88

Luers, A. L., Mastrandrea, M. D., Hayhoe, K. and Frumhoff, P. C. (2007) *How to Avoid Dangerous Climate Change: A Target for U.S. Emissions*, Union of Concerned Scientists, Cambridge, MA

Monbiot, G. (2007) *Heat: How to Stop the Planet from Burning*, South End Press, Cambridge, MA

Pacala, S. and Socolow, R. (2004) 'Stabilization wedges: Solving the climate problem for the next 50 years with current technologies', *Science*, vol 305, pp968–972

Rabe, B. G. (2002) *Greenhouse & Statehouse: The Evolving State Government Role in Climate Change*, Pew Center on Global Climate Change, Arlington, VA

Shaheen, S. (2008) 'Summary of expert interviews & regional workshops on land use, transportation & climate change', Transportation Sustainability Research Center, Berkeley

United States Environmental Protection Agency (US EPA) (2007) '2007 U.S. Greenhouse Gas Inventory Report', Washington DC www.epa.gov/climatechange/emissions/usinventoryreport.html, accessed 10 September 2007

Wheeler, S. M. (2008) 'State and municipal climate change plans: The first generation', *Journal of the American Planning Association*, vol 74, no 4, pp481–496

Wilson, J. (2007) 'State Air Board official resigns', *Los Angeles Times*, 3 July, http://articles.latimes.com/2007/jul/03/local/me-airboard3, accessed 8 September 2008

11

Climate Change and Australian Urban Resilience:
The Limits of Ecological Modernization as an Adaptive Strategy

Jason Byrne, Brendan Gleeson, Michael Howes and Wendy Steele

Introduction

Climate change in Australia has elicited a polarized response. As elsewhere, most Australians are transfixed by the looming threats and mind-numbing scale of likely changes. Some simply deny the risk, while a few have begun to make small changes at the household scale (e.g. energy efficient light-bulbs or appliances, green power, public transportation, recycling, solar hot water) (see Slocum, 2004, for a Canadian comparison). But very few signs suggest that Australian society has begun to take the urgent action required if we are to stave off catastrophic climate change (Low, 2008). The likely consequences for the world's driest continent are dire indeed: prolonged drought and episodic rainfall, heightened storm intensity, increased flooding, extreme heatwaves and frequent bushfires, severe coastal erosion, widespread insect-borne diseases (e.g. dengue fever, malaria and Ross River virus), failing food-bowls, climate refugees, unprecedented species extirpation and, ultimately, the

need to abandon some settled areas (Allen Consulting Group, 2005; Australian Greenhouse Office, 2006; Buckley, 2007; Local Government Association of Queensland, 2007).

Australians are slowly awakening from the dream of unlimited prosperity to face the reality of ecological limits. We are confronted by an enormous challenge: how to adapt our settlements, agricultural systems and infrastructure to the coming changes. Urban areas will arguably feel the impacts of climate change the worst (Branz Ltd, 2007; Gill et al, 2007). As one of the most urbanized countries in the world, with over 90 per cent of the population concentrated in towns and cities (Gleeson and Low, 2000; Australian Bureau of Statistics, 2008) Australia has much to lose from climate change. Major cities are located close to or on the coast and many are highly vulnerable to climate change impacts (e.g. sea level rise, flooding and storm damage). While Australian planning has finally begun to take the first steps towards adapting our settlements to the future crisis, two crucial questions confront us: 'Are our actions too few or too late?' and 'How well can spatial planning respond to climate change challenges?' (Bulkeley, 2006; Lyth, 2006).

In this chapter we outline some Australian planning responses to climate change, focusing on the resilience of Australian cities and their ability to respond to anticipated climate change impacts.[1] We use the theoretical lens of ecological modernization, which proposes that economic growth can be decoupled from environmental harm via a technical, institutional and philosophical transformation. We consider how Australian planners have thus far addressed the challenges of adapting Australia's cities to cope with anticipated climate change impacts, acknowledging that many adaptive responses also have mitigative functions (e.g. decentralizing electricity generation via photovoltaic panels can lessen peak demand, lower emissions and reduce infrastructure vulnerability) (Hamin and Gurran, 2008).

We begin by briefly examining urban coordination at the Commonwealth level and the impact of past federal policy on Australia's response to climate change. Next we consider the implications of the Commonwealth Government commissioned Garnaut report (Garnaut, 2008), which establishes the case for and efficacy of national emissions reduction targets, and subsequent policies for Australian planning. We then examine current federal, state and local responses, using South East Queensland (Australia's fastest growing region) – and the Gold Coast specifically as a case study. Recognizing that adaptation will necessitate social change as well as policy and technological responses, and that climate change impacts will have social and environmental justice consequences, we discuss how Australian planners are tackling a range of climate change issues (e.g. climate responsive housing, more efficient transport systems, and infrastructure protection). We then consider why Australian planning systems have enabled or constrained adaptive responses to climate change. Our discussion illustrates how the nested scale of loosely coordinated policy responses actually plays out 'on-the-ground'. We conclude by considering what the Australian experience may offer the rest of the world.

Historical context

Contrary to the common image of a sprawling rural nation, Australia is, and has long been, one of the world's most urbanized societies. Suburbanization occurred early – from the late 19th century – and set the pattern for urban development and – for most people – everyday life from that point onwards. By 1900, around 50 per cent of Australians owned their homes, compared with just 10 per cent in the United Kingdom.

The suburban experience dominated national development and life during the 20th century and shows no sign of loosening its grip in the new millennium. The Australian suburban experience shares features with its counterparts in other developed nations – for example, from the mid 20th century, a low density, car-based urban form and a social ecology marked by a high degree of 'familism'. And yet it also demonstrates unique qualities which help to explain the political and institutional responses to recently manifesting ecological threats, notably climate change. As the American scholar, Bruegmann (2005), points out, Australia appears to be the only developed country where the political left developed a strong attachment to – and advocacy for – suburbanization. For instance, leading late 20th century Australian urban scholars such as Hugh Stretton and Patrick Troy, drew upon social justice perspectives in support of suburbanization, which they argued delivered to the working class both material wealth and the resources for a good life. Australia also developed relatively stronger planning systems and mechanisms than did other settler societies, especially the USA, to guide suburban growth and ensure that it was equitable as well as timely (Bruegmann, 2005).

The Australian environmental movement grew out of this suburban experience but soon turned its back on its birthplace (Davidson, 2006). In recent decades, green critique has tended to cast suburbia in increasingly dystopian terms – a sprawling landscape of waste and natural destruction – tending to undermine and to some extent confuse progressive politics. Suburbanites largely ignored the increasingly shrill declamations of environmentalism and, in the past decade, a powerful new politics of 'aspirational suburbia' was encouraged and drawn upon by a conservative national government seeking to subordinate ecological concerns to the imperative of economic growth. For the Howard national government (1996–2007), the suburban constituency was assumed to value material welfare and personal improvement over ecological and shared social concerns. During this time, the 'Commonwealth' – as Australia's national government is known – withdrew from any active participation in urban policy, eschewing any concern for the planning of urban development and active shaping of (sub)urban consumption to achieve sustainability. National coordination of urban change was rejected in favour of a technocratic faith in improved resource efficiency which assumed markets and industries would deliver in a context of seemingly unfettered growth. In the latter years of the Howard government, the problems of uncoordinated urban growth came home to roost in the form of failing housing markets, infrastructure deficits and congested transport systems. Environmental concern, especially about climate change, swelled in suburbia. In 2007, the Howard government was defeated in a national election dominated by a mood of anxiety, even anger, in Australia's suburban heartlands.

The patchy record of urban coordination at the Commonwealth level in the last three decades is partly explained by the influence of ecological modernization (EM), or at least its weak variants, on national thinking. In the face of mounting evidence of environmental failure, EM has buttressed an embedded faith in technological innovation and market adjustment as superior alternatives to active coordination of economic growth and urban development. The newly elected Rudd national government (2007) has re-entered the field of national urban policy, but thus far only weakly, emphasizing, for example, deficiencies in urban transport and energy infrastructure. Within Australia, urban environmental policy remains largely undeveloped.

Policy environment – the ecological modernization frame

EM is a label that has been attached to a paradigm advocating social change towards sustainability. The term is not widely used outside academia but the ideas to which it refers have been highly influential in shaping many environmental policies, plans and management systems through its links to sustainable development (Hajer, 1995; Weale, 1998; Mol and Sonnenfeld, 2000; Grant and Papadakis, 2004). The core argument of EM is that although democracy, the state and the market have gone astray, they can be restructured in a way that will make them sustainable (Christoff, 1996; Mol and Spargen, 2000; Dryzek, 2005; Howes, 2005). EM argues that economic growth can be decoupled from raw material throughput, energy use and waste generation by applying new technology and redesigning institutions (Berger, et al, 2001; Dryzek, 2005; Howes, 2005).

Ecological modernization presumes that economic and environmental goals need not be mutually exclusive (Gouldson and Murphy, 1997; Curran, 2001). Well designed interventions by government are assumed not to hinder economic growth but instead to stimulate new and more efficient industries (Blowers, 1997; Weale, 1998; Mol and Sonnenfeld, 2000). Industry reduces its costs through increased technological efficiency and both the environment and community benefit from less pollution and waste. Governments continue their regulatory roles but are also recast as facilitators assisting industry to become more sustainable. New policies are directed towards correcting market failures by improving information on the impacts of actions, imposing green tax regimes to internalize negative externalities, and pricing ecological goods and services to reflect their 'true' value (Costanza et al, 1997; Lundqvist, 2000). Proponents strongly emphasize retaining the key institutions of modernity (science, technology, the market, industry and the state) but embedding ecologically-reformed economic practices within them (Berger et al, 2001).

Ecological modernization ranges from the original weak 'techno-corporatist' approach – focusing mainly on technological change, to the strong 'reflexive' approach that encourages a political transition to an 'ecological democracy' (Christoff, 1996; Dryzek, 2005). Many variants fall between these two extremes but adopt some elements from each (Fisher and Freudenburg, 2001). The differences between the strong and weak versions of EM can be illustrated by reviewing five core themes in EM programmes for social change:

1 Technological innovation to foster efficient resource use and reduce damage (which plays a substantial role in both the strong and weak versions);
2 Providing economic imperatives as incentives for firms to improve their environmental performance (in both strong and weak versions);
3 Political and institutional change rendering policy making more open and flexible (modest in the weak version, substantial in the strong);
4 Transforming the role of social movements so they act as both watchdog and/or partner in decision making (in the strong version);
5 Discursive changes recasting environmental issues as opportunities to improve outcomes for the environment, business and the community (an essential aspect of the strong version).

These themes recur in various guises throughout a number of analyses (see Berger et al, 2001, pp58–59 and Welford and Hills, 2004, p325, for original material). However, only recently has the EM lens been directly applied to analysing policy and planning responses to climate change.

Australia's national response to climate change

Weaker versions of EM underpinned the Australian government's policy commitment to 'Ecologically Sustainable Development' in the early 1990s (Howes, 2005). They also supported the initially positive response to the 1992 'Framework Convention on Climate Change', where the Keating Labor government believed that technological innovation alone would effect simple and rapid cuts to emissions with limited economic cost (Bulkeley, 2001; Christoff, 2005). The transition to the conservative Howard government in 1996 coincided with some disillusionment with this stance and led to a more restrained policy response until 2007. For example, voluntary programmes for energy conservation were established, a modest mandatory renewable energy target was set for the energy sector and, to placate the coal industry, funds were provided to research carbon capture and storage. In 2001, Australia joined the US in abandoning the Kyoto Protocol and under pressure from growing public concern sought to establish the Asia–Pacific Partnership on Clean Development and Climate in 2006 as an alternative.

Following its election in 2007 and subsequent signing of the Kyoto Protocol, the Rudd Labor government embarked upon a year-long policy process to establish an emissions trading system. After it appointed Ross Garnaut (2008) to undertake the analysis and formulate discussion papers for public comment, the government released its final white paper in late 2008, committing Australia to cutting greenhouse gas (GHG) emissions by 5–15 per cent by 2020 and 60 per cent by 2050 (measured against a 2000 baseline) by:

• introducing a national emissions trading scheme covering 1000 of the major GHG emitters (that are responsible for 75 per cent of total national emissions) by 2010;
• establishing a €250 million fund to support research, development and deployment of carbon capture and storage technology;
• providing extra funds to help the energy sector and energy intensive export exposed firms adjust;
• funding further research into climate change by universities and the Commonwealth Scientific and Industrial Research Organisation (CSIRO);

- setting a target to generate 20 per cent of all energy from renewable sources by 2020;
- facilitating a climate adaptation strategy that includes €6.5 billion for a national 'Water for the Future' plan;
- providing a €100 million 'International Forest Carbon Initiative' to assist developing countries protect forests as carbon sinks;
- establishing domestic subsidies for household investments in solar hot water services, photovoltaic panels and rainwater tanks (initiatives that have been supported by the federal, state and local levels of government) (Australian Government, 2008).

This set of policies clearly exhibits the weak ecological modernization assumptions on which they are founded, with a strong focus on technical innovation and research funding. The emission trading system promotes an economic incentive for producers and consumers to change their behaviour, but has no regulations to effect this response. Although the policy-making process has been extensive and open to public scrutiny, the ability of environmental groups to effectively influence the final targets has been curtailed by powerful energy and export sectors. There is little evidence of stronger versions of EM and there has been no move to undertake a major restructuring of government, nor has the role of social movements been transformed (although just getting the emissions trading system on the policy agenda was a major victory). And there is little evidence of a major shift in policy discourses – the major focus has been upon costs for industry rather than benefits to society or the environment.

Despite this set of policies, a coherent national strategy for climate change adaptation, engaging all levels of government, has yet to emerge. To implement the above-described national framework, the Commonwealth government has relied upon the cooperation of state government heads through the Council of Australian Governments (COAG). The intent is to:

- improve government's ability to predict impacts;

- fund risk assessment programmes;
- fund programmes to develop adaptation strategies;
- educate decision makers and businesses;
- develop policy tools for state and local levels of government.

To paraphrase the Commonwealth government, adaptive responses include; 'bearing the loss, sharing the loss, modifying the threat, preventing impacts, changing uses, changing location, research, and education' (Allen Consulting Group, 2005, pp103–104). A key initiative of this strategy has been developing a national centre for climate change adaptation. According to COAG, the centre will: 'synthesise knowledge, coordinate and commission research, activities, broker research partnerships and provide information for decision makers in a form relevant to their sectoral or regional need'(Council of Australian Governments, 2007, p7). The primary aims then are to 'build adaptive capacity' and to 'reduce vulnerability'.

National responsibilities

The Commonwealth government (hereafter Commonwealth) has taken the lead on identifying likely impacts of climate change and assessing vulnerabilities at a regional level – charging the CSIRO and Bureau of Meteorology (BOM) with this task. And the Commonwealth has also taken responsibility for developing a national digital elevation model for better predicting impacts and has developed a national water initiative for coordinating information on stream-flow and water availability. Local councils will soon be able to consult regional impact atlases that illustrate the likely climate change impacts in their area.

But most Commonwealth actions have been directed towards funding state and local governments to undertake risk assessment or implement adaptive responses. Although the Commonwealth has identified €9 billion for its climate change response, it has thus far only set aside €7.4 million over four years for this purpose (Australian Greenhouse Office, 2005).

Local governments are eligible for €25,000 grants to undertake risk assessments – a paltry sum considering the scope and scale of the work required. Councils may also apply for a Community Water Grant to promote initiatives such as installing solar panels on school buildings and low flow showerheads, rainwater tanks and dual flush toilets in residential buildings. Another initiative has been to provide rebates (up to half the cost) to households for installing household photovoltaic power – which amounts to around €4000. And further rebates are available for solar hot water systems (about €500). Yet on-the-ground outcomes are scattered.

The Commonwealth has been mired in debates around the relative merits of mitigation (especially the national emissions trading scheme). The global economic crisis has hampered the Commonwealth's responses as the Rudd government has feared a voter backlash against spending on the environment. Investigative studies, policy reports, education strategies and economic levers (e.g. grants to state and local governments) have comprised the main actions, leading some CSIRO climate scientists, who estimate that a million homes are already at risk (Bardon, 2008), to plead for the Commonwealth government to override state and local governments to prevent development in future flood-prone areas and upon vulnerable coastlines through a national building code.

State government responses and responsibilities

State governments (hereafter State) meanwhile have scrambled to respond to the existing impacts of climate change – especially the prolonged drought that has gripped most of the continent (the worst on record). Too numerous to detail here, some states have adopted similar strategies that merit closer attention, as do some unique responses. However, many strategies now identified as 'adaptation responses' were already enshrined in policy documents, strategies and plans – typically under the rubric of sustainability or smart growth – suggesting that the catalogue of state responses is as much rhetoric as action.

Queensland (Qld.), Victoria (Vic.) and Western Australia (WA), for example, have commissioned desalination plants – WA's is powered by wind energy but Queensland's is reliant upon coal power. Queensland has linked dwindling water resources through an integrated delivery grid and New South Wales (NSW) is adopting a similar approach. WA's adaptation response has largely been to gather more information and monitor the pace and scale of change, an approach also taken by Tasmania (Tas.) and the Northern Territory (NT). But some responses are unique. The NSW government has enacted carbon trading legislation, Tasmania has undertaken an audit of government emissions and the Australian Capital Territory (ACT) now requires full disclosure of building energy efficiency at the time of sale. These and other state government adaptation responses are summarized in Table 11.1.

Queensland's response

Queensland's response typifies that of many other states. Queensland has set aside €215 million for climate change mitigation and adaptation – €150 million of which is reserved for a climate fund to generate annual funding for climate projects. The state released a climate change strategy in 2007 called 'Climate Smart 2050', which – like other Australian states – sets a target of reducing carbon emissions by 60 per cent of 2000 levels by the year 2050. The state has been investing heavily in clean coal and carbon capture technology (€150 million) as most electricity is generated from coal and Queensland exports enormous quantities of coal globally (currently earning the state €7.5 billion per annum) (Queensland Government, 2007). But solar, wind, geothermal and biomass technologies are also being funded – though currently to a lesser extent.

The Climate Smart strategy has established a number of adaptation measures including a ban on broadscale vegetation clearing and financial incentives for households to minimize water consumption and improve energy efficiency. For instance, until recently a Home Waterwise

Table 11.1 State government responses to climate change

	Queensland	New South Wales	Australian Capital Territory	Victoria
Adaptation strategy	Climate Smart 2050	Greenhouse Plan 2005	Weathering the Change 2007–2025	Our Environment Our Future Action Plan
Climate fund	€151 million	€171 million	None	None
Legislation	Vegetation Management Act (1999)	• Carbon Rights (1998) • Native Vegetation Act (2003)	None	None
Research initiatives	• Clean coal • Geothermal • Fuel cell • Cloud seeding • Livestock	• Bushfires • Water • Biodiversity • Weeds and pests • Geosequestration	• Urban impacts	• Carbon storage • Agricultural adaptations • Impact modelling
Environmental measures	• Tree clearing ban • Reafforestation • New national parks • Wildlife corridors	• Tree clearing ban • Tree plantations • Farm forestry • Wildlife corridors	• Tree clearing ban • Reafforestation • Bushfire abatement • Homeowner incentives • Wildlife corridors	• Vegetation mapping • Reafforestation • New national parks • Wildlife corridors • Catchment management
Education programme	• Awareness campaign • Behavioural change • School curriculum	• Awareness campaign • Behavioural change • Farmer training • School curriculum	• Awareness campaign • Behavioural change	• Awareness campaign • Behavioural change
Water resources	• Water efficiency rebates • Water restrictions • Overconsumption fines • Rainwater tank rebate • Desalination • New dams • Water grid • Wastewater recycling	• Water efficiency standards • Water restrictions • Water grid	• Water efficiency	• River protection legislation • Water restrictions
Energy measures	• Energy efficiency standards • Renewable energy targets • Clean coal • Electric water heater ban • Solar feed in tariff • Compact fluorescent bulb giveaway • Natural gas rebate • Smart meters	• Energy efficiency standards • Light emitting diode traffic signals • Mandatory minimum energy performance standards • Methane capture and re-use • Carbon trading	• Energy efficiency standards • Methane capture and re-use • Solar feed in tariff • Solar hot water rebate • Energy efficient streetlights • Mandate 4 star rating for new buildings	• Energy efficiency standards • Renewable energy targets • Clean coal • Smart meters • Mandatory minimum energy performance standards
Transport changes	• Public transit upgrades • Fuel efficient vehicle registration discount	• Public transit upgrades • Clean car consumer guide	• Public transit upgrades • Fuel efficient vehicle registration discount	• Public transit upgrades • Government hybrid vehicles • Modal shift education

Tasmania	South Australia	Western Australia	Northern Territory
Framework for Action on Climate Change	Greenhouse Strategy 2007–2025	Making Decisions for the Future (2007)	Strategy under development
None	None	None	None
Climate Change State Action Act (2008)	Climate Change and Greenhouse Emissions Reduction Act (2007)	• Carbon Rights Act (2003) • Tree Plantations Agreement Act (2003)	None
• Biosequestration • Agriculture • Antarctic impacts	• Dryland salinity	• Emissions reduction • Geothermal • Biodiesel • Methane capture • Biosequestration	None
• Forest management	• Reafforestation • Conservation strategy • Forecast habitat impacts	• Monitoring fisheries impacts • Conservation strategy	• Vegetation management
• Awareness campaign • Behavioural change • School curriculum • Training programmes	• Awareness campaign • Behavioural change • School curriculum • Training programmes	• Awareness campaign	• Behavioural change
• Water efficiency • Rainwater tanks • Review of water allocation plans • Water restrictions	• Water efficiency • Water restrictions	• Desalination • Wastewater recycling • Consumption targets	• Water efficiency
• 100% renewable energy target • Energy efficiency • Solar feed in tariff • Light emitting diode traffic signals	• Energy efficiency standards • Solar feed in tariff • Renewable energy targets • Solar hot water for new homes • Smart meters • Solar panel installations on government buildings	• Energy efficiency standards • Renewable energy targets • Gas-fired power-station • Clean coal • New government-funded wind-farms • Solar hot water rebate	• Energy efficiency incentives
• Public transit upgrades • Government electric vehicles	• Alternative fuel bus fleet • Government alternative-fuel vehicles	• Public transit upgrades • Modal shift education programme	None

Table 11.1 *continued*

	Queensland	New South Wales	Australian Capital Territory	Victoria
	• Mandate ethanol fuel blend • Cycle network upgrades • Modal shift education programme • Alternative fuel bus fleet	• Alternative fuel ferries • Transit integration	• Alternative fuel bus fleet • Cycle network upgrades • Free bus travel for bike riders • Replace government fleet vehicles	• Alternative fuel bus fleet
Agricultural responses	• Managing livestock and soil emissions • Clearing offsets • Promoting farm forestry	• Managing livestock and soil emissions • Farmer training • Promoting farm forestry	None	• Ecosystem services market valuation • Promoting 'healthy soils' • Promoting vegetation protection
Health programme	Researching health impacts	Promoting hospital co-energy generation	None	• Developing a heat-wave emergency plan • 'Greening' hospitals and aged-care
Commercial/ Industrial	• Energy efficiency audits • New energy efficiency standards	Voluntary greenhouse rating scheme	Use national building code	• New energy efficiency standards
Housing	• Demonstration sustainability houses • Requirements for water and energy efficiency • Phase out of electric hot water systems and rebates for gas a solar	• Retrofit state-owned housing with energy and water efficient fixtures • Education campaign on air-conditioner use	• Mandatory disclosure of energy efficiency at time of sale • Mandatory 5 star rating for all new housing	• Retrofit state-owned housing with energy and water efficient fixtures • Mandatory 5 star rating for all new housing • Gas-boosted solar hot water for all new houses • Introducing mandatory disclosure of energy efficiency at time of sale
Land use planning responses	• Draft South East Queensland Regional Plan (2009-2031) • State planning policy under development • State Coastal Management Plan • Coastal vulnerability assessment • Storm gauge and tide buoys being installed • Disaster management plan under review	• NSW metropolitan strategy • Streamlining approvals process for low-emissions technology • Introducing planning guidelines promoting walking and cycling • Promoting a hydrogen economy • Developing infrastructure to facilitate waste to energy technology	• Fostering integrated land use and transport planning • Regional vulnerability assessment • Assessing climate change impacts in urban areas • Undertaking social impact analysis • Draft water sensitive-urban design guidelines	• Melbourne 2030 strategic plan • Draft Victorian coastal strategy • Reviewing development standards for river protection • Green-wedge management plans • New urban growth boundary • Transit oriented development

Tasmania	South Australia	Western Australia	Northern Territory
• Modal shift education programme • Alternative fuel bus fleet • Fuel efficient vehicle registration discount • Driver education	• Modal shift education programme	• Alternative fuel bus fleet • Cycle network upgrades • Subsidy for LPG fuel conversion subsidy • Investigating regenerative braking in trains • Investigating vehicle emissions testing	
• Researching sustainable yields • Food impact report • Climate-smart farms • Emissions measurement tools	• Seed conservation • Trialling new crops • Protecting viable agricultural land • Support tools for farm management	• Researching alternative agriculture • Promoting farm forestry	• Drought tolerant crops • Monitoring and controlling pests and weeds
None	None	Reviewing potential impacts	None
None	• New energy efficiency standards • Foster micro-wind turbine technology	None	None
• Energy efficiency audits • Means-tested insulation upgrade rebates • Mandatory minimum energy efficiency standards	• System to rate residential sustainability	• Introducing 5 star rating for all new housing • Water efficiency requirements	None
• Introducing climate change impact statements into decision-making • Assessing coastal risks • Including climate change in regional planning • Promoting transit integrated planning • Restricting development in floodplains • Hazard management and response planning	• Reviewing vulnerability of critical infrastructure • Reviewing hazards and emergency planning • Mapping coastal vulnerability • Fostering transit-oriented development • Removing development approval requirements for solar panel installation	• State coastal planning strategy • Network city strategy • Promoting transit-oriented development • Strengthening residential design code to include water sensitive urban design and passive solar design	• Avoiding building in storm surge zones • Improved cyclone-resistance building codes • Guidelines for climate-oriented design

Scheme was in place. For a mere €10, a licensed plumber would visit a home and check for leaks, install a low-flow showerhead and water pressure regulator, and recommend water conservation measures. This scheme cost the state €9 million per annum but unfortunately has now been axed in the face of the global financial crisis and dwindling state coffers.

The Queensland government continues to subsidize domestic rainwater tanks, solar hot water systems and energy efficient appliances (e.g. front-loading washing machines). A million compact fluorescent light-bulbs have been given away to households as part of an education campaign, and the Sustainable Housing Code requires that all new housing is equipped with energy efficient lighting, efficient hot water (e.g. solar, gas or heat exchange) and water conserving devices. Demonstration sustainability houses have been constructed across the state to showcase these technologies; sub-tropical design incorporating passive ventilation is a strong feature (South East Queensland is expected to experience up to 30 days a year over 35 degrees Celsius by 2070 compared with the current 3) (Local Government Association of Queensland, 2007; Australian Greenhouse Office, 2008).

The state government has also recently activated a state-of-the-art water recycling plant to recycle wastewater for commercial and domestic consumption and has established an electricity buy-back scheme where surplus power is purchased back from domestic photovoltaic installations at three times the domestic tariff (44 cents per kilowatt hour). The state has committed €117 million to an integrated regional cycling network and is expending large sums on light rail projects to improve public transport patronage. Transit oriented development and 'smart growth' (e.g. compact urban forms and densification) are now enshrined as key land use planning strategies, such as the South East Queensland (SEQ) Regional Plan.

Just five years ago, the South East Queensland Regional Plan (2005–2026) contained only five references to climate change and GHGs – most of them referring to mitigation. The newly released Draft South East Queensland Regional Plan (2009–2031) has its

primary regional policy now dedicated to climate change and sustainability; adaptation features prominently. The plan is a statutory document requiring local government town planning schemes to comply with its provisions. These provisions now include the development of an SEQ Regional Climate Change Management Plan, 'protection from climate hazards and protection of food supplies', 'compact urban form', transit equity, passive solar design and ventilation, requirements for risk analysis, extensive tree planting and better protection of – and access to – open space, to name just a few (The State of Queensland, 2008).

Local government responses

The response at the local level has also been mixed. Once again, it is beyond the scope of this chapter to document all local government responses, but some examples will illustrate the range of actions being taken – some duplicate national and state efforts.

For example, some Victorian councils (e.g. Wellington Shire) have attempted to place a moratorium on coastal development (later withdrawn due to resident and property developer opposition). The Melbourne City Council has constructed Australia's first six-star green building with passive ventilation and cooling, solar power, natural lighting and vegetation walls (Council House 2). Some New South Wales councils have begun to designate areas specifically for habitat conservation, anticipating the likely impacts that climate change will have on biodiversity (Hamin and Gurran, 2008). Wyong Council (NSW) is establishing a buy-back fund for properties damaged by rising sea levels and the Greater Taree Council is determining its legal liability for past approvals of coast-front developments. The Ku-ring-gai Council has commenced harvesting stormwater to irrigate its parks and greenspaces. In Western Australia, the City of Melville has a grey-water re-use programme for similar purposes. South Australia's Port Adelaide–Enfield Council has commissioned a flood risk study and its District

Council of Yorke Peninsular recently refused a coastal development on the grounds of anticipated sea level rises. Finally, the Darwin City Council in the Northern Territory has developed an environmental management strategy to protect natural resources vulnerable to climate change impacts (SMEC Australia, 2007).

These examples show the wide variety of adaptive responses being taken at the local level. Our empirical focus on South East Queensland and the Gold Coast City Council demonstrates that local, regional and state coordination of some responses is beginning to occur, albeit in an ad-hoc and incremental fashion.

Climate change adaptation on the Gold Coast

South East Queensland (SEQ) is the fastest-growing region in Australia (Department of Infrastructure and Planning, 2008). In the next two decades the region's population is expected to increase by almost 2 million, from the current level of around 2.8 million – generating demand for over 700,000 new dwellings. This explosive growth is being driven by sea-change migration to the area's attractive landscapes, expanding economy and relaxed lifestyle associated with its sub-tropical climate. Although the region spans an area of 22,890km^2, the population is heavily concentrated in coastal towns and cities including Brisbane (Queensland's capital), and the Gold Coast (Australia's sixth largest city). In reality, the largely suburban settlements from Noosa in the north to Coolangatta in the south are becoming a 200km long conurbation (Spearitt, 2008).

A variety of landscapes characterize the region, including rainforest-covered extinct shield volcanoes, broad sandy beaches punctuated by headlands, extensive estuaries, mangrove-fringed rivers, and a wide coastal plain – some of which is still cultivated for sugar cane. Parts of the built environment are extremely susceptible to flooding; Brisbane and the Gold Coast have both sustained heavy damage in the past. And the Gold Coast which straddles a narrow coastal shoreline and broad floodplain is especially vulnerable (see Figure 11.1) (Godber, 2005).

Australia's glitzy tourist destination, the Gold Coast, is a strange amalgam of extravagant consumption (e.g. super-yachts, skyscraper apartments, canal-front mansions and 'ocean view jacuzzis'), tourism (swanky hotels, caravan parks, theme parks and ecotourism), beach culture (gold lame bikinis and world-renowned surf-breaks) and nature-oriented suburban living (e.g. family parks, shopping malls and palm-tasselled backyard pools). Some parts of the city are not dissimilar to Spain's Costa del Sol, Florida's Fort Lauderdale or Mexico's Cancún; others have more in common with Hawaii or Costa Rica. As the most biodiverse city in Australia, and the fastest growing settlement in the region, the city's climate change responses warrant closer inspection.

The Gold Coast City Council (GCCC) is pursuing a variety of mitigation and adaptation measures. As part of its commitment to the Cities for Climate Protection programme, Council has announced an ambitious goal to be carbon neutral by 2020 (GCCC, 2001, 2007). Although its adaptation responses have been widely praised, many were in fact pre-existing and linked to sustainability objectives (e.g. Local Agenda 21). For example, GCCC has a longstanding partnership with the Queensland State Government for a beach re-nourishment programme to replenish beaches suffering erosion. And its town planning scheme has pre-existing planning policies and guidelines to manage coastal and flood-prone development (e.g. dune revegetation, water sensitive urban design and mosquito control). For instance, new beachfront houses have long been required to install a protective rock wall behind the coastal dunes – able to withstand a 1 in 50 year storm event (i.e. large cyclone). Yet GCCC often cites this as a climate change adaptation.

But some of GCCC's climate change responses are new. It has backed sustainable housing demonstration projects (three thus far – see Figure 11.2), is undertaking water and energy efficiency initiatives for its buildings, and is installing light emitting diodes in the city's traffic lights to reduce electricity consumption.

Photo: Jason Byrne

Figure 11.1 Gold Coast built environment

Its swimming pools will soon be heated using solar heating and its beachfront barbecue facilities will be upgraded to energy efficient appliances. GCCC is raising the wall of the city's main dam – the Hinze Dam – to increase its storage capacity and to mitigate downstream flooding, and now has a policy requiring an additional 27cm of building floor clearance above the state's 1 in 100 year ARI flood level (based on CSIRO modelling). In conjunction with the state government, GCCC is establishing a Sustainable Housing Code that will cover all new houses built on the Gold Coast and has established a series of wildlife corridors to preserve ecological connectivity between the coast and the hinterland rainforests. Finally, the new desalination plant (funded partly by GCCC and partly by the state) is also an adaptive measure – though a dubious one given its coal-fired power-source.

Planning systems issues and adaptive strategies

Why then is planning for climate change in Australia so fragmented, and why have the land use planning responses been so late? Australian systems of governance and the various planning systems operating within the country are partly responsible. Unlike the United Kingdom or New Zealand, the Australian Commonwealth government has no constitutionally defined planning powers or obligations. Neither, however, is it prevented or hindered by the constitution from undertaking urban policy. Instead, like the United States, planning has largely been a 'state concern'. In Australia, the seven states and two territories have promulgated different types of planning legislation. Although there are some similarities between these sub-national systems, no two are the same (for a comprehensive review see Gurran, 2007).

Photo: Jason Byrne

Figure 11.2 Sustainable house, Gold Coast

Most Australian states and territories have metropolitan planning instruments, regional planning instruments, state planning policies of various forms and building codes. With the exception of Queensland, Australian planning systems are largely prescriptive in nature. Enabling legislation typically provides for land use planning and management through a town planning scheme, usually administered by local government (a level of government constituted by the state and lacking its own constitutional basis). These schemes zone land parcels for prescribed uses. These land uses are controlled through a zoning and development table setting out development standards (e.g. boundary setbacks, building heights, parking requirements etc.). Queensland, however, like New Zealand and some US states, has a performance-based planning system. Land use planning schemes in Queensland establish desired environmental outcomes and performance objectives that land and property developers use to guide their developments and comply with the intent of the scheme (Baker et al, 2006). Until recent amendments were made to the Integrated Planning Act (1997), very few forms of development were prohibited in Queensland.

Despite the style of planning system adopted by each state, the effect of these arrangements has been to divide up responsibilities in ways that can often hamper efficient responses to problems like climate change. State governments are dependent upon the Commonwealth for funding, but ardently defend their interests and protect their autonomy. If the Commonwealth desires unified action it must broker deals with the states or cajole them into complying with national interests. And actions across state borders – which pay no heed to river catchments, conurbations or biogeochemical processes – can be difficult to achieve. As can be seen from our discussion of national, state and local responses to climate change, efforts can be duplicated and responsibilities hard to determine – with subsequent buck-passing across all three levels.

From the local to the global: Lessons from the Australian experience

The legacy of the Howard national government was urban policy that was left largely to the states and territories, even though its own policy settings (e.g. immigration) profoundly shaped the course of metropolitan development. At the state level, urban policy in recent decades has fixed on the ideal of 'urban consolidation' (compaction) as a means to achieve efficiency and sustainability in the urban system (see Green and Handley, Chapter 4). Consolidation has been the leading principle of metropolitan scale planning through the 1990s and beyond, exemplified, for example, in the new draft South East Queensland Regional Plan 2009–2031, which seeks to contain the metropolitan footprint of Brisbane and its connected urban sub-regions (Ipswich, Gold Coast and Sunshine Coast) and to increase the share of new housing supply in brownfield redevelopment areas.

It can be argued that the long fixation on urban form, especially density, in Australian metropolitan planning has been to the detriment of critical urban structural issues, including the fundamental layout of land uses and the distribution and ranking of urban centres (Troy, 1996). Australia's large, complex metropolitan regions have developed increasingly problematical structures due to over-centralization of investment, economic activity and travel behaviour. Arguably, the fixation on consolidation (i.e. densification) as a dominant urban policy setting reflects faith in a weak version of ecological modernization (EM). The idea that growth can be improved, not curbed, and made sustainable through technological adjustments to the built environment can be read as a variant of EM strategy.

The consolidation policy setting has only opposed outward growth, not growth per se, as reflected in the increasing massification and intensification of the metropolitan environment via densification. Yet hand in hand with densification, residents of Australian cities like the Gold Coast continue to consume more energy and generate higher greenhouse emissions (Australian Conservation Foundation, 2007). Such trends are also evident in the United States and Europe. Increasing evidence of policy failure suggests the need for a more critical and directive view to urban growth and climate change that addresses the fundamental problems of overconsumption and overproduction in household and commercial sectors – not just technological tinkering with the symptoms.

A shift to a new, decentralized urban structure in metro regions seems desirous as a means to fundamentally lowering energy consumption by facilitating the localization of economic activity, food production and waste management. Ultimately, however, there appears no simple 'spatial fix' for problems like greenhouse emissions. Planning will need to work in concert with political economic policies that reshape the fundamental causes of overconsumption and overproduction. Consistent with stronger versions of EM, this will require major shifts in policy discourse as well as a fundamental rethink of prevailing governance processes. Challenging questions about how to distribute the burden of environmental adjustments within Australian and global human settlements must also be raised.

Conclusion: After the horse has bolted…

Fortunately it is not too late to do something about climate change, and Australian planners have begun to tackle the many challenges associated with mitigation and adaptation. Most Australian states either have or are in the process of developing climate change adaptation strategies. Many are pursuing diverse actions such as fostering research and monitoring, educating their constituents, promoting transit oriented development, and providing incentives for household-scale adaptive responses (e.g. water conservation and energy efficiency). Some have taken very progressive actions such as developing energy efficiency ratings for buildings and mandating their disclosure at the time of sale, promoting the retention and enhancement of urban forests (ACT), linking water resources into an integrated grid (Qld.), mandating rainwater tank use (SA), and phasing out electric hot water systems (Qld., NSW, ACT). But few have taken the regulatory pathway, preferring instead to use incentives, rebates, education and demonstration projects to effect change.

Arguably an area with greater potential for climate change adaptation is in land use planning. The planning responses taken thus far include risk analysis and minimization, education, monitoring, reporting, new codes and standards, new assessment methods, improved coordination, and integrated emergency response / disaster planning. But as we have seen, strategies such as urban consolidation or building desalination plants can have paradoxical and perverse consequences – that is, reducing carbon sinks and habitat patches and increasing energy use (Hamin and Gurran, 2008). While reducing lot sizes has been a goal of most metropolitan plans over the past decade – allowing for more efficient use of infrastructure – it may actually have promoted less efficient and more vulnerable urban landscapes. The 'affluenza' that gripped Australian society during the economic boom has meant bigger houses, smaller gardens, larger and more appliances (e.g. flat screen TVs and clothes dryers), consumption oriented lifestyles and less greenspace. We may actually be worse off as a result of consolidation.

Ecological modernization – the idea of decoupling economic growth from environmental harm – has been the primary strategy for climate change adaptation and mitigation in Australia thus far. And it has been weak, not strong ecological modernization ideas that have driven climate change strategies at all levels of government. Energy conservation, transit-oriented development and urban consolidation are all technologies that embody ideas of eco-efficiency. But in many ways unfettered growth and economic prosperity without large-scale lifestyle changes or the pain of seriously reworking the way we interact with Australian socio-political and biogeochemical systems is a path to disaster.

Noticeably absent from Australian adaptive responses are initiatives such as identifying areas that may have to be abandoned and areas where new development will be prohibited, retrofitting existing building stock to bolster resilience to climate change impacts, and developing uniform climate-change building codes for higher intensity cyclones, heatwaves, storm surges, flooding and so on. Problematically, there is an assumption across the Australian adaptation literature that buildings have a 20 to 50 year lifespan. In reality, many Australian cities have buildings much older than that. Adaptation must therefore ensure that existing buildings are retrofitted with insulation, water- and energy-efficient devices and upgraded to meet new building codes (e.g. replacing roofs that are susceptible to hail damage). The option of strategically abandoning some areas or buying back the most vulnerable sites and relocating communities must also be seriously considered.

Perhaps most concerning about the way that Australian state governments have addressed climate change adaptation thus far has been their total neglect of environmental and social justice issues. Impoverished and ethno-racially marginalized communities have the most to lose from climate change. While wealthy communities located on the coast or next to estuaries and canals will undoubtedly suffer – at least they will have the resources to escape or rebuild. Remote Aboriginal settlements and immigrant and impoverished communities confined to older housing stocks that lack thermal efficiency, relegated to areas far from social and community services and lacking public transportation, will be seriously impacted by climate change. Heatwaves, increased costs of electricity and water, flooding, storm damage and spreading vector-borne diseases will take a heavy toll on these communities. They will disproportionately carry the burden unless governments expand adaptation measures to include affordable housing, healthcare, better access to transportation and urban greenspaces, and act to curb conspicuous consumption and profligate growth. Future research must address these issues.

Note

1   Resilience refers to the ability of cities to respond
    to, adapt to and recover from stress and
    catastrophic events related to climate change (e.g.
    violent storms, heatwaves, large scale flooding and
    epidemics). Resilient cities will 'bounce back' from
    such events without suffering high death-rates
    and/or long-term damage to critical infrastruc-
    ture, the integrity of life-sustaining systems and
    social institutions (Alberti and Marzluff, 2004;
    Campanella, 2006; Commonwealth Scientific and
    Industrial Research Organisation et al, 2007).

References

Alberti, M. and Marzluff, J. M. (2004) 'Ecological
    resilience in urban ecosystems: Linking urban
    patterns to human and ecological functions', *Urban
    Ecosystems*, vol 7, no 3, pp241–265
Allen Consulting Group (2005) 'Climate change risk
    and vulnerability: Promoting an efficient response
    in Australia', Australian Greenhouse Office,
    Department of the Environment and Heritage,
    Canberra
Australian Bureau of Statistics (2008) 'Australian social
    trends, 2008', www.abs.gov.au/AUSSTATS/
    abs@.nsf/Lookup/4102.0Chapter3002008,
    accessed 15 December 2008
Australian Conservation Foundation (2007)
    'Consumption atlas', www.acfonline.org.au/
    consumptionatlas/, accessed 12 December 2008
Australian Government (2008) 'Carbon Pollution
    Reduction Scheme: Australia's low carbon future',
    Commonwealth of Australia, Attorney General's
    Department, Canberra
Australian Greenhouse Office (2005) 'National
    Climate Change Adaptation Program', Department
    of Environment and Heritage, Canberra
Australian Greenhouse Office (2006) 'Adaptation
    planning for climate change', *Australian Planner*, vol
    43, no 2, pp8–9
Australian Greenhouse Office (2008) 'Australia's
    settlements and infrastructure: Impacts of climate
    change', www.greenhouse.gov.au/impacts/
    settlements/html, accessed 1 August 2008
Baker, D. C., Sipe, N. G. and Gleeson, B. (2006)
    'Performance-based planning: Perspectives from
    the United States, Australia and New Zealand',
    Journal of Planning Education and Research, vol 25,
    no 4, pp396–409

Bardon, J. (2008) 'Call for climate change planning
    code', *Australian Broadcasting Corporation*,
    www.abc.net.au/news/stories/2008/10/27/
    2402714.htm, accessed 27 October, 2008
Berger, G., Flynn, A., Hines, F. and Johns, R. (2001)
    'Ecological modernization as a basis for environ-
    mental policy: Current environmental discourse
    and policy and the implications on environmental
    supply chain management', *Innovation,* vol 14, no
    1, pp55–72
Blowers, A. (1997) 'Environmental policy: Ecological
    modernisation or risk society?' *Urban Studies*, vol
    34, nos 5–6, pp845–871
Branz Ltd (2007) 'An assessment of the need to adapt
    buildings for the unavoidable consequences of
    climate change', Australian Greenhouse Office,
    Department of the Environment and Water
    Resources, Canberra
Bruegmann, R. (2005) *Sprawl: A Compact History*,
    University of Chicago Press, Chicago
Buckley, R. (ed) (2007) *Climate Response: Issues, Costs
    and Liabilities in Adapting to Climate Change in
    Australia*, Griffith University, Gold Coast
Bulkeley, H. (2001) 'No regrets?: Economy and
    environment in Australia's domestic climate
    change policy process', *Global Environmental
    Change*, vol 11, no 2, pp155–169
Bulkeley, H. (2006) 'A changing climate for spatial
    planning', *Planning Theory and Practice*, vol 7, no 2,
    pp203–214
Campanella, T. J. (2006) 'Urban resilience and the
    recovery of New Orleans', *Journal of the American
    Planning Association*, vol 72, no 2, pp141–146
Christoff, P. (1996) 'Ecological modernisation,
    ecological modernities', *Environmental Politics*, vol
    5, no 3, pp476–500
Christoff, P. (2005) 'Policy autism or double-edged
    dismissiveness?: Australia's climate policy under the
    Howard government', *Global Change, Peace and
    Security*, vol 17, no 1, pp29–44
Commonwealth Scientific and Industrial Research
    Organisation, Arizona State University and
    Stockholm University (2007) *Urban Resilience:
    Research Prospectus. A Resilience Alliance Initiative for
    Transitioning Urban Systems towards Sustainable
    Futures*, Commonwealth Scientific and Industrial
    Research Organisation, Canberra
Costanza, R., d'Arge, R., de Groot, R., Farber, S.,
    Grasso, M., Hannon, B., Limburg, K., Naeem, S.,
    O'Neill, R.V., Paruelo, J., Raskin, R. G., Sutton, P.
    and van den Belt, M. (1997) 'The value of the
    world's ecosystem services and natural capital',
    Nature, vol 387, no 6630, pp253–260

Council of Australian Governments (2007) 'National Climate Change Adaptation Framework', Australian Government

Curran, G. (2001) 'The third way and ecological modernisation', *Contemporary Politics*, vol 7, no 1, pp41–55

Davidson, A. (2006) 'Stuck in a cul-de-sac? Suburban history and urban sustainability in Australia', *Urban Policy and Research*, vol 24, no 2, pp201–216

Department of Infrastructure and Planning (2008) 'South East Queensland', www.dip.qld.gov.au/seq, accessed 15 December 2008

Dryzek, J. (2005) *The Politics of the Earth: Environmental Discourses*, Oxford University Press, Oxford

Fisher, D. R. and Freudenburg, W. R. (2001) 'Ecological modernization and its critics: Assessing the past and looking towards the future', *Society and Natural Resources*, vol 14, no 8, pp701–709

Garnaut, R. (2008) *The Garnaut Climate Change Review – Financial Report*, Cambridge University Press, Melbourne

Gill, S., Handley, J., Ennos, R. and Pauleit, S. (2007) 'Adapting cities for climate change: The role of green infrastructure', *Built Environment*, vol 33, no 1, pp115–133

Gleeson, B. and Low, N. (2000) *Australian Urban Planning: New Challenges, New Agendas*, Allen and Unwin, St Leonards, NSW

Godber, A. (2005) 'Urban floodplain land-use – acceptable risk?' *The Australian Journal of Emergency Management*, vol 20, no 3, pp22–40

Gold Coast City Council (2001) 'The Gold Coast 2010 Cities for Climate Protection Program: 2001 Action Plan', Gold Coast City Council, Gold Coast

Gold Coast City Council (2007) 'Carbon neutral by 2020: Gold Coast City Council responding to climate change', Gold Coast City Council, Gold Coast

Gouldson, A. and Murphy, J. (1997) 'Ecological modernization: Restructuring industrial economics', in M. Jacobs (ed) *Greening the Millennium? The New Politics of the Environment*, Blackwell Publishers, Oxford, pp74–86

Grant, R. and Papadakis, E. (2004) 'Challenges for global environmental diplomacy in Australia and the European Union', *Australian Journal of International Affairs*, vol 58, no 2, pp279–292

Gurran, N. (2007) *Australian Urban Land Use Planning: Introducing Statutory Planning Practice in New South Wales*, Sydney University Press, Sydney

Hajer, M. (1995) *The Politics of Environmental Discourse: Ecological Modernization and the Policy Process*, Clarendon Press, Oxford

Hamin, E. M. and Gurran, N. (2008) 'Urban form and climate change: Balancing adaptation and mitigation in the U.S. and Australia', *Habitat International*, doi:10.1016/j.habitatint.2008.10.005, online

Howes, M. (2005) *Politics and the Environment: Risk and the Role of Government and Industry*, Earthscan, London

Local Government Association of Queensland (2007) 'Adapting to climate change: A Queensland local government guide', Local Government Association of Queensland, Brisbane

Low, N. (2008) 'In praise of public planning in an era of climate change', *Urban Policy and Research*, vol 26, no 2, pp141–144

Lundqvist, L. (2000) 'Capacity building or social construction? Explaining Sweden's shift towards ecological modernisation', *Geoforum*, vol 31, no 1, pp21–32

Lyth, A. (2006) 'Climate proofing Australian urban planning: Working towards successful adaptation', *Australian Planner*, vol 43, no 2, pp12–15

Mol, A. P. J. and Sonnenfeld, D. A. (2000) 'Ecological modernisation around the world: An introduction', *Environmental Politics*, vol 9, no 1, pp3–14

Mol, A. P. J. and Spargen, G. (2000) 'Ecological modernisation', *Environmental Politics*, vol 9, no 1, pp17–49

Queensland Government (2007) 'Climate Smart 2050: Queensland climate change strategy 2007', State Government of Queensland

Slocum, R. (2004) 'Polar bears and energy efficient lightbulbs: Strategies to bring climate change home', *Environment and Planning D: Society and Space*, vol 22, no 3, pp413–438

SMEC Australia (2007) 'Climate change adaptation actions for local government', Australian Greenhouse Office, Department of the Environment and Water Resources, Canberra

Spearitt, P. (2008) 'The water crisis in South East Queensland', in P. Troy (ed) *Troubled Waters: Confronting the Water Crisis in Australia's Cities*, Australian National University E-Press, Canberra

The State of Queensland (2008) 'Draft South East Queensland Regional Plan', Department of Infrastructure and Planning, Brisbane

Troy, P. (1996) *The Perils of Urban Consolidation: A Discussion of Australian Housing and Urban Development Policies*, The Federation Press, Sydney

Weale, A. (1998) 'The politics of ecological modernization', in J. Dryzek and D. Schlosberg (eds)

Debating the Earth: The Environmental Politics Reader, Oxford University Press, Oxford, pp301–318

Welford, R. and Hills, P. (2004) 'Ecological modernisation, environmental policy and innovation priorities for the Asia Pacific Region', *International Journal of Environment and Sustainable Development*, vol 2, no 3, pp324–340

12

Beyond a Technical Response:
New Growth-management Experiments in Canada

Pamela Robinson

Introduction

The image of 'Canada' often conjures up a country of mountains, trees, tundra and charismatic megafauna. Despite Canada's many natural assets, Canada is also a country of (sub)urban development. Most recent population data indicates that 80 per cent of Canadians live in an urban centre of 10,000 people or more (Statistics Canada, 2008). But the densities of these urban areas vary across large Canadian cities. For example, the population density per square kilometre in Montreal is 4438.7; in Calgary 1360.2 and in Vancouver 5039.0. These variations are further amplified when the contrast between the urban core and suburban periphery is assessed. The Greater Golden Horseshoe region is Canada's most populous metropolitan area (see Figure 12.1) covering 33,500km². With a population of 8.1 million people it is the region where 25 per cent of Canadians make their home but the region's densities range from 24.4 people per square kilometre in the town of Kawartha Lakes to the high of 3972.4 in Toronto.

These data become important when exploring the terrain of climate change response in Canada. The Government of Canada (also known as the federal government) is the level of government that negotiates Canada's international commitments with regard to greenhouse gas (GHG) emissions. But the federal government does not hold sole jurisdiction or control over the activities that constitute the most significant contributions to Canada's GHG emissions. The provincial and territorial governments have jurisdiction over some GHG-emitting activities and local or municipal government does as well.

In Canada, municipal governments fall under the jurisdiction of provincial governments or territories that define, supervise and regulate which powers municipalities will receive and which activities they will engage in. Generally speaking, these municipalities:

- exert at least partial control over land use through zoning and official plan documents;
- issue building permits and development approvals;
- control parking supply and prices;
- are responsible for roads and public transit;
- oversee parks and recreation services;
- play a regulatory and management role in power and gas utilities (Federation of Canadian Municipalities, 2008a).

This 'to-do' list of municipal functions, suggests that despite provincial governments, municipalities have clear potential to contribute to GHG emissions reduction. More specifically, using 1990 emissions data it is estimated that Canadian municipalities have direct control, indirect control or influence over approximately 52 per cent of domestic emissions. Direct control over emissions comes from the municipal governments' use of fuels and electricity in its operations, methane gas capture, greening activities and urban forestry. Indirect control over emissions comes from institutions and enterprises over which the municipality has indirect control through directorships, funding, shared facilities, and so on. Influence over emissions results from activities that are at least partly controlled or influenced by municipal government laws, taxes or regulation (Municipalities Issue Table, 1998). But how is this control exercised? For example, in the case of the Greater Toronto Area (GTA), recently released data indicate that residential vehicles are responsible for 21 per cent of the 60 million tonnes of CO_2e emitted on an annual basis; residential homes 23 per cent, non-residential buildings 31 per cent, commercial and public vehicles 15 per cent and waste 5 per cent (Toronto City Summit Alliance, 2008, p6). Municipal GHG emission reduction success in Canada has come mostly from technical 'fixes', such as landfill gas capture and building and energy and water retrofits, which account for a relatively small proportion of the overall emissions, rather than a substantive change in urban lifestyles (Robinson, 2006).

To respond effectively to the growing need to further reduce GHG emissions, Canadian municipalities must therefore move beyond the technological fix (see Byrne et al, Chapter 11). The role of urban planners here is significant: they 'should strive to build new social and political capital in support of growth management efforts so that the foundation for future emissions reductions is strong rather than low density and automobile dependent' (Robinson, 2006).

This chapter will first assess in greater depth the extent to which Canadian municipalities have responded to climate change and then consider two emerging case studies where new approaches to growth management are being experimented that have the potential to lead to emissions reductions. It concludes with recommendations for future research related to urban growth management in Canadian communities.

Diagnosis of our current state of affairs

As the imminent and distant threats of global climate change become more readily apparent, governments have responded in a variety of manners. While international negotiations continue, nation states have few tangible and measurable success stories to report in terms of effective response to climate change mitigation and adaptation. In contrast, local governments in the industrialized world have made noteworthy progress toward meaningful response to this global problem (Robinson and Gore, 2005; Gore and Robinson, forthcoming). Toronto and Vancouver among others were early entrants to climate response and continue to implement projects that result in measurable emissions reductions. However, as noted above, a closer investigation of the nature of this municipal success reveals that it is based upon largely technical, project-specific initiatives (e.g. land fill gas capture, building energy retrofits, etc.). Furthermore, larger systemic change resulting in new patterns of (sub)urban form that reduces automobile dependency and thus fossil fuel consumption remains elusive in part due to weak decision-making regimes and half-hearted efforts at citizen engagement (Robinson, 2006).

Canada was an early leader in the call to reduce GHG emissions as the host of the Toronto Conference on the Changing Atmosphere in 1988. Local government in Canada recognized the importance of climate change, with Vancouver and Toronto developing early response plans. Now, some 20 years later, local government in Canada is the only level of government that can claim any real progress in terms of emissions reductions. There are 166 Canadian municipalities registered with the Federation of Canadian Municipalities (FCM) and International Council for Local Environmental Initiatives (ICLEI) 'Partners for Climate Protection' (PCP) programme. This represents 65 per cent of Canadians, or approximately 21 million people (Gore and Robinson, forthcoming).

The PCP programme advocated a 'five milestone' process to help direct municipal response:

1 creating a greenhouse gas emissions inventory and forecast;
2 setting an emissions reductions plan;
3 developing a local action plan;
4 implementation of the action plan;
5 monitoring progress and reporting results.

Table 12.1 illustrates the distribution of climate change response across Canada's 20 largest municipalities.

These 20 municipalities represent 76 per cent of the Canadian population. These data reveal that the early municipal leaders in Canada are also the ones that have accomplished and implemented the most. FCM distinguishes between emissions generated by local government operations (corporate) and those from activities occurring within municipal boundaries (community). This distinction is important because the success accomplished to date by Canadian municipalities comes largely from corporate emissions while growth management contributions are accounted for under the community emissions. Further success is anticipated from projects funded by FCM's Green Municipal Funds, a programme designed to stimulate investment in municipal infrastructure that reduces environmental impact including

Table 12.1 Twenty largest Canadian municipalities and participation in FCM's Partners for Climate Protection (PCP) Programme

Municipality	Province	Population 2006	PCP member date joined
1. Toronto	Ont.	2,503,281	1990
2. Montreal	Que.	1,620,693	September 1998
3. Calgary	Alta.	988,193	1994
4. Ottawa	Ont.	812,129	February 1997
5. Edmonton	Alta.	730,372	1995
6. Mississauga	Ont.	668,549	December 1999
7. Winnipeg	Man.	633,451	November 2006
8. Vancouver	B.C.	578,041	1995
9. Hamilton	Ont.	504,559	November 1996
10. Quebec	Que.	491,142	January 1997
11. Brampton	Ont.	433,806	NO
12. Surrey	B.C.	394,976	July 1996
13. Halifax	N.S.	372,679	February 1997
14. Laval	Que.	368,709	March 1997
15. London	Ont.	352,395	November 1994
16. Markham	Ont.	261,573	February 2007
17. Gatineau	Que.	242,124	NO
18. Vaughn	Ont.	238,866	NO
19. Longueuil	Que.	229,330	NO
20. Windsor	Ont.	216,473	December 2002

Source: Adapted from Gore (2008).

decreasing GHG emissions (FCM, 2008b). Hence, GHG reduction success has come from the completion of discrete projects. What remains elusive is progress in tackling the larger, more systemic causes of GHG emissions, largely urban sprawl.

Systemic barriers to municipal response therefore present some of the most significant obstacles to further emission reductions in Canada (Robinson, 2006). The foremost challenge land-use planners face in Canada with regard to making a meaningful contribution to GHG reduction is to find the right combination of financial tools, intergovernmental coordination/cooperation, appropriately positioned public policy and a more active and engaged polity and civil society that will limit urban sprawl. Scholars suggest that the presence of a strong planning framework is a necessary prerequisite to implementing sustainable development principles, among which climate change is often cited (Parkinson and Roseland, 2002; Portney, 2002; Berke and Manta Conroy, 2004). Canadian planning's long standing tradition of regional planning and the paired use of comprehensive plans with implementing zoning byelaws would suggest that more success in combating sprawl should have been achieved.

Bulkeley (2006) accurately pinpoints the tension in land use planning that exists between the need to develop longer time horizons within which to take into consideration issues of climate protection, and the pressure and need to act swiftly to put such strategies in place. Tomalty and Alexander (2005) point to the absence of political will, the absence of policy frameworks allowing for planning innovation, a lack of interest on behalf of the development community in non-traditional designs, the financial impacts of municipal taxation and development charges and consumer preference for low-density urban development as barriers to the implementation of growth management policies. However, the ongoing encroachment of low density, automobile-dependent, single land use development into the countryside in Canada reminds us that a strong policy process is not sufficient and that a more integrated series of policy and programme responses are required.

Others remind us that civil society is an important resource for planners when developing a vision for change (Kenworthy and Newman, 1999; Roseland, 2005). Yet, the project-focused and technologically specific nature of Canadian local governments' success in emissions reductions leaves little room for the public's input or engagement. In Chapter 23 Claire Haggett argues that renewable energy projects in the UK have provided opportunities for public engagement at the site or project-specific stages of planning. In contrast, the Canadian projects have not resulted in these opportunities, in part because the emissions reductions have resulted from strategic procurement (e.g. green fleets) or the installation of on-site technology (landfill gas capture): activities which typically do not involve public input. The consumer preference for low-density urban development, when compared to Canadian public opinion polling suggesting that a significant proportion of the Canadian public is deeply concerned with climate change issues, suggests that the public suffers from cognitive dissonance. This could be in part attributed to a lack of meaningful opportunities for the public to engage in articulating a new vision for change with regard to climate and energy related challenges.

Many academic and practitioner planners have clearly argued and delineated the specific nature of land-use and transport-related policy change that is necessary to respond to climate change. Our current state of affairs suggests that while we know what to do, we lack the ability to actually do it. One explanation for this 'implementation deficit' (see Bulkeley, 2006) could be that our current system of urban governance, defined as the process by which local governments manage their relationships with higher levels of government, civil society and the private sector, serves as a barrier to effective implementation.

A recent Canadian investigation of barriers to smart growth implementation illustrates many disfunctionalities in the process of decision making at the local government level and the relationships Canadian local governments have with higher orders of government, civil society and the private sector (see Tomalty and Alexander, 2005). In rapidly growing communities across Canada there are many battles between developers and citizens groups over protecting greenfields from suburban encroachment. The negative influence of private sector developers working against local efforts to curb sprawl signals the systemic nature of the urban governance problems faced by Canadian municipalities when attempting to respond with specific policy-derived tools to climate change issues.

To conclude, the current paradigm of Canadian local response to emissions reductions efforts can be described as a 'work in progress' with room for more municipalities to become involved and more progress to be made toward true emissions reductions being achieved. Perhaps more significantly, the broader, more integrative challenge of tackling the further expansion of low-density, automobile-dependent suburban sprawl is paramount. With this diagnosis comes the recognition that within our current paradigm of response, new governance, civic engagement and legislative frameworks are needed to support effective growth management. In short, despite success in emissions reductions being achieved through technological approaches, a new paradigm of effective growth management is needed in order for Canadian local government to further reduce GHG emissions.

What's next?

In the 20 years that have passed since the release of the World Commission on Environment and Development report 'Our Common Future' (1987), our understanding of sustainability issues has evolved from first generation approaches to second generation ones (Robinson, 2008). This distinction between first- and second-generation approaches also applies to the specific urban sustainability challenge of climate change. In this context, first generation approaches are those where emissions reductions are achieved through technological approaches and second generation efforts are those derived through effective growth management practices.

In the broader case of urban sustainability new forms of state–society relationships and organizational management that take a holistic approach to sustainability are required. With the recognition that our current institutional responses and governing processes have produced our current state of unsustainability comes the realization that these same institutions will face challenges if they are solely relied upon to design and implement the processes needed to deliver more sustainable futures (Robinson, 2008). This recognition that we need new institutions to guide our response to urban sustainability and, within that framework, climate change response, serves as an indicator of the significant scope and scale of the paradigmatic change. Our largest challenge is to mobilize emissions response while simultaneously expanding the adaptive capacities of our communities to respond to the inevitable uncertainties that climate change will bring.

New opportunities to respond to growth management

The remaining part of this chapter will explore two growth management initiatives that offer promise in terms of reigning in urban sprawl. The two case studies presented here offer new approaches in the Canadian context to responding to the governance, civic engagement and planning framework deficits earlier mentioned. They were also selected because they represent growth management interventions at two different scales: the provincial level and the local government level. The first case explored will be one of policy alignment between new legislation in the Province of Ontario, and the second is the City of Vancouver's (British Columbia) EcoDensity Efforts.

Growth management in Ontario

In the Province of Ontario some progress has been made in terms of growth management. For the first time in over 50 years, the foundation has been laid for an integrated growth management strategy. Through the introduction and passage of a series of growth-management related legislation, the Province is forcing regional and local governments to reconsider how and where they allow new growth to occur. While it is early days in terms of implementation, these legislative interventions offer the potential to reduce the number of single-family homes built at densities that do not support transit.

Table 12.2 provides an overview of significant policies and programmes introduced by the Government of Ontario with the goal of containing the rapid rates of urban sprawl that the Greater Golden Horseshoe region is confronting. This series of planning interventions by the Province is noteworthy because the elements contain consistent themes of smart growth with policy interventions to protect prime agricultural land from suburban sprawl, and new areas specifically identified for intensification, settlement and employment, with accompanying plans for new and/or revitalized infrastructure and transportation. The Province of Ontario estimates that without the implementation of the Growth Plan, there would be a 45 per cent increase in carbon dioxide emissions (Ministry of Public Infrastructure Renewal, 2006, p27). This integration and coordination between programmes and policies offers the best prospect for containing suburban growth thus far in the history of Ontario. But is it enough?

Reviews are mixed. Cherise Burda (2008) offers a strong and detailed critique of the linkages between the new growth management strategy and Ontario's GHG emissions reduction strategy. This research finds weaknesses in the lack of policy integration between growth management efforts and the Province's climate change goals. It suggests that the targets in the Places to Grow growth management plan are not ambitious enough and fail to include new legal mechanisms to withdraw already granted suburban development rights. It consequently signals that there are no formal requirements for local government to develop plans to reduce energy use.

While the use of 'greenbelts' as growth management boundaries is not unique to Ontario, the significance of Ontario's Greenbelt comes from the political leadership demonstrated by the provincial government in developing the programme of related legislation and plan-making. Because the Ontario land use planning legislative framework requires planners to consult the public in their plan-making exercises, the Greenbelt also expands opportunities for the public to become engaged in the challenges of responding to growth-management issues and projects.

While Carter-Whitney is optimistic about the Greenbelt's potential impact, stating that 'Ontario's Greenbelt is positioned to be the most successful and most useful Greenbelt in the world' (2008, p1), other scholars identify weaknesses with Ontario's Greenbelt Act and Greenbelt Plan (Fung and Conway, 2007): the boundary decision-making process has been called more 'political science' than natural science; local governments are resistant to further planning direction being imposed by the Province; there are significant concerns about leapfrog development occurring beyond the border and the costs of housing increasing within the border and to its south; and the agricultural community is divided about the impact of the new legislation on their land values and farm viability. In spite of vociferous critique from pockets of the agricultural community and developers whose land is no longer developable, the Province of Ontario initiated a 'Growing the Greenbelt' consultation series in the spring of 2008, further signalling their serious intent to protect agricultural land, curb sprawl and improve ecological integrity. One of the weaknesses of the Ontario planning system is that once already permitted, suburban land development rights are difficult to withdraw. Further, there is no consolidated, rigorous data assembled that inventories the number of hectares of now-designated Greenbelt land with pre-existing development rights. So, although there is a Greenbelt, no one really knows how much of it, as a whole, is actually still developable.

As regional and local government response to this new growth management regime unfolds a new challenge is being presented. The Places to Grow Act requires municipal conformity to

Table 12.2 A selection of growth-management related policy initiatives in Ontario

Legislation/Policy	Description
The Ministry of Public Infrastructure and Renewal *Created October 2003*	A body that combines the Smart Growth Secretariat with the Superbuild Corporation (highway extension programme) with the intention of integrating land use and infrastructure planning (www.pir.gov.on.ca).
The Greenbelt Act *Introduced October 2004*	Legislation that establishes a 240,000-hectare greenbelt in the Greater Golden Horseshoe (GGH) area, within which urban development will not be permitted (www.e-laws.gov.on.ca/html/statutes/english/elaws_statutes_05g01_e.htm).
The Greenbelt Plan *Effective December 2004*	The specific goals and policies (www.mah.gov.on.ca/Page189.aspx) intended to enact the Greenbelt Act.
The Places to Grow Act *Revised March 2005*	A legal framework (www.e-laws.gov.on.ca/html/statutes/english/elaws_statutes_05p13_e.htm) to coordinate planning and decision making for long-term growth and infrastructure renewal in Ontario. It gives the Province the power to designate geographical growth areas and to require municipalities to bring their official plans into conformity with the growth plan for their area (www.placestogrow.ca/index.php?lang=eng).
Provincial Policy Statement (PPS) *Revised March 2005*	Revised planning rules (www.mah.gov.on.ca/Page1485.aspx) that allow development only in areas where it can be sustained and supported by infrastructure. These include new policies to support intensification, more transit-friendly land use patterns, stronger direction on land use policies for improved air quality and alternative and renewable energy.
Strong Communities Act *Effective March 2005*	Legislation (www.mah.gov.on.ca/Page1433.aspx) that requires planning decisions on applications subject to the new PPS 'shall be consistent with' the new policies. It allows more time and opportunity for public scrutiny in the planning process.
Greater Golden Horseshoe (GGH) Growth Plan *Released June 2006*	A plan (www.placestogrow.ca/images/pdfs/FPLAN-ENG-WEB-ALL.pdf) to delineate and set policy for where and how growth/development can occur in the GGH, including the identification of intensification nodes, built-up areas, settlement lands, greenfield areas and employment lands.
Bill 104 *Enacted April 2006*	The creation of the Greater Toronto Transportation Authority (http://www.ontla.on.ca/web/bills/bills_detail.do?locale=en&BillID=414&isCurrent=false&detailPage=bills_detail_about) (now Metrolinx) (http://www.metrolinx.com/), which replaced the Go Transit Act (2001) (www.e-laws.gov.on.ca/html/statutes/english/elaws_statutes_01g23_e.htm). This new body will consolidate the previous patchwork approach to public transport.
Bill 51 *Effective January 2007*	The Planning and Conservation Land Statute Law Amendment Act (www.mah.gov.on.ca/Page211.aspx), establishing local appeal bodies to hear appeals as an alternative to the Ontario Municipal Board (OMB) (www.omb.gov.on.ca/).
Move Ontario *Announced June 2007*	A $17.5 billion capital investment in public transit infrastructure to serve the Greater Toronto Area and Hamilton (GTAH) beginning in 2008 (www.premier.gov.on.ca/news/Product.asp?ProductID=1384).
Go Green Ontario *Released August 2007*	Ontario's Action Plan on Climate Change (www.gogreenontario.ca/plan.php) that includes short-, medium-, and long-term targets to reduce Ontario's GHG emissions through MoveOntario 2020, renewable energy sources, land-use and growing green communities and creating 'green' jobs through the Next Generation Jobs Fund (www.ontario-canada.com/ontcan/en/progserv_ngjf_en.jsp).

Source: Adapted from Burda (2008).

be implemented by 2009 but the new regional transportation plan was being concurrently developed by Metrolinx (formerly the Greater Toronto Transportation Authority – see Table 12.2). The concept presented in the Metrolinx plan (2008) offers a sequentially more complex and far-reaching version of a new regional transportation network. However, this plan is being developed toward the end of the conformity window required by Places to Grow. This overlap in timing presents a new challenge in terms of maximizing capacity to effectively control suburban growth. If the Ontario Places to Grow growth plan contains specific boundaries for growth with associated population growth targets, how can planning for land-use intensification be at its most effective in the absence of a Regional Transportation Plan? This case of 'many growth management balls in the air' is a signal that while the Province is serious about curbing sprawl perhaps it needs to consider how and when the elements of this complex approach might best fit together to produce a mutually reinforcing system. Furthermore the issues raised here signal the challenge that governments face when attempting to fundamentally change old approaches: the shift to a new paradigm can be slowed or impeded by the entrenchment of the old paradigm.

This brief overview of key elements of Ontario growth management policy and legislation reveal that there is significant political investment being made in a strong foundation for managing suburban growth in southern Ontario. This growth management framework is an important addition to the Ontario land-use planning legislative toolkit, thus expanding the potential capacity of local government to reduce sprawl. Because this framework originates with the Province of Ontario, all regional and local governments must 'have regard' for it in their local planning and as such it alleviates some of the requirement for expanded political will at the local scale to support growth management. This 'top–down' approach could be perceived to be heavy-handed and perhaps serving as an instigating cause of local resistance. However, since Canadian confederation in 1867, the Provinces have provided the directives for land use planning and as such, local governments must respond and this historical relationship – top-heavy as it may seem – is the context in which local government has had to function since its inception. In Ontario the land use planning process prescribes how and when the public must be consulted, thus expanding opportunities for citizen participation in growth management decision making, but the extent to which local authorities have capitalized upon this prescribed intervention as a means of engaging the public in dialogue about behavioural change options remains unexplored. To conclude, this investment is the most significant growth-management effort made thus far in Canada. However, it will take at least ten years to assess whether this investment will yield significant returns in terms of GHG emissions reductions.

Ecodensity: Vancouver

The second case study this chapter will explore is the City of Vancouver's Eco-Density initiative. EcoDensity is: 'an acknowledgement that high quality and strategically located density can make Vancouver more sustainable, liveable and affordable' (City of Vancouver, 2008a, City of Vancouver, 2008c). The City of Vancouver is widely regarded as a liveable city but the municipal government recognizes that this liveability is threatened by increasing housing costs, the environmental impacts of population growth, and the strain that new development can put on existing services if not managed effectively (City of Vancouver, 2008b). According to the City, only 11 per cent of the City's land area is used for multi-unit dwellings and single family homes account for over half of the land area in the City (City of Vancouver, 2008b).

The City Council first approved the EcoDensity process and terms of reference in 2006 (City of Vancouver, 2008d). This approval led to the implementation of a public consultation process with a series of public meetings, on-line consultations, workshops, an EcoDensity Fair and a speakers' series. The Draft EcoDensity Charter was referred to Council a year later,

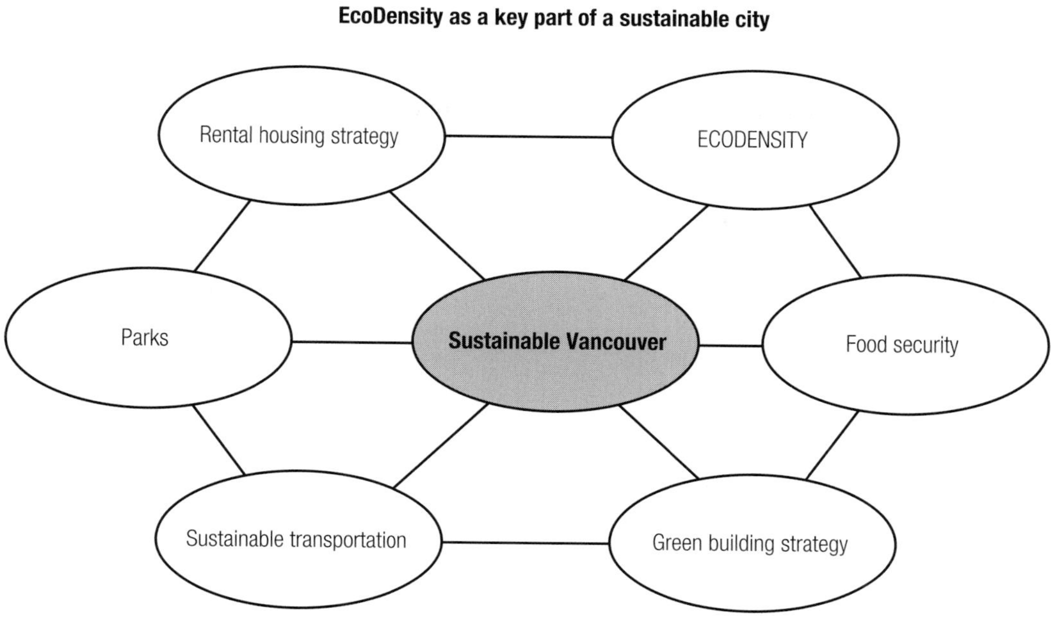

EcoDensity as a key part of a sustainable city

Source: Toderian et al (2008, p3)

Figure 12.1 EcoDensity in context

which led to a series of consultative activities through 2008.

The EcoDensity project involves increased density applied across a variety of urban forms (e.g. key centres and nodes – neighbourhood and transit; in existing low-density residential areas) while ensuring that this increased density is delivered through the use of state-of-the-art green building design and high-calibre architectural design that enhances the character of the neighbourhoods. The City is linking its EcoDensity initiatives to EcoStructure programmes which contain three elements: transportation, community amenities and green systems (City of Vancouver, 2008b).

The City is 'selling' the EcoDensity concept as a means of delivering the following outcomes that contribute to sustainability and liveability in the City:

- make walking, transit and cycling easier for more people;
- take advantage of existing infrastructure;
- allow for new green systems that reduce

and improve use of energy, water and materials;
- introduce urban agriculture to reduce 'food miles';
- create more complete communities by having housing diversity within walking distance of shops and services, and accessible to transit. (City of Vancouver, 2008b, p7)

Vancouver intends that the EcoDensity process will result in three outputs:

1 EcoDensity Charter: Principles to guide future decisions on planning and development to achieve EcoDensity in Vancouver.
2 EcoDensity Toolkit: Resources to help professionals and interested citizens understand the environmental implications of their development choices and to encourage the implementation of greener and more affordable housing choices.
3 Action Plan: A series of shorter and longer term actions to implement EcoDensity in Vancouver. (City of Vancouver, 2008b, p38).

Vancouver's explicit efforts to 'sell' the density concept are distinctive in Canada. Although many municipalities, local and regional, have included efforts to intensify land-uses in their planning efforts, Vancouver is distinctive for putting the density issue at the forefront and tackling it head on. The City's efforts to engage the public in a forward looking dialogue about density are also noteworthy. While it is common for local and regional governments in Canada to seek public input (often through traditional consultative measures such as public meetings) on strategic plans (called official plans or official community plans) it is not common for municipal government to engage its citizens in dialogue and conversation about specific planning issues (e.g. density) outside the context of specific development plans. The City of Vancouver's efforts here are distinctive and perhaps indicative of what it will take to shift the public's mindset about further increasing density. The City's investment in this programme is curious because the City of Vancouver has already received international recognition for its intensification efforts (e.g. downtown Vancouver, South East False Creek) yet it still opted to invest in this significant campaign to 'sell' density to its residents. This process of engagement raises the question of whether these predevelopment project conversations will make it easier, harder or both for developers to proceed with new high-density developments. From a broader perspective, the EcoDensity project serves as a reminder that developers and local governments, despite their previous success with building livable, higher density developments, still confront the challenge that the public is not easily convinced that higher density results in positive outcomes at the community or individual householder scale.

The EcoDensity initiative is not without its detractors. As with many other urban densification policies, there can be local resistance to the perceived environmental, economic and social impacts. Other critiques of the project centre around two major themes:

1   The impacts of the intensification will be felt more by working-class citizens than more wealthy ones.

2   While the City has consulted widely on the idea of EcoDensity, its consultation process for the first priority areas has not been as extensive or inclusive as some detractors might like.

These critiques further reinforce the earlier raised point that for effective growth management to take place, the process by which we achieve it is important. The case of EcoDensity is a reminder that a fundamental component of an effective growth management policy development process is citizen participation. Equitable planning processes must consider whose voices are being heard and whose voices are absent or marginalized. Similarly, it is imperative to consider who bears the burden of implementation. In international discussions about emissions reduction regimes, countries in the global south routinely remind their neighbours to the north that it was their patterns of industrialization that rapidly increased the anthropocentric contribution to climate change and there should a be concordant bearing of the responsibility for action and change by these same countries. Parallels can be drawn with the Vancouver EcoDensity critique. Affluent, large single family homeowners traditionally have higher GHG emissions yet some Vancouverites feel that it is the residents of the smaller, more dense, working class neighbourhoods who are being asked to shoulder the increased densities.

Conclusion

Municipal governments are important institutions in addressing climate change. To date, they have successfully implemented a range of technical fixes to reduce emissions within city operations and have demonstrated ongoing commitment to further reduce GHG emissions. Despite the potential of growth management to simultaneously reduce costs transportation and GHG emissions and other air pollutants, local governments are still in the early stages of developing this approach.

This chapter has suggested that effective response to growth management in Canadian communities is dependent upon new policy frameworks, strong political will and meaningful opportunities to include the public in decision making. The Ontario growth management strategy and the City of Vancouver's EcoDensity efforts are the two most promising growth management efforts currently underway in a country that continues to wrestle with the challenge of curbing public demand for single family, low density, automobile dependent homes. Although these initiatives are in the early stages of implementation, they both represent innovative efforts to accommodate urban population growth in a manner that has the potential to reduce GHG emissions by creating an urban form that reduces transportation and building energy use. Ongoing study of these efforts will shed light on the vexing North American problem of suburban sprawl and its contribution to global climate change. These cases give rise to a range of important research questions that include:

- Toronto and Vancouver, along with Montreal are considered Canada's most 'metropolitan' cities in terms of citizens having an urban mindset. Yet, planners across scales face challenges when seeking to increase the densities of new developments. If this 'sell' is hard in these metropolitan centres, what are the prospects like for other Canadian cities and ultimately Canadian greenhouse gas emissions reductions?
- The North American political appetite for command and control type policies in the context of environmental politics is weak as is evidenced by Canadian and American responses to the ratification and implementation of the Kyoto Protocol. One common thread linking these case studies is the planners' need to be deliberate and interventionist in their efforts to manage growth in a way that can contribute to reducing emissions. Is this tension insurmountable or do these cases suggest alternative but complementary first-steps toward progress?
- Canadian urbanists often draw from European urban examples when making the case that increased urban density need not

come at the expense of beautiful public spaces, access to sunlight, and liveable scale. But, critics are quick to suggest that 'what works in Amsterdam/London/Paris/Copenhagen can't work here'. Is this really the case? What can Canadian spatial planners learn from our European counterparts with regard to new ways to achieve higher densities?

To close, these case studies and subsequent questions to guide further research serve as a reminder that spatial planners have an important role to play in mediating the tension between short-term political goals and long-term urban settlements that are sustainable, desirable and affordable in the short, medium and long term.

References

Berke, P. R. and Manta Conroy, M. (2004) 'What makes a good sustainable development plan? An analysis of factors that influence principles of sustainable development', *Environment and Planning A*, vol 36, pp1381–1396

Bulkeley, H. (2006) 'A changing climate for spatial planning?', *Planning Theory and Practice*, vol 7, no 2, pp203–214

Burda, C. (2008) *Getting Tough on Sprawl: Solutions to Meet Ontario's Climate Change Targets*, Pembina Institute, http://ontario.pembina.org/pub/1612

Carter-Whitney, M. (2008) 'Ontario's Greenbelt in an international context: Comparing Ontario's Greenbelt to its counterparts in Europe and North America', www.cielap.org/pub/pub_internationalgreenbelt.html, accessed 10 July 2008

City of Vancouver (2008a) 'What is EcoDensity', www.vancouver-ecodensity.ca/content.php?id=39, accessed 2 May 2008

City of Vancouver (2008b) 'EcoDensity Primer', www.vancouver-ecodensity.ca/content.php?id=2, accessed 2 May 2008

City of Vancouver (2008c) 'EcoDensity Primer: An introduction to building communities that are green, liveable and affordable in Vancouver', www.vancouver-ecodensity.ca/content.php?id=2, accessed 2 May 2008

City of Vancouver (2008d) 'Detailed history: EcoDensity background', www.vancouver-ecodensity.ca/content.php?id=38&sid=0&pid=0&language=en, accessed 10 July 2008

City of Vancouver (2008e) 'Policy Report (Urban structure): Revised charter and initial actions', www.vancouver-ecodensity.ca/content.php?id=42, accessed 30 May 2008

Federation of Canadian Municipalities (2008a) 'Partners for Climate Protection', www.sustainablecommunities.fcm.ca/partners-for-climate-protection/, accessed 30 May 2008

Federation of Canadian Municipalities (2008b) 'The Green Municipal Fund', www.sustainablecommunities.fcm.ca/GMF/, accessed 10 July 2008

Fung, F. and Conway, T. (2007) 'Greenbelts as an environmental planning tool: A case study of southern Ontario, Canada', *Journal of Environmental Policy & Planning*, vol 9, no 2, pp101–117

Gore, C. (2008) 'Local governments and climate change advocacy in Canada: Reflections on network (in)effectiveness or the politics of indifference', paper prepared for the Canadian Political Science Association Conference, Vancouver, 4–6 June

Gore, C. and Robinson, P. (in press) 'Local government response to climate change: Our last, best hope?', in H. Selin and S. VanDeveer (eds) *Changing Climates in North American Politics*, MIT Press, Cambridge, MA

Government of Ontario (2007) 'Go Green: Ontario's action plan on climate change', Government of Ontario, Toronto

Metrolinx (2008) 'Preliminary directions and concepts: Development of a regional transportation plan for the Greater Toronto and Hamilton Area, Metrolinx, Toronto, ON

Ministry of Public Infrastructure Renewal (2006) *Growth Plan for the Greater Golden Horseshoe,* MPIR, Toronto

Municipalities Issue Table (1998) *Foundation Paper,* National Climate Change Programme, Government of Canada, Ottawa

Newman P. and Kenworthy, J. R. (1999) *Sustainability and Cities: Overcoming Automobile Dependence*, Island Press, Washington DC

Parkinson, S. and Roseland, M. (2002) 'Leaders of the pack: An analysis of the Canadian Sustainable Communities 2000 municipal competition', *Local Environment*, vol 7, no 4, pp411–429

Portney, K. (2002) 'Taking sustainable cities seriously: A comparative analysis of twenty-four U.S. cities', *Local Environment*, vol 7, no 4, pp363–380

Robinson, P. (2006) 'Canadian municipal response to climate change: Measurable progress and persistent challenges for planners', *Planning Theory and Practice Interface*, vol 7, no 2, pp218–223

Robinson, P. (2008) 'Urban sustainability in Canada: The global–local connection', in C. Gore and P. Stoett (eds) *Environmental Challenges & Opportunities: Local-Global Perspectives on Canadian Issues*, Emond Montgomery Publications, Toronto

Robinson, P. and Gore, C. (2005) 'Barriers to Canadian municipal response to climate change', *Canadian Journal of Urban Research*, vol 14, no 1, Supplement, pp102–120

Roseland, M. (2005) *Toward Sustainable Communities: Resources for Citizens and Their Governments*, New Society Publishers, Gabriola Island, BC

Statistics Canada (2008) 'Population and dwelling counts: 2006 Census', *The Daily*, 13 March, www.statcan.ca/Daily/English/070313/d070313a.htm, accessed 26 June 2008

Toderian, B., Howard, R. and Kuhlman, T. (2008) 'EcoDensity: Revised charter and initial actions', City of Vancouver Policy Report to City Council, www.vancouver-ecodensity.ca/webupload/File/01%20Covering%20Council%20Report_FINAL.pdf, accessed 26 February 2009

Tomalty, R. and Alexander, D. (2005) 'Smart growth in Canada: Implementation of a planning concept', Canada Mortgage and Housing Research Paper, www.smartgrowth.ca/research/Smart%20Growth%20in%20Canada-Implantation%20of%20a%20Planning%20Concept.pdf, accessed 15 January 2008

Toronto City Summit Alliance (2008) 'Greening Greater Toronto', http://64.34.53.199/pdf/Report.pdf, accessed 24 June 2008

World Commission on Environment and Development (1987) *Our Common Future*, Oxford University Press, Oxford

13

Planning for Offshore Wind Energy in Northern Europe

Stephen Jay

Renewables stepping offshore

The major expansion of renewable forms of energy is now an accepted part of the strategy for reducing the use of fossil fuels and the associated emission of greenhouse gases (GHGs) (European Commission, 1997; Department of Trade and Industry (DTI), 2007). The relative speed and extent of renewables development in some countries over the last two decades has been impressive (Coenraads and Voogt, 2006; International Energy Agency (IEA), 2008), such as the uptake of wind energy in some north European countries. Although the regulatory means of bringing this about has varied between different countries, planning bodies have generally been closely involved in the advent of modern renewables, and have brought their influence to bear on this changing pattern of energy production (e.g. Kellett, 2003; Khan, 2003; Toke, 2005). It is partly because schemes have been diffused that many planning authorities have found themselves engaging closely with the development of renewables. In some countries, such as the UK, planning authorities have found a new role in dealing with renewables because of their small scale compared to conventional energy schemes, which have tradi-

tionally been dealt with by central government.

The most recent and perhaps surprising stage in the renewables revolution has been that the wind energy industry has stepped offshore. The possibility of exploiting marine renewable sources of energy, especially wind, tidal and wave power, has been mooted for some time, but it is only in the last few years that significant progress has been made. Specifically, the large scale capture of offshore wind resources has now become technically and economically feasible. Moreover, the enormity of the resources available for many maritime nations is finally being appreciated, with the UK being described, for example, as the 'Saudi Arabia of wind energy' (Boyle, 2006, p26) because of its vast marine hinterland with consistently strong winds blowing across it.

In fact, some pioneer offshore wind farms were developed during the 1990s. The first was built in 1991 near Vindeby in Denmark, followed by a few others in Denmark, Sweden, the Netherlands and the UK; these all consisted of a small number of turbines (up to 11) in shallow waters close to shorelines, with an output of just a few megawatts.[1] However, 2002 marked a new phase of development, when Denmark again led the way by switching on the Horns Rev wind

Source: Reproduced courtesy of Vattenfall Group

Figure 13.1 Horns Rev offshore wind farm, Denmark

farm, comprising 80 turbines 14–20km from the coast (Figure 13.1). Since then, the UK, Ireland, the Netherlands and Sweden have also developed schemes with outputs of the order of 100 megawatts. They are also planning future projects, as are a number of other European countries, the US, Canada and China (Offshore Center Danmark, undated). Some of these schemes are on a quite unprecedented scale; for example: the UK has recently given consent for the London Array development in the outer Thames estuary, which will comprise about 270 turbines with an output of 1000 megawatts. This is a hitherto undreamt-of magnitude for wind farms, and will place it amongst the ranks of the major electricity generators.

Stepping offshore therefore appears to represent the coming of age of wind power. A number of governments that have set themselves tough targets for renewable energy production over the coming years are looking to offshore wind to make a major contribution to realizing their ambitions, and are providing significant regulatory and financial support for the sector, as

illustrated below. The European Union is also giving a strong lead to member states with the potential to exploit offshore wind (Commission of the European Communities (CEC), 2008). Furthermore, a vibrant industry has now emerged that is being relied upon to deliver this secure and low-carbon solution to our energy needs. Against a background of liberalized and competitive energy markets, energy companies are vying with each other for the opportunity to move into offshore areas that have been selected for development.

Gaining consent for offshore wind farms

Proposals for new renewable energy schemes on land generally come under normal arrangements for infrastructure planning. Typically, proposals are dealt with under the planning system for the area, administered by local planning bodies which therefore make key decisions about individual schemes in the context of the many other demands for the use of space. In some cases, central government authorities have a more decisive role, especially for larger schemes, though even here, local planning bodies will usually have a major say. Whatever the arrangements, planning bodies are increasingly being called upon to facilitate the development of renewables in their area.

The situation offshore is quite different. The territory of coastal planning bodies usually ends at or near the shoreline, which effectively means that planning systems as a whole do not extend out to sea. Instead, marine activities tend to be overseen by centralized government departments which exercise powers over specific sectors, such as navigation, ports, dredging and fishing. These powers have been accrued in a piecemeal fashion, sometimes in the wake of various international agreements covering the use of the seas, and adapted over time to meet new demands. So the pattern that has grown up is one of fragmented, complex and sometimes overlapping controls, exercised by different government bodies with narrowly defined responsibilities (Kay and Alder, 2005). This situation is amply illustrated by the case of offshore wind farms, where developers have to negotiate a complex process of gaining a range of consents that were not originally designed to deal with marine renewables, and which may take a period of several years to complete. In the UK, for example, two different consent routes are available under separate pieces of legislation covering energy and navigation, and additional authorizations must also be gained on safety and environmental matters. Accordingly three or more government departments may be involved in the planning process (Jay, 2008). The consents regime is further complicated by the different arrangements that may apply either side of the limit of territorial waters (generally 12 nautical miles from the coast). Beyond this limit, nations do not exercise full sovereignty, but they may establish the right to exploit certain resources such as energy (by declaring an 'exclusive economic zone', for example); however, regula-tory controls have to be modified for this very different legal environment.

Within this scenario, there is relatively little opportunity for planning bodies to play a formal part in the development of offshore wind farms (except in the few countries where planning authorities do have jurisdiction over coastal waters, as illustrated in the case of Sweden below). In general, the contribution of planning is restricted to two aspects. Firstly, nearby planning authorities are centrally involved in the onshore elements of a wind farm, such as cabling and other connections to the existing electricity network, for which planning consent may be required. Although onshore infrastructure is secondary to the main wind farm development, it still requires careful integration to its surroundings. The possibility is now also being raised of combining the electricity supply from neighbouring offshore wind farms into more coherent and efficient grid connections, making effective onshore coordination all the more critical.

Secondly, planning authorities that look out to the offshore sites will be consulted for their views by the central government bodies dealing with the principal wind farm applications. This gives them a potentially significant role, but in practice their opinions may carry relatively little weight in the overall process. For instance, the experience of UK planning authorities along the coast that have been consulted over the recent phase of offshore wind farms is that their input, though based on a detailed understanding of the overall setting of the proposed wind farms, was marginal and sometimes disregarded in the decision-making process. This raises the possibility of the loss of insights that may be best expressed through the mechanisms of planning, not least regarding the integration of wind farms into complex patterns of coastal activity, which local planning authorities are well-placed to articulate and handle. There are, arguably, examples of poor decision making regarding the siting of offshore wind farms where the well-argued cases of local planning authorities have been overridden (Jay, 2008).

Enter spatial planning?

This situation is not static, however, as significant changes are beginning to take place in the way that marine activities are governed. It is now increasingly recognized that the sectoral framework that has evolved for regulating marine activities, within which offshore wind farm development has so far taken place, is inadequate. This is not only because it complicates life for users of the sea, but also because it leaves the seas open to harm from poorly coordinated activities (Department for Environment, Food and Rural Affairs (Defra), 2002; Elliott et al, 2006). Some steps are therefore being taken for a more integrated approach to managing the seas and improving the marine environment (European Parliament and Council, 2008). Within this context, calls are being made specifically from a planning perspective for the creation of a strategic, spatial planning system for the marine environment (Tyldesley, 2004); this is now being implemented in some countries, especially those where the development of offshore wind energy is being pursued most vigorously. This opens the way for the more comprehensive application of the principles and practices of planning to the expansion of offshore renewables, and indeed, this is one of the stated intentions of marine planning (Defra, 2007). In fact, there is already evidence from some countries of a more positive incorporation of the contribution of spatial planning into offshore wind farm development.

This said, there has been little investigation to date on the relationship between spatial planning and the development of offshore wind energy. This is in contrast to the burgeoning literature addressing the role of planning in the growth of land-based wind farms and renewables in general (see Ellis et al, 2007 for a summary). Equivalent offshore studies are only just emerging (Gray et al, 2005; Ellis et al, 2007; Firestone and Kempton, 2007), with the UK being a key focus of research (Jay, 2008). In this chapter, attention is turned instead to three other European countries where offshore wind is being given a seat at the top table of energy production and where spatial planning principles are playing an important role in shaping its implementation. Sweden, Denmark and Germany have been chosen because they represent a progression of strategic outlook on planning for offshore wind, as summarized in the headings below. Also, they share common marine borders that raise additional transboundary questions that are likely to take on growing importance with the intended expansion of offshore wind

Sweden: Local offshore wind farms

Between 1998 and 2002, three small offshore wind farms were completed in Swedish waters. They are all located in shallow, sheltered waters close to islands in the southern Baltic Sea. A larger wind farm has recently started operating in the Öresund, the narrow strait shared with Denmark. Other medium-sized projects are in the pipeline, and also a very major scheme further out to sea, near the marine border with Germany.

Sweden is unusual in marine management terms in that its terrestrial planning system extends significantly into the sea. Counties and municipalities (the two levels of local government) have planning jurisdiction as far as the limits of territorial waters. This means that the authorization of wind farms at sea (all of which have been located within territorial waters so far) follows essentially the same procedures as those on land and local authorities have a central role in their development. However, extra permits are needed offshore, considerably lengthening the time taken for projects to be implemented.

In addition to the involvement of local authorities, a strategic lead is being given by a government agency responsible for promoting the development of wind power. The Swedish Energy Agency (SEA) aims to achieve a fourfold increase in wind energy by 2020, with the intention that a third of the final output should come from offshore. The agency is facilitating this expansion by various measures, such as aiming to streamline authorization procedures and supporting wind energy in more challenging locations, including offshore, through research and financial subsidies (SEA, 2008). It has also set out principles of 'national interest' that favour the development of wind energy. Following on from this, broad criteria have been established for selecting suitable sites for windfarms, such as a minimum average wind speed. Exclusion criteria have also been defined, including areas with less capacity than 10 megawatts and offshore areas deeper than 30 metres. Other no-go areas are set down in a national environmental code, including coastal areas and archipelagos, and areas of outdoor recreation and nature conservation. Areas of national interest for shipping and fishing are also prohibited. Despite these wide-ranging exclusions, the agency has been able to identify about 420 areas of national interest for wind power on land and sea, which are now being put to the local authorities.

The agency has taken a partnership approach to achieving its goals, entering into negotiation with the local authorities about possible sites: 'these are the criteria we send out to the county administration, they send it out to the municipalities, then we get objections back; we work it out together and send it out again and so on, because we want to work together with the municipalities, so it gets a little more accepted' (Swedish Energy Agency officer, interview 25 August 2008). This perhaps reflects the high degree of autonomy of local authorities in Sweden and the sometimes resistant attitude that they have had to wind farms (Khan, 2003). Ultimately, the authorization of wind farms at land and sea is in the hands of local authorities. As far as larger schemes are concerned, including all offshore, both counties and municipalities must give their consent. Consultation with official bodies and the public plays an important part in the procedures (SEA, 2008), as illustrated by the consents process:

- The applicant carries out a formal process of consultation with all affected parties, including government bodies, local communities and other stakeholders. The results of this exercise form part of the application.

- An environmental impact statement is prepared and included in the application.
- Written application is made to the county, which carries out its own consultation.
- A decision is made, which is open to appeal by any party.
- An application must also be made to the municipality for a building permit.
- For this to be granted, the municipality's strategic-level plan may need to be modified to provide backing for the wind farm.
- A detailed development plan will be drawn up for the scheme, which is also subject to consultation and open to appeal.

The agency is also encouraging local authorities to be proactive in seeking possible sites, saying 'don't sit around and wait for development to come to you, start planning, look at the wind map, where do you think you can plan for a wind device in your municipality, and talk to your neighbours so you don't put them on the border with another municipality or country, work together' (Swedish Energy Agency officer, interview 25 August 2008). Along with this, attention has also been given to considering carefully the environmental consequences of wind farms, such as their effects upon the land/seascape and finding the optimum means of accommodating them within their settings (Palmberger and Jakélius, 1998).

One of the authorities that have been most active in supporting the expansion of offshore wind energy is the Administrative Board of Kalmar County, in the south-east of the country. The county includes Kalmarsund, a strait between the mainland and the island of Öland, where two of the earliest schemes are located and another is planned. The board has drawn up policy in collaboration with its municipalities and a neighbouring county in support of offshore wind energy, taking into account the environmental conditions of the area and setting out preferred options for wind farm development (Länsstryrelsen Kalmar Län, 2003). For instance, one of the key issues is the visual appearance of the wind farm from land and the consequence of this for coastal areas where tourism is important. This has led to a preference for turbines being arranged so they form simple lines rather than cluttered 'forests' when seen from particular vantage points.

In Sweden, therefore, the local planning authorities are taking a leading role in the development of offshore wind energy, in marked contrast to nations where planning authorities along the coast have no more than a marginal influence, such as the UK. This is partly because of the unusual jurisdiction of Swedish coastal planning authorities over territorial waters and also, perhaps, the robust control they exercise over development in their areas in general. However, there is also evidence that they are working collaboratively with the lead government agency for wind energy, and are bringing their spatial planning skills to bear on finding the best means of incorporating wind farms into their wider settings, both on land and at sea.

Denmark: Strategic mapping for offshore wind

As mentioned above, the world's first offshore wind farm was built in Danish waters, and Denmark has maintained its lead in exploiting offshore wind energy (though the UK is steadily taking over this position). The nation's pioneering status was a natural progression from its onshore industry, which expanded rapidly in the late 1980s, as the plentiful shallow, sheltered waters around the country's long coastline presented an obvious opportunity for the extension of wind power. Pilot projects were developed in the early 1990s with government encouragement and support. This was followed in 1997 by a government-led action plan for expansion, resulting in further schemes close to the coastline and two large demonstration projects several kilometres from the coast (including the world's first large-scale offshore wind farm). There are now eight offshore wind farms of varying sizes in operation, dotted around the coasts of Jutland and the islands (Danish Energy Authority (DEA), 2005). Two other large projects further out to sea are also

coming on line. This growth has been characterized by a gradual building of experience and confidence, and also by attention being given to environmental issues and public acceptability, with studies being carried out jointly by the industry and government agencies (Dong Energy et al, 2006).

In addition, there has been a political consensus in favour of offshore wind energy. The most recent expression of this is the 2005 energy strategy in which a commitment was made to meeting 30 per cent of national gross energy demand from renewable sources by 2025 (Ministry of Transport and Energy, 2005). Offshore wind is expected to make a major contribution to this target, for which an inter-departmental committee has now drawn up a strategy (DEA, 2007). However, there has been a shift to a more market-oriented approach to the sector, so that energy companies must now tender for particular projects, or indeed propose their own sites, rather than sites being allocated by government.

In contrast to Sweden, coastal municipalities in Denmark have no jurisdiction beyond the shoreline, so the primary administrative responsibility for offshore wind energy falls to the DEA (which now comes under a recently formed Ministry of Climate and Energy). This agency implements government renewables policy and considers itself to be the planning authority for the sea with regard to offshore wind. The agency led the development of the recent strategic study for expansion. This consisted primarily of a mapping exercise for identifying potential sites for offshore wind energy, by which layers were built up to show areas of constraint and opportunity. This took into account:

- technical considerations, such as wind speeds, water depth, distance to shore;
- existing electricity grid conditions and the need for reinforcement;
- existing areas of constraint, such as shipping lanes, military zones, nature protection areas and mineral extraction areas;
- possible future constraints;
- environmental considerations, such as effects upon landscape and wildlife.

Figure 13.2 Potential offshore wind farm sites in Danish waters

This exercise led to 23 sites of 44km² each being identified as suitable for wind energy development, widely scattered through Danish waters (circled areas in Figure 13.2). This tended to focus simply on finding potential sites for this one activity within the limitations imposed by other activities rather than an integrated approach. Nonetheless, an emphasis on dialogue with interested parties allowed some degree of negotiation: 'at some point you need to prioritize, would this area be more suitable for fishing or wind power or natural protection? At some stage we got to the point where we needed to argue our case for what would be the best use' (DEA officer, interview 21 August 2008). Interestingly, sites already developed for offshore wind would not necessarily have been chosen by this process. Indeed, it has become all the more

important to consider other marine activities carefully because of the growing pressures on Danish waters, especially the designation of new nature protection areas and the rise in shipping traffic (DEA, 2007). The 23 sites combined have a total possible capacity of 4600 megawatts, or more than 8 per cent of national energy consumption. Two of them have been put out for tender with the intention of achieving an additional installed capacity of 400 megawatts by 2012.

This strategy does not prevent energy companies from taking their own initiatives in proposing sites for wind farms. An 'open door' option is also available, which allows companies to apply for permission to build wind farms on sites that they choose for themselves. These are assessed against similar criteria to those used in the strategic mapping exercise. There has been some activity on the part of companies proposing small-scale projects on this basis, with test sites being encouraged especially near places such as industrial areas or ports, where the sense of impact is likely to be relatively minor.

The DEA is also central to consent procedures for individual projects, and offers a 'one stop shop' to applicants. For example, once a concession is granted to a successful tender, a licence is given to allow preliminary investigations and environmental studies, followed by consent for construction, with conditions, and a licence to produce electricity. This is a much more straightforward arrangement than in other countries developing offshore wind energy, where authorizations tend to be convoluted and split between different arms of government. The agency can thus draw together and weigh up the many interests expressed through consultation in a relatively coordinated manner.

The emphasis on consultation and gaining consensus when deciding on individual schemes extends to the involvement of coastal municipalities. Local municipalities are treated as important consultees from an early stage, and their views carry considerable weight and may be decisive, especially for projects close to the shore where issues such as noise and visual impact may be significant. For one particular proposal, 'the project didn't happen because there was great public resistance, and because of this, the municipality also spoke against it'. On the other hand, 'for Horns Rev II, we asked all the municipalities along the coast that would be able to see the park from the shore and the city of Esbjerg, a construction port; it brings a lot of activity for them, so the view of this park was positive' (DEA officer, , interview 21 August 2008).

Denmark therefore shares certain characteristics with Sweden in its planning for offshore wind energy: the importance given to local planning perspectives and the pivotal role of a dedicated government agency working in collaboration with interested parties. In addition, a more strategic level of planning has been established. Although this has largely been a sectoral exercise rather than a comprehensive spatial planning approach to the use of Danish waters,

there are indications that more traditional activities are having to adjust to this new use of marine space. 'The location of large scale wind farms offshore has triggered a need to balance the many and varying sea-use interests. Overall, the sea around us has gradually become the object of planning' (DEA, 2007, p3).

Germany: Marine spatial planning for offshore wind

Germany has two marine areas under its jurisdiction, separated from each other by Schleswig-Holstein: a wedge of the North Sea, stretching as far as a maritime border with the UK, and a narrow section of the southern Baltic Sea. A clear distinction is made in the administration of both of these areas by the 12 nautical mile limit of territorial waters: territorial waters come under the authority of the coastal states (the Länder), whereas the seas beyond have been designated as an exclusive economic zone (EEZ) under the control of the federal government.

Although Germany has been at the forefront of developing wind energy capacity on land since the early 1990s, it has only recently sought to follow the example of its northern neighbours in extending the capture of wind energy to sea areas. A federal offshore wind energy strategy was adopted in 2002, setting out a pathway for this venture (Federal Ministry for the Environment, Nature Conservation and Nuclear Safety (FMENCNS), 2002). This strategy is set in the context of challenging national targets for renewable energy, which have most recently been set at the production of 25–30 per cent of total energy needs from renewable sources by 2020. There is also an expectation of a slow-down in building new land-based wind farms in the coming years. Offshore is therefore the new frontier in the quest for clean sources of energy; it is hoped that a massive 25–30,000 megawatts of installed capacity can be attained by 2030. Despite these aspirations, however, by the end of 2008 no progress had been made on the construction of offshore wind farms, apart from work on a small pilot project.

Because of the administrative split of German waters, different regimes are in place for developing wind farms in the territorial waters and the EEZ. However, the main focus of attention for offshore wind energy is the EEZ, partly because the nearshore territorial waters include large areas of nature conservation importance which are considered too sensitive for development. The main body which deals with wind farm applications in the EEZ is the Federal Maritime and Hydrographic Agency. The agency acts under marine legislation which is designed to favour the expansion of wind energy. Crucially, this requires the agency to approve a project so long as neither navigation nor the marine environment is adversely affected, regardless of other considerations. The agency (along with the Waterways and Shipping Directorate) handles the necessary consents for a project and carries out cross-departmental and other stakeholder consultation on behalf of an applicant (Bundesamt für Seeschifffahrt und Hydrographie (BSH), 2008a).

The effect of this developer-friendly regime has been to encourage a rush of speculative applications from energy companies keen to stake a claim in the EEZ. Under the legislation, the agency is obliged to offer sites on a first-come, first-serve basis, and companies are free to lodge applications at relatively low cost to themselves and with no cast-iron commitments. The result is that large sections of the EEZ have been partitioned between competing developers, with about two dozen projects (of up to 80 turbines each) now approved, and a host of others under consideration (BSH, 2008b). Companies remain unwilling, however, to realize the schemes for which they have authorization.

This very unsatisfactory situation has led to a rethink of the best way to allocate sites for offshore wind farms. The approach that is developing instead is a comprehensive system of spatial planning for German waters. 'Now it is really chaotic, we are not very happy about this. We hope that with the spatial plan coming into force that maybe this will be stopped' (Federal Maritime and Hydrographic Agency officer, interview 28 August 2008). This should place offshore wind energy in the context of all other marine activities, so that interests as a whole can be balanced, conflicts reduced and synergies found. The agency's remit has now been modified to allow broader spatial planning issues to be taken into account, rather than just navigation and environmental issues, when assessing applications.

The system has come into force through an amendment to Germany's founding planning legislation, to extend existing competencies for terrestrial planning to the EEZ. The Maritime and Hydrographic Agency is now responsible for preparing a marine plan and environmental report for the EEZ, which will provide the definitive framework for offshore wind energy. Similarly, for territorial waters, the coastal Länder have extended their planning competencies out to sea and are expanding their existing land use plans accordingly.

A draft plan for the EEZ has now been completed. This sets out what are likely to be three priority areas for offshore wind energy, where effort will be concentrated on implementing actual schemes. Any conflicting uses will be excluded from these areas. Similar 'suitability areas' are being designated by the Länder in the territorial waters. Interestingly, a measure of cooperation between different interests has been a feature of plan making. For example, shipping authorities have supported the designation of a wind energy area in between shipping lanes as a means of reinforcing traffic separation. Design criteria for wind farms have also been set out in the EEZ plan, such as limiting the maximum height of turbines to 125m on visual grounds. The hope is clearly that this planning exercise will give a much stronger lead on the expected parameters of offshore wind farms.

Other measures are also being taken to encourage the implementation of offshore wind energy. Most importantly, under new legislation, grid connections are now the responsibility of the grid operators rather than the wind energy companies, which should lead to a more coherent pattern of cabling, enabling, for instance, an offshore transformer to be built to collect power from several neighbouring wind farms. The federal government is also financing a test site of

Figure 13.3 The Kreigers Flak projects, Baltic Sea

large turbines with the aim of kick-starting the industry into action.

So amongst the nations described in this chapter, Germany has gone the furthest in applying the principles of strategic planning to offshore wind energy development, in that offshore wind energy is being set in the context of the multiple demands being placed upon the marine environment. This is in line with the notion of marine spatial planning which is being widely advocated as a more integrated approach to marine management (CEC, 2006; Douvere and Ehler, in press). In Germany,

however, it is specifically the difficulty of implementing offshore wind energy that has provoked the formation of a marine planning system. The problems encountered are perhaps a reflection of the precipitous manner in which the federal government has sought to implement offshore wind energy, in contrast to Denmark and Sweden's more gradual gaining of experience, but this has, arguably, resulted in the most progressive approach to planning for offshore wind energy adopted so far by the three nations.

Where the waters meet

As offshore wind farms extend further from coastlines, transnational issues inevitably arise. This is already apparent in the case of Sweden, Denmark and Germany, as they share some restricted sea areas with ample potential for wind farms. For example, in the Baltic Sea, their marine boundaries meet in a shallow area called Kriegers Flak. There are proposals for three schemes here, one in each national segment. Kriegers Flak 1, 2 and 3 are at various stages of development, with a possible total of about 300 turbines covering an area of at least 100km² (Figure 13.3). These projects were conceived with little discussion between the authorities of the three countries apart from formal cross-border consultations, and certainly with no plan in mind for the marine area as a whole. The strategic approach to offshore wind energy that is emerging within these nations has not yet been scaled up to this wider region, which is a shortfall given the integral nature of this marine space and of existing activities and conditions across it. At present, there is a danger of development proliferating with little consideration for the wider consequences; one can imagine a scenario in which one country's options might be limited by a neighbouring country's initiatives, if, for example, a wildlife habitat has already been compromised by those initiatives.

Having said this, some wider thinking has been given to grid issues for Kriegers Flak, with the possibility of projects sharing a connection to land, so that electricity would be exported from one or two sectors to the country where the connection is made (CEC, 2008). More generally, the possibility of developing a transnational grid becomes more feasible as wind farms are built in border areas. Again, this calls for coordinated action and attention to wider spatial implications.

It should also be noted that the prospect of widespread offshore wind energy has led to an important intergovernmental initiative. The German, Danish and Swedish governments made an agreement in 2007 to cooperate more closely in wind energy deployment in the North and Baltic Seas (Governments of the Kingdom of Denmark, Federal Republic of Germany and Kingdom of Sweden, 2007). Initially, this envisages joint research on the environmental impacts of wind farms, and reflects a desire to expand offshore wind energy without causing damage to the wider marine environment. Norway, the Netherlands and the UK have also joined the discussions taking place in this context.

Conclusions

The imperative to tackle climate change is a major driver for the current development of offshore wind energy, and has become central to the discourse of energy policy makers and the industry alike. The lead in developing this form of renewables is being taken by nations that have the potential to extend their already well-established land-based wind industry into easily exploitable waters off their coastlines, especially in northern Europe. However, this takes the development of renewables into not just a new physical environment, but also the choppy waters of marine administration, with its renowned departmental fragmentation and lack of cross-sectoral frameworks. With the exception of countries like Sweden, there has been little place for spatial planning in the shaping of marine activities until recently.

Nonetheless, the pressure to develop offshore wind energy, and the need to do this with due regard to other marine activities and to protecting the marine environment, has begun to focus the attention of government bodies responsible for offshore wind energy on more strategic and integrated approaches that owe much to the practices of spatial planning. Offshore wind is proving to be a catalyst for the expansion of the boundaries of spatial planning into previously prohibited territory (CEC, 2008). The new contributions to marine governance that are emerging include, as illustrated above, the extension of plan-making to the sea, strategic mapping for individual sectors like offshore wind energy, and comprehensive spatial planning for all marine activities. These offer the possibility of coordinating marine activities more successfully, incorporating environmental considerations more sensitively into marine development, integrating schemes more carefully into their surroundings and dealing more holistically with the land–sea interface.

Underlying these efforts are important institutional features that have been key, at least at the national level in the examples described above, to progress being made towards the establishment of offshore wind energy. Both the formation of broad strategies and the detailed planning of individual schemes are characterized by strong central government leads through dedicated agencies, genuine collaboration between central and local levels of government and meaningful public and stakeholder involvement in which differing perspectives are captured and shaped through the process of engagement with the wider vision. Consensual processes of decision making are therefore playing a key role in the expansion of this potentially important technology for reducing reliance on fossil fuels and also in the formation of more integrated strategies for the use of marine space as a whole.

Acknowledgements

Thanks are due to the following people who gave invaluable help for these accounts: Fredrik Dahlström of the Swedish Energy Agency, Ingegärd Widerström of Kolmar County, Sweden, Steffen Neilsen and Mette Buch of the Danish Energy Authority, and Nico Nolte and Carolin Abromeit of the German Federal Maritime and Hydrographic Agency.

Note

1   The output of power stations is usually expressed in megawatts. Large conventional power stations, burning coal, for example, typically have a capacity of 500 to 2000 megawatts. Wind farms have had a much lower capacity, perhaps as little as a couple of megawatts for a scheme of just a few turbines. However, the development of more powerful turbines and the trend to much larger wind farms is changing this pattern, so that much larger scale wind farms are now being developed with capacities of several hundred megawatts or more.

References

Boyle, G. (2006) 'UK offshore wind potential', *Refocus*, July/August, pp26–29

Bundesamt für Seeschifffahrt und Hydrographie (BSH) (2008a) 'Wind farms', www.bsh.de/en/Marine_uses/Industry/Wind_farms/index.jsp, accessed 15 November 2008

BSH (2008b) 'Contis maps', www.bsh.de/en/Marine_uses/Industry/CONTIS_maps/, accessed 15 November 2008

Commission of the European Communities (CEC) (2006) 'Towards a future maritime policy for the Union: A European vision for the oceans and seas', Green Paper, COM 275, CEC, Brussels, http://ec.europa.eu/maritimeaffairs/pdf/com_2006_0275_en_part2.pdf, accessed 15 November 2008

CEC (2008) 'Offshore wind energy: Action needed to deliver on the energy policy objectives for 2020 and beyond', Commission of the European Communities, Brussels, http://ec.europa.eu/energy/strategies/2008/doc/2008_11_ser2/offshore_wind_communication.pdf, accessed 15 November 2008

Coenraads, R. and Voogt, M. (2006) 'Promotion of renewable electricity in the European Union', *Energy and Environment*, vol 17, no 6, pp835–848

Danish Energy Authority (DEA) (2005) 'Offshore wind power: Danish experiences and solutions', DEA, Copenhagen, www.ens.dk/graphics/Publikationer/Havvindmoeller/uk_vindmoeller_okt05/pdf/havvindmoellerapp_GB-udg.pdf, accessed 15 November 2008

DEA (2007) 'Future offshore wind power sites 2025', report of the Committee for Future Offshore Wind Power Sites, Danish Energy Authority, Copenhagen, www.ens.dk/graphics/Publikationer/Havvindmoeller/Fremtidens_%20havvindm_UKsummery_aug07.pdf, accessed 15 November 2008

Department for Environment, Food and Rural Affairs (Defra) (2002) *Safeguarding Our Seas: A Strategy for the Conservation and Sustainable Development of our Marine Environment*, Defra, London, www.defra.gov.uk/Environment/water/marine/uk/stewardship/index.htm, accessed 15 November 2008

Defra (2007) *A Sea Change: A Marine Bill White Paper*, Cm 7047, Department for Environment, Food & Rural Affairs, London, www.defra.gov.uk/marine/pdf/legislation/marinebill-whitepaper07.pdf, accessed 15 November 2008

Department of Trade and Industry (DTI) (2007) 'Meeting the energy challenge', A White Paper on Energy, Cm 7124, The Stationery Office, Norwich, www.berr.gov.uk/files/file39387.pdf, accessed 15 November 2008

Dong Energy, Vattenfall, Danish Energy Authority and Danish Forest and Nature Agency (2006) 'Danish offshore wind: Key environmental issues', DEA, Copenhagen, www.ens.dk/graphics/ Publikationer/Havvindmoeller/ havvindmoellebog_nov_2006_skrm.pdf, accessed 15 November 2008

Douvere, F. and Ehler, C. (2009) 'New perspectives on ecosystem-based management: Initial findings from European experience with marine spatial planning', *Journal of Environmental Management*, vol 90, no 1, pp77–88

Elliott, M., Boyes, S. and Burdon, D. (2006) 'Integrated marine management and administration for an island state: The case for a new marine agency for the UK', *Marine Pollution Bulletin*, vol 52, pp469–474

Ellis, G., Barry, J. and Robinson, C. (2007) 'Many ways to say "no" – different ways to say "yes": Applying Q-methodology to understand public acceptance of wind farm proposals', *Journal of Environmental Planning and Management*, vol 50, no 4, pp517–551

European Commission (1997) 'Energy for the future: Renewable sources of energy', White Paper for a Community Strategy and Action Plan, Communication from the Commission COM(97) 599 final, European Commission, Brussels, 26 November, http://ec.europa.eu/energy/ library/599fi_en.pdf, accessed 15 November 2008

European Parliament and Council (2008) 'Directive 2008/56/EC of the European Parliament and of the Council of 17 June 2008 establishing a framework for community action in the field of marine environmental policy (Marine Strategy Framework Directive)', *Official Journal of the European Communities*, vol L164, pp19–40, http://eur-lex.europa.eu/LexUriServ/ LexUriServ.do?uri=OJ:L:2008:164:0019:0040: EN:PDF, accessed 15 November 2008

Federal Ministry for the Environment, Nature Conservation and Nuclear Safety (FMENCNS) (2002) 'Strategy of the German government on the use of off-shore wind energy in the context of its National Sustainability Strategy', unpublished document, www.bmu.de/files/pdfs/allgemein/ application/pdf/offshore.pdf, accessed 15 November 2008

Firestone, J. and Kempton, W. (2007) 'Public opinion about large offshore wind power: Underlying factors', *Energy Policy*, vol 35, no 3, pp1584–1598

Governments of the Kingdom of Denmark, Federal Republic of Germany and Kingdom of Sweden (2007) 'Joint declaration on cooperation in the field of research on offshore wind energy deployment', signed at Berlin 4 December 2007, unpublished document, www.eow2007.info/ uploads/media/Joint_Declaration.pdf, accessed 15 November 2008

Gray, T., Haggett, C. and Bell, D. (2005), 'Offshore wind farms and commercial fisheries in the UK: A study in stakeholder consultation', *Ethics Place and Environment*, vol 8, no 2, pp127–140

International Energy Agency (2008) *Renewables Information 2008*, IEA, Paris

Jay, S. (2008) *At the Margins of Planning: Offshore Wind Farms in the UK*, Ashgate, Aldershot

Kay, R. and Alder, J. (2005) *Coastal Planning and Management*, 2nd edition, Taylor and Francis, Abingdon

Kellett, J. (2003) 'Renewable Energy and the UK Planning System', *Planning Practice and Research*, vol 18, no 4, pp307–315

Khan, J. (2003) 'Wind power planning in three Swedish municipalities', *Journal of Environmental Planning and Management*, vol 46, no 4, pp563–581

Länsstryrelsen Kalmar Län (2003) 'Samordnad policy för lokalisering av havsbaserad vindkraft i södra Kalmarsund', unpublished document, Länsstryrelsen Kalmar Län, Kalmar

Ministry of Transport and Energy (2005) 'A visionary Danish energy policy: Summary', unpublished document, www.ens.dk/graphics/Energipolitik/ dansk_energipolitik/Energistrategi 2025/Presentation_Summery_190107_Final_ UK.pdf, accessed 15 November 2008

Offshore Center Danmark (undated) 'Offshore wind farms', www.offshorecenter.dk/offshore windfarms.asp, accessed 15 November 2008

Palmberger, B. and Jakélius, S. (1998) *Vindkraft i Harmoni*, Statens Energyimyndighet, Eskilstuna, Sweden

Swedish Energy Agency (SEA) (2008) 'Wind power: Building and connecting large wind turbines', Swedish Energy Agency, Eskilstuna, www.swedishenergyagency.se, accessed 15 November 2008

Toke, D. (2005) 'Explaining wind power planning outcomes: Some findings from a study in England and Wales', *Energy Policy*, vol 33, pp1527–1539

Tyldesley, D. (2004) 'Making the case for marine spatial planning in Scotland', report commissioned by RSPB Scotland and RTPI in Scotland, unpublished document, www.rtpi.org.uk/download/ 639/Making-the-Case-for-Marine-Spatial-Planning-in-Scotland.pdf, accessed 15 November 2008

14

Sustainable Construction and Design in UK Planning

Yvonne Rydin

Introduction

Tackling the climate change agenda demands a strategic approach but it also requires action at the level of the individual development proposal. This chapter looks at the emerging policy framework for promoting sustainable construction and design in developments that come forward to the planning system for planning permission. It focuses on the individual building or development site, rather than policies for new eco-towns or existing urban areas. As well as considering emergent planning practice, it considers the challenges that this throws up for the planning system.

Construction activity in the UK has long been criticized for the poor energy efficiency of its buildings. In the housebuilding sector this has been linked to a lack of innovation in building practices and a relatively low skill base, itself a function of a high reliance on self-employment and labour-only subcontracting within the labour force. In the non-domestic sector, building innovation has been more apparent with the emergence of steel-framed sheds for retail, indus-

trial and warehouse developments and innovative forms of concrete-with-cladding construction, often with considerable glazing, for office and other commercial premises. However, neither of these forms prioritize energy efficiency either. As a result, the rise of the climate change agenda within planning debates has meant that the UK planning system has found itself in advance of changes occurring within the development sector, with the limited exception of flagship sustainability projects.

The following chapter begins by considering the policy framework that now exists to promote sustainable construction and design through the planning system in the UK. It goes on to discuss certain key aspects of the institutional context for policy change that, despite this emergent framework, can inhibit actual progress on the ground in developing more sustainable buildings and estates. The nature of the challenge this situation poses for the planning system is then analysed, before a final section concludes on the way forward for embedding sustainable construction and design within planning decision-making and development practice.

The emerging planning policy framework[1]

2006 was a turning point for the way that the British planning system responded to the climate change agenda. In the pre-Budget statement in the autumn of that year, the future Prime Minister, Gordon Brown, announced the intention that all new housebuilding should be zero-carbon by 2016 and, in the meantime, that zero-carbon houses would be exempt from stamp duty (the tax payable on purchase of a property). By December 2006, these announcements were wrapped up in a package under the heading Towards Greener Building (Department of Communities and Local Government (CLG), 2006a).

Part of this package was the Code for Sustainable Homes (CLG, 2006b). This is a government-endorsed rating system for new housing, which builds upon the Building Research Establishment's Eco-Homes assessment. The Code for Sustainable Homes uses a six-star ranking system, with zero-carbon development achieving six stars. The number of stars awarded depends on the points accrued under nine headings (see Box 14.1). As from 1 May 2008, assessment of all new dwellings against the Code is mandatory, although this does not imply the necessity of meeting any particular standard. Homes built with English Partnership or Housing Corporation funding do have to meet standards set against the Code but this does not apply to market housing.

Towards Greener Building proposed a staged process by which all residential development would achieve the six star ranking. Central to this would be progressive improvements to the Building Regulations, particularly those elements dealing with energy and water efficiency, although the Code itself deals with a wider range of issues. The Building Regulations have already been subject to change, with a new version containing more stringent energy efficiency measures in Part L taking effect from April 2006. These increase energy efficiency standards by 40 per cent over 2002 levels. But perhaps even more significant than the increase

in expected efficiency standards was the shift in approach embodied in these new Regulations.

Rather than prescriptively requiring certain construction elements (such as walls or windows) to reach specified energy efficiency standards, the new Regulations set a required standard for energy consumption and associated carbon emissions and then let the building designer choose elements so as to reach those standards. This approach required considerable modelling of the energy consumption associated with a particular building design and of the carbon emissions associated with that energy consumption. At the heart of this modelling is the Standard Assessment Procedure or SAP.

A parallel version of the Code for Sustainable Homes for non-domestic development is due for issue in summer 2008, again replacing the Building Research Establishment's Environmental Assessment Method for rating a variety of non-domestic buildings, known as BREEAM (www.breeam.org). And in the March 2008 Budget, the Chancellor announced that, subject to consultation with industry, all new non-domestic buildings should be zero-carbon by 2019, saving an estimated 75 $MtCO_2$ over 30 years.

Beyond such assessment, ranking and regulation of individual developments, the local planning system has also been increasingly concerned to promote sustainable construction

---

Box 14.1 The code for sustainable homes

Energy/CO_2 emissions Water efficiency	Minimum standards set at each level
Building materials Surface water run-off Waste (domestic and construction)	Minimum standards set at entry level only
Pollution Health and well-being (including noise, daylight and lifetime adaptability) Management (such as a Home Users Guide) Ecology	No minimum standards set

and design through policy and guidance. While there has been national policy guidance for a while to the effect that sustainable development is at the core of the planning system (CLG, 2005), it is only more recently that the advice has become specifically about how to generate more sustainable buildings and developments, as opposed to urban forms and settlements.

During the early 2000s, this left a gap into which some local planning authorities moved. The early years of the new millennium saw a number of innovative authorities begin to develop green building guides and guidance on sustainable construction options. Mention could be made of the London Boroughs of Enfield and Camden in this regard. But probably the local authority whose name is most associated with action on sustainable construction and design is the London Borough of Merton. The so-called Merton Rule was devised by a planner in the borough to encourage the incorporation of on-site renewable energy generation into new developments. It simply stated that 10 per cent of the development or building's energy needs should be met from on-site renewables. Solar panels for heating water, photovoltaics for generating electricity, wind turbines and bio-mass combined heat and power plants all qualified as on-site renewables.

The aim was not just to reduce the reliance on fossil fuels for space and water heating at the margins. Rather, by consistently requiring developers to incorporate such technologies, the hope was that a threshold would be breached. Developers would come to see such installations as a routine and expected element of their development proposals. And, cumulatively, this would contribute to the capacity of the renewables technology market to deliver products and to do so at lower cost.

The Royal Town Planning Institute gave LB Merton an award for innovative policy making. But, more significantly, the Merton Rule has become very popular with other local planning authorities. There is a dedicated website www.themertonrule.org.uk and some 170 local authorities have signed up to using the Rule or are actively considering doing so. Perhaps the clearest indication of the success of the Merton

Rule is that the Greater London Authority sought to out-Merton other local authorities by incorporating into the 2007 amendments to the London Plan, a policy for 20 per cent of energy needs to be met by on-site renewables. The Government has also given support for the incorporation of on-site renewable energy generation and the spread of the Merton Rule through guidance in Planning Policy Statement 22 (CLG, 2004).

But beyond the Merton Rule, there is now a wealth of Supplementary Planning Guidance (SPG) giving advice to planning applicants on how to incorporate measures into their developments that will both mitigate carbon emissions and ensure that the development is adapted to anticipated climate change, as well as covering other aspects of sustainability such as biodiversity. Box 14.2 provides some indication of the breadth of coverage of such SPGs.

National planning guidance is now encouraging local planning authorities to develop such guidance. In particular, the Planning Policy Statement supplement on climate change

---

Box 14.2 Typical coverage of SPG on sustainable construction and design

- Use of brownfield sites
- Density of development
- Location of development with regard to public transport
- Energy use
- Energy generation
- Waste including recycling
- Adaptability to climate change
- Flooding prevention
- Materials (e.g. low embodied energy)
- Water use
- Water pollution
- Air pollution
- Noise
- Micro-climate
- Indoor comfort
- Inclusive design
- Secure design
- Open space and biodiversity

emphasizes the need to reduce carbon emissions associated with new development and makes it clear that local planning authorities should be testing their planning strategies against ambitious carbon targets. While emphasizing the importance of the location of new development, it also explicitly recommends that decentralized, renewable and low-carbon energy supplies should be planned into new developments (CLG, 2007a).

Finally, there is the prospect of encouraging changes to the existing built environment by removing the requirement for certain developments to engage with the planning system altogether. Following a review of the regime of permitted development rights for householders (CLG, 2007b), England is following Scotland in removing the need for installation of microgeneration appliances to most domestic buildings to obtain prior planning permission with effect from April 2008. This seeks to ensure that the bureaucracy of development control does not hinder individuals from taking action to improve the sustainability of existing dwellings.

The institutional context for sustainable construction and design policy

This level of policy activity shows considerable initiative on the part of the planning system, both nationally and locally. It would suggest that significant change in the built environment can be expected to mitigate and adapt to climate change. However, to conclude that this is indeed going to be the case, one would need to be sure about both the content and implementation of this policy framework. In this section, the influence of the institutional context on the content of the policies will be examined. This suggests that the good intentions of policy for sustainable construction and design may flounder in the uncoordinated involvement of many different organizations at central government level.

At the UK national level, responsibility for the issues underpinning sustainable construction and design are split between different govern-

ment departments. The Department of Business, Enterprise and Regulatory Reform (BERR, formerly the Department of Trade and Industry) looks at this issue from the perspective of industry, covering the construction and development industry but also the energy industry. As such they are responsible for developing a Sustainable Construction Strategy (Department of Trade and Industry (DTI), 2006). There are a range of initiatives that BERR undertakes as part of this approach including the Microgeneration Strategy and the Low Carbon Buildings Programme; the latter has a budget of some £80 million to encourage the installation of microgeneration technologies, including the much-derided householder projects stream which has proved highly problematic to implement effectively.

BERR also takes the lead on links with the building materials industry; this sector has an important role to play in bringing forward products with higher sustainability performance and enabling development to meet inter alia new energy efficiency standards. In this task, the work of BERR interfaces with that of the European Commission, where there has been considerable activity on standardizing the measurement of the environmental performance of building materials. And an important, though relatively unrecognized issue is that BERR is also responsible for the details of the SAP, which is the basis of much modelling of energy needs in buildings.

Meanwhile, CLG has overall responsibility for the planning system, across the process of making planning documents and decision making on applications for permission to develop, and the Building Regulations (the regulations of buildings standards). In addition CLG has responsibility for housing policy and this includes matters such as the Housing Information Packs (HIPs) that are required on the sale of a house and include an energy performance rating. But while all these different responsibilities come under the CLG umbrella, the organizational divisions between different sections within the same department can often result in policy making within 'silos'. Lowe and Oreszczyn (2008) have pointed to confusions between the Building Regulations and the Code for Sustainable Homes, for example, because these are dealt with by different teams in CLG with different advisors.

Then there is the Department for Food, Environment and Rural Affairs (Defra) which has responsibility, implemented through the Environment Agency, for water issues. These include both adaptation to enhanced flood risks and water resource management, encouraging greater water efficiency within new developments in parts of the country vulnerable to water stress, such as south-east England. Defra is also the sponsoring department for Natural England, the agency which oversees biodiversity protection and enhancement, also important aspects of sustainability. Perhaps slightly surprisingly, Defra is also responsible for overseeing the implementation of the EU Directive on the Energy Performance of Buildings, with its requirements for energy performance certification for all buildings from 2008, overlapping awkwardly with the CLG remit on HIPs.

And finally, there is the Treasury, which has an overall remit for government finances and which, therefore, provides funding for various schemes to incentivize (or not) aspects of sustainable construction. The money for the BERR Low Carbon Scheme, the 2006 stamp duty reforms and the 2008 announcement that commercial microgeneration installations will not trigger an increase in rateable value (and hence the property-based business tax) are positive examples of the Treasury's stance. The continuing differential between VAT on new build (which is zero rated) and refurbishment (which attracts VAT at 15 per cent) is an unfortunate counter-example.

The Government has struggled to ensure that these multiple departments, and divisions within departments, are working together on a common basis to promote sustainable construction and design. In part, this is a matter of the detail of policy, affecting how they are implemented. For example, the stamp duty exemption for zero-carbon housing is reported to have had a very low take-up because of the different definitions of zero-carbon being used by the Treasury and CLG, with the former refusing to

consider off-site renewables as relevant to the carbon reduction potentially associated with new buildings (Lowe and Oreszczyn, 2008). Attempts have been made to resolve the confusion over the definition of zero-carbon through a report by the UK Green Building Council (2008).

Meshing together different policies at the level of definition and wording takes considerable time and resources, both within government departments and in involving external stakeholders. Examples of the kinds of activity that have been set up to think through policy in this are the Building Regulations Ministerial Round Table, the 2016 Task Force to encourage the shift towards zero-carbon housebuilding and the Construction Products and Regulations Impact Team looking at building materials.

But the lack of coordinated policy making on this topic has also been due to differential priority being given to sustainable construction and different perspectives on how change should be achieved. From the BERR (and perhaps the Treasury) point of view, sustainable construction can be driven forward from within the sector, through corporate social responsibility. This has promoted more of an emphasis on voluntary initiatives. The tendency from the Environment Agency meanwhile has been to push for more stringent applications of current regulations, certainly with regard to building in floodplains and developments that threaten biodiversity.

CLG's approach is based on the twin elements of regulatory control and plan-making that lie at the heart of the planning system. The development control system offers the prospect of planning committees and officers rejecting applications that do not meet sustainability criteria. It also offers the potential for considerable negotiation on development details spurred by the existence of regulatory control. Plan making, particularly under the new regime of spatial planning, demands a commitment to stakeholder engagement and joint working in order to develop plans and strategies. This can then facilitate the implementation of those plans and strategies as their key elements are incorporated into the investment strategies and decision making of the stakeholders themselves.

Thus in engagements between government departments, there is a clash of cultures which has to be worked with and overcome if coordinated policy action to deliver sustainable construction and design is to be achieved.

The challenges posed for the planning system

The mix of stakeholder-led strategic planning and negotiation-based regulation that has been used to describe the CLG's overall policy approach also describes the everyday reality of planning practice in local planning authorities. Focusing at this level of practice suggests a number of challenges in achieving more sustainable new development.

First, it is clearly necessary to build a local commitment to achieving sustainability through influencing construction and design practices. This has to occur across the local politicians and officers (planning-related and otherwise) involved in shaping such local practices. Without this, policies for sustainable construction and design cannot become embedded within local planning documents.

But even with such policy statements in place, the political and professional will to implement them in specific instances has to be applied. There is little point having a Merton Rule in place, if it is not used in practice. The problem here is that any specific planning application raises a complex mix of costs and benefits. These have to be weighed against each other and it is not within the culture of development control practice to consistently prioritize only one planning consideration among the mix. Some might argue that this is both a strength and a defining feature of decision making with the planning system. However, if progress is to be made on reducing carbon emissions from new development, then this will require a degree of prioritization of this goal and associated planning tools that might deliver on the goal.

The way that sustainable development has been defined within planning policy guidance as encompassing economic, social and environmental dimensions already lends itself to trade-offs between the environmental and other dimensions. This is particularly the case where it is assumed that social benefits can be obtained through permitting market-led economic development. It will require a strong local coalition of values in favour of sustainable construction to ensure policies are consistently implemented. The hope is often expressed that design solutions will be developed that deliver reduced carbon emissions alongside improved townscape quality, increased social well-being, development marketability and so on. However, while such win–win outcomes are clearly desirable, sometimes difficult decisions will need to be made and climate protection may be sacrificed to other goals.

The composite nature of sustainable construction and design also makes this difficult. As Box 14.2 above illustrates, there are many different aspects of a development that contribute to its overall level of sustainability. It is tempting to prioritize one element – carbon emissions is the obvious target – but this is a thinner interpretation of sustainable urban development than has been promoted so far. Those concerned with adaptation to climate change, water security and biodiversity would be rightly concerned at the lost opportunity that the sustainable construction and design concept offers if climate change were always to be emphasized at the cost of other sustainability concerns.

One way of testing out if local commitment to implementing sustainability policies is effective is through monitoring. Unfortunately monitoring is poorly developed within the planning system. Under the spatial planning reforms, there is now a requirement for each local planning authority to produce an Annual Monitoring Report, which will report on progress in producing the documents set out in the Local Development Scheme and also the progress on those policies contained in the planning documents.

However, the Government's required Core Output Indicators (CLG, 2008a) only include two indicators that are of direct relevance to sustainable construction and design. These are number of applications permitted contrary to the advice of the Environment Agency on flooding or water quality grounds and the renewable energy capacity installed in new developments. It is up to local authorities to develop more comprehensive indicator frameworks to monitor the achievement of sustainability in new development.

Even with such indicator sets, it is likely that much non-compliant development will go undetected. Core Output Indicators and other local indicators will probably be monitored through a desk exercise, using the information in planning application files. The prospect of actually monitoring the extent to which developments as built conform with the planning permission, including all conditions and details of the S106 agreement, are heavily curtailed by available resources. The same applies with the monitoring of compliance with the Building Regulations. This often happens on a sampling basis and concern has been reported over the limited effectiveness of enhanced Building Regulations if there is no attention to how they are implemented in practice (Lowe and Oreszczyn, 2008).

Even if problems of non-compliance with enhanced sustainability standards set out in planning permissions are identified, enforcement action on such breaches is required if sustainable construction and design policies are to be taken seriously. Enforcement is another under-resourced area of planning practice. Government statistics show that in 2006/7 local authorities in England issued about 5500 enforcement notices but it is the contravention notices (where planning conditions have not been complied with) that are more relevant to sustainable construction concerns, with only just over 1000 of such notices issued in 2006/7.

Clearly there will need to be enhanced resourcing of both monitoring and enforcement if there is to be any certainty that the fine words of planning policy are actually changing development practice.

Beyond these issues of planning practice, there are a range of technical concerns that may inhibit progress on sustainable construction and design. This is, to a large extent, a rapidly evolving area: as new technologies are being applied in new situations, a number of issues are coming to the fore and more may be expected in the coming years. Concerns that have yet to be resolved include (Rydin, forthcoming):

- whether the SAP (that is the basis of the modelling of energy needs in a building under the building regulations) is actually a good indicator of energy consumption post-occupation;
- what measure to use as the basis for calculating requirements for on-site energy generation through renewables technology;
- how the increased energy efficiency of buildings through changes to the built fabric impact on the efficiency, effectiveness and viability of on-site microgeneration using renewable energy sources;
- a similar concern with the impact of increased energy efficiency on Combined Heat and Power (CHP) plants;
- the overall carbon emissions associated with biomass CHP plants given that the biomass often has to be processed and transported considerable distances;
- the effectiveness in terms of energy generation and carbon reduction of the different types of microgeneration installations.

Listing such technical concerns highlights that this is an area where the boundaries of planning expertise and competence are being stretched. Not only are these issues that have not traditionally been within the remit of planning school curricula, they tend to involve expertise outside the design and social science disciplines that the planning profession has been based upon. This raises the important issue of planners' knowledge with regard to sustainable construction and design.

One way to address this would be to suggest that planners require training to be able to discuss all these aspects with developers and their advisers. This would fit within the traditional 'knowledge gap' approach to such situations. However, in the context of everyday planning practice, it is worth asking what kind of knowledge, and how much, planners actually require.

Two aspects of current planning practice, particularly at the development control stage, stand out. First, planners are under considerable time pressure because of the central government expectation that minor planning applications will be decided within 8 weeks and major ones within 13 weeks. This limits the extent to which they can go into any specific aspect of the development proposal, including probing the depths of consultants' reports on different dimensions of sustainability.

Second, the lengthening list of aspects of the proposal that need consideration threatens to overload the cognitive capacity of any planner to engage with all these aspects. The pile of paperwork facing a development control case worker is already considerable. Providing more training on these different aspects may help with things like the understanding of reports, but it is not realistic to expect a planner to be trained up to the level of the different experts who provide such reports.

There is thus a difficult balance to be maintained between enhancing planners' understanding of sustainable construction and design issues to the point at which they can effectively exercise the planning skill of synthesizing material on the diverse dimensions of a development proposal, on the one hand, and not putting unrealistic pressures on the capacity of a planner to absorb knowledge on sustainable construction and design, given their practical working context.

The way forward

In the light of this discussion, what are the prospects for achieving more sustainable urban development and how should planning practice change?

The Towards Greener Building document has triggered a serious shift in the policy agenda, with new forms of urban development becoming a priority within the planning system. The distinctive feature of this agenda is that it is as concerned with the development control and building regulation stages as with policy formulation, if not more so. There is a real need to consider how sustainability can be promoted through the development plan and supported by

detailed SPG. There is also a need to consider the practice of negotiation and regulation on specific planning applications and to monitor outcomes against relevant indicators.

While it is neither realistic to expect planners to become experts on all the different dimensions of sustainable construction and design, nor necessary within the context of their everyday practice, there clearly needs to be learning occurring within the planning system about what sustainable construction and design actually means and how to achieve it. This is likely to involve engagement within learning networks, providing access to a range of expertise on a need-to-know basis, rather than on the basis of individual knowledge acquisition (Rydin et al, 2007).

The key features of such learning networks (sometimes also referred to as communities of practice (Wenger, 1998)) are that they are formed around practical problems that are commonly agreed to be a priority. Within such networks, members learn from each other. As they do so, they change their working practices and routines but they also change their professional identity, aligning themselves with the new ways of working. This can feed through, over time, into changed priorities and goals for the organization within which the professionals are working. In this way culture change and learning are integrally related.

All learning requires feedback and this means that planning organizations, particularly local planning authorities, will need to develop mechanisms for monitoring and reporting on successes and failures within their own experience of trying to implement sustainable construction and design. This in-house form of learning can complement the learning-from-others arising from involvement in networks on this issue. However, monitoring can also be a straight-jacket inhibiting learning. Organizations need to be allowed to fail occasionally in order to learn how to succeed. Over-zealous and premature monitoring can work against learning from failure.

All these forms of learning are important in a broader sense because the success of this agenda will depend to a large extent on its impacts being visible and achieving the desired goals. Effective energy efficiency measures within buildings will reduce energy costs and this will reinforce the demand for innovative urban developments, incorporating CHP and high levels of insulation. But if these measures fail to deliver or have undesirable side-effects (such as excessive internal moisture) then this will undermine the agenda. The planning system has a responsibility to learn how to implement sustainable construction and design effectively to maintain the momentum of the greener building initiative.

There is though a final word of caution that needs to be sounded. While fostering new development that will mitigate climate change, be adapted to climate change impacts and deliver other sustainability benefits is important, it is nowhere near enough on its own. It is often cited that at least half of carbon emissions are associated with the built stock and this is put forward as a strong argument for the greener building agenda. While this is true, there are three other aspects to consider.

First, given the relatively low rates of turnover in the urban stock, even under the conditions of increasing housebuilding envisaged by the Government, most of the existing building stock will still be with us in 2020, even 2050. It has been suggested that 75 per cent of the dwellings in 2050 has already been built (Power, 2008). In this case, it is not enough just to focus on new build, although this can provide an important model for how urban development should look and operate. It will be necessary to tackle the existing building stock, both residential and commercial, in order to raise the energy efficiency levels at least. In 2006 only 33 per cent of all dwellings had 150mm or more of loft insulation; in the private rented sector the figure fell as low as 21 per cent (CLG, 2008b). Much of the non-prime commercial stock was built in times when energy efficiency was not a concern. This takes the greener building agenda beyond the concern with new build to think about the building stock and indeed the built environment as a whole.

Planning has a role to play here too but it will be one that reconsiders how urban regeneration is currently framed. There may be scope for returning to some of the ideas of the 1970s concerned with housing improvement and bringing together housing and planning policy in a new way. In commercial areas, town centre management needs to engage with the sustainability agenda, and the planning system can bring this into its place-making activities.

The other two aspects can be more briefly dealt with as other chapters in this volume deal with them at more length. The first of these concerns the fact that transport is the fastest growing source of carbon emissions and planning urgently needs to consider whether it is doing enough to reverse this trend. The second relates to the importance of dealing with the energy and carbon efficiency of appliances used within buildings. As energy efficiency of the built form increases, this means that a greater proportion of energy will be used in appliances within buildings; and this applies to commercial as well as residential buildings. It is likely that decarbonizing the generation of electricity will be necessary to radically alter overall carbon emissions associated with our use of buildings and this will means tackling centralized as well as decentralized electricity generation.

Sustainable construction and design may seem a minor add-on to the broader policy initiatives aimed at mainstreaming the climate agenda, but this chapter has argued that it has a vital role to play in ensuring that the grand ambitions of this agenda are actually implemented. Without this we will still be living and working in the same old kind of buildings and urban development will have missed out on the opportunity to provide us with a truly sustainable urban future.

Note

1 This chapter was written before the departmental reorganizations of 2008 and 2009 that resulted in the creation of the Department of Energy and Climate Change and the reorganization of BERR into the Department of Business, Innovation and Skills.

References

CLG (2005) 'Planning Policy Statement 1: Delivering sustainable development', Department of Communities and Local Government, www.communities.gov.uk/publications/planningandbuilding/planningpolicystatement1, accessed 20 January 2009

CLG (2006a) 'Towards greener building', Department of Communities and Local Government, www.communities.gov.uk/archived/publications/planningandbuilding/buildinggreener, accessed 20 January 2009

CLG (2006b) 'Code for Sustainable Homes: A step-change in sustainable home building practice', Department of Communities and Local Government, www.planningportal.gov.uk/uploads/code_for_sust_homes.pdf, accessed 20 January 2009

CLG (2007a) 'Planning Policy Statement: Planning and climate change', supplement to Planning Policy Statement 1, Department of Communities and Local Government, www.communities.gov.uk/planningandbuilding/planning/planningpolicyguidance/planningpolicystatements/planningpolicystatements/ppsclimatechange/, accessed 20 January 2009

CLG (2007b) 'Changes to Permitted Development: Consultation Paper 1: Permitted development rights for householders', Department of Communities and Local Government, www.communities.gov.uk/archived/publications/planningandbuilding/changespermitted, accessed 20 January 2009

CLG (2008a) 'Regional spatial strategy and local development framework: Core output indicators', Update 2/2008, Department of Communities and Local Government, www.communities.gov.uk/publications/planningandbuilding/coreoutputindicators2, accessed 20 January 2009

CLG (2008b) 'Housing and Planning Key Facts' Department of Communities and Local Government, www.communities.gov.uk/publications/corporate/statistics/keyfactsaugust2008, accessed 20 January 2009

Department of Communities and Local Government (CLG) (2004) 'Planning Policy Statement 22: Renewable energy', www.communities.gov.uk/planningandbuilding/planning/planningpolicyguidance/planningpolicystatements/planningpolicystatements/pps22/, accessed 20 January 2009

Department of Trade and Industry (2006) 'Review of sustainable construction', DTI, www.berr.gov.uk/files/file34979.pdf, accessed 20 January 2009

Lowe, R. and Oreszczyn, T. (2008) Regulatory standards and barriers to improved performance for housing', State of Science Review undertaken for the Government Office of Science's Foresight project on Sustainable Energy Management and the Built Environment

Power, A. (2008) 'Does demolition or refurbishment of old and inefficient homes help to increase our environmental, social and economic viability?' State of Science Review undertaken for the Government Office of Science's Foresight project on Sustainable Energy Management and the Built Environment

Rydin, Y. (forthcoming) 'Planning and the technological society: Discussing the London Plan', *International Journal of Urban and Regional Research*

Rydin, Y., Amjad, U. and Whitaker, M. (2007) 'Environmentally sustainable construction: Knowledge and learning in London planning departments', *Planning Theory & Practice*, vol 8, no 3, pp363–380

UK Green Building Council (2008) 'The definition of zero carbon', report by the Zero Carbon Task Group, May, UK Green Building Council, www.ukgbc.org/site/resources/showResourceDetails?id=180, accessed 20 January 2009

Wenger, E. (1998) *Communities of Practice: Learning, Meaning, and Identity*, Cambridge University Press, Cambridge

15

Making Space for Water:
Spatial Planning and Water Management in the Netherlands

Jochem de Vries and Maarten Wolsink

Introduction

In Al Gore's documentary 'An Inconvenient Truth' the Netherlands is portrayed as one of the major potential victims of climate change. In a dramatic animation large portions of country are flooded and become part of the North Sea. It is obvious that climate change has potentially disastrous implications for a country of which 35 per cent lies below sea level. Even today 65 per cent of the country would run the risk of regular flooding if there were no protective measures such as dikes and storm surge barriers. It is clear that water management is the core issue of climate change in the Netherlands (WRR, 2006).

Nevertheless, on the one hand Gore's documentary overdramatizes the Dutch situation which, unlike some low deltas in the developing world is in a relatively favourable position to counter the effects of climate change. As protecting the country against flooding is literally nearly as old as the land and keeping the Dutch feet dry is enshrined in the constitution, political support for protective measures is almost self-evident. Furthermore, being one of the wealthiest countries in the world, the country is capable of taking necessary protective measures. However, it would be a mistake to see the challenges that climate change poses as the same old story. Particularly in relation to spatial planning, the issues raised by climate change require nothing short of a revolutionary change of policy (De Vries, 2006).

At present innovative policies have been developed, but this is by no means a straightforward success story. The Dutch case shows that the transition towards climate proof water management and spatial planning requires changes in many different aspects of policy making. Furthermore, because water management and spatial planning are strongly embedded in the existing institutional and spatial context, successfully changing the existing practices in water management and spatial planning is not only a matter of changing policies, but more importantly a matter of institutional changes.

In the first section we describe the historical roots of water management in the Netherlands. It will become clear that water management developed into an almost isolated sector within government and society. Subsequently we discuss the challenges that the Netherlands faces in relation to climate change. Adapting the physical environment to the needs of the water system is apparently the key issue. Therefore, in section four we turn to the relationship between water management and spatial planning and more specifically to the – already in place and planned – policy innovations that contribute to 'spatializing' water management. In the last section we reflect on the policy and institutional changes.

Water management in isolation

'God made the world and the Dutch made the Netherlands' as the saying goes. The 'battle against water' has been a unifying force in the Netherlands for centuries. It is of considerable symbolic value for Dutch culture in general and its planning culture in particular. The 'hydrological hypothesis' is often used to explain the traditionally large societal acceptance in the Netherlands of government intervention in general and in the physical environment in particular (Faludi, 2005). As a result of these physical circumstances, regional water boards emerged based on private initiative to manage the 'common pool resource' of safety from flooding (Dolfing, 2000). In fact they are the oldest democratic institutions in the Netherlands, as the establishment of many of these boards predated the establishment of municipal corporations (Faludi, 2005). From the 13th century onwards water boards were established on an ad hoc basis. Every time a new problem arose – such as the need to build a dike – a new corporation was created. These water boards were created by the directly affected inhabitants. As a result 'interest–taxation–representation' became the guiding principle for water boards. The amount of taxes and the degree of participation in decision making were determined by the extent to which someone benefited from the services provided by the water board (Van Steen and Pellenbarg, 2004).

From the 19th century onwards, steps were taken to reduce the number of local water boards – which amounted to several thousand, but a drastic reduction had to wait until the second half of the 20th century. In 1950 the number of water boards was still 2500. By 2003 this number was reduced to 48 and a further reduction to 25 is likely in the near future. In accordance with the principle 'interest–taxation–representation' the water boards are decentralized functional governments. They raise their own taxes, which have to cover their activities, and have their own democratically elected council. Most water boards are responsible for protection against flooding, water quantity, the maintenance of regional waterways and, since the 1970s, water quality. The water board council is directly elected by people with direct interest (such as land owners, property owners, inhabitants and users) in the activities of the water board (Dicke, 2001). Low turn out at elections is a sign of a lack of interest or unawareness by the general public. In the past different debates have been started on the question of whether the water boards are outdated and should be abolished. This, however, has never materialized (Wiering and Immink, 2006).

At the national level the Ministry of Transport, Public Works and Water Management and its executive branch Rijkswaterstaat, is the main player in the field of water management. Rijkswaterstaat was founded in 1798 when the Netherlands became a unitary state under French rule. Before 1795 the Netherlands was a confederation of provinces and water management was the subject of cooperation between provinces. Civil engineers traditionally dominate the professional culture of Rijkswaterstaat. Its central role in protecting the country from flooding combined with its status as one of the first and strongest unified state institutions, makes the Ministry of Transport, Public Works and Water Management and Rijkswaterstaat a powerful actor. As a result it is sometimes seen as being a state within the state (Faludi, 2005).

In addition to a water sector that is largely organized in powerful single sector agencies, other principles of dealing with water issues are also institutionalized. Firstly, a dominant perspective – even paradigm (Wiering and Immink, 2006) – is the striving for unambiguous safety from flooding everywhere in the country and a strong reliance on science in general and engineering in particular. An approach developed in which predefined generic norms about the risk of flooding (1:10,000 year, 1:4000 year etc.) form the point of departure for designing water infrastructure such as dams and dikes. After establishing the specific norm, it can be calculated how strong and high a barrier should be. While this approach suggests a high degree of precision, it has a tendency of neglecting uncertainties (Roth and Warner, 2007).

Secondly, the engineering perspective contributed to the idea that the water system could be moulded to the needs of society. Massive land reclamation in large single projects started in the 17th century with, for example, the Beemster polder which is now a UNESCO cultural heritage site. With the steam engine, larger projects could be realized, such as the Haarlemmermeerpolder, which nowadays is the home of the international airport Schiphol. In the 20th century the Zuiderzeewerken and the IJsselmeerpolders turned large parts of the large internal sea into a lake and parts of it into land. Adapting the water system to the needs of society also prevailed in river management, where the course of the main rivers was changed to adapt to the needs of shipping, leading to straightening and narrowing of the riverbed. This process of 'normalization of rivers' led to a significant loss of space for water (Van Heezik, 2007).

A third area where the water system was adapted to the needs of society is the relationship between water management and other sectors of society, such as agriculture and spatial planning. The nearly complete control over the water system led to a situation in which sectors simply demanded a specific water regime and the water authorities made sure these demands were met. For example groundwater tables can be regulated with great precision and according to the type of land use – housing or agriculture – the ideal level is chosen. One of the effects of this relationship between the water sector and other sectors is to create clear spatially divided spheres of influence. In precisely defined areas – such as the coastal zone, dikes, water extraction areas – water management has been the dominant activity to the extent that it has overridden other activities that are deemed incompatible with it.

In the rest of the country spatial planning plays the dominant role and the water system has to adapt to the needs of land use as determined by the planning system.

A fourth characteristic of the way in which water management is governed concerns the style of policy, which is also heavily influenced by an engineering ethos. Project planning – as opposed to strategic planning (Faludi, 1987) – or project management – as opposed to process management (De Bruijn and Ten Heuvelhof, 2004) – in which fixed targets and command and control are key characteristics, suits water engineers well. This means that many initiatives in the water sector have a blueprint character and, as a result, a fixed end state and strict timetable. The famous example of this approach is the Delta Plan (Meijerink, 2005). This huge endeavour was drafted after the flood disaster of 1953 in which nearly 2000 people lost their lives. This plan included the building of dams by which the estuaries in the south-west of the Netherlands were cut off from the sea and as a result turned into fresh water lakes (Pols et al, 2007).

In the 1970s the first challenges to the water sector's traditional approaches took place. The large-scale enhancement of the main river dikes, following the Delta Plan, by Rijkswaterstaat was challenged because of its ecological and landscape impact. The discussion and collaborative planning process introduced a different approach which was directed at preserving local identity. However, this solution was never implemented (Wiering and Driessen, 2001; Wolsink, 2003). More successful public protest concerned the last large land reclamation of the Zuiderzeewerken, the Markerwaard. A coalition of environmentalists, inhabitants and water sports enthusiasts organized a pressure group, which developed a counter plan. It was a mature plan, based on sound expertise. This was the first time that Rijkswaterstaat lost its grip on the agenda-setting process in the field of water management (Dicke, 2001). The Markerwaard has never been realized.

In the same period protest grew against the implementation of the Delta Plan. A coalition of fishery lobbyists and environmental groups opposed the idea of blocking the East-Scheldt, which would have turned the estuary into a fresh water lake. This involved an ingenious design of the dam, which could remain open except in emergencies (Dicke, 2001). Today the East-Scheldt estuary is a national park. The significance of the East-Scheldt dam episode is that safety was not the only and overriding rationale for decision making. Ecological criteria for the first time played an important role. Therefore water management was no longer the exclusive domain of engineers: biologists, economists and spatial planners were equally entitled to have their say.

Since 1989 integrated water management has been the official policy. Indeed, the East-Scheldt case can be viewed as the cradle of integrated water management (Saeijs, 1991; Disco, 2002). From an integrated water management perspective 'water policies' should be framed on the basis of geographically coherent water systems. It includes surface and ground water, riverbeds, banks and technical infrastructure. It is about both water quantity and quality. Integrated water management brings hydrological relations to the fore instead of dealing with individual water bodies from a single function perspective. The approach aims to not only internally integrate between different water management areas, but also externally integrate between water management and other policy areas such as spatial planning (Dicke, 2001).

Integrated water management marked the first step in the direction of integrating water management with spatial planning. Nevertheless the debate about 'spatializing' water management gained particular momentum when the level in the main rivers Rhine and Meuse in 1993 and 1995 reached record levels. In one instance nearly 200,000 people had to be evacuated in the face of possible floods. In the same period, the challenge of climate change adaptation was added to the political agenda. Together this led to the emergence of a new discourse on water management – living with water – which at this moment exists alongside the traditional discourse – keeping our feet dry – that is based on a technocratic paradigm of guaranteeing safety (Wiering and Immink, 2006).

Dealing with climate change
ARK policy strategy

In response to the Intergovernmental Panel on Climate Change (IPCC) 2001 report, in which a strategy for adaptation was added to the existing climate change mitigation agenda, several reports were issued relating to the inventories of vulnerabilities to climate change by, for example, the Dutch national meteorological institute (KNMI, 2006). When the national white paper on spatial development – Nota Ruimte – (VROM, 2004) was discussed in the First Chamber of the Parliament (the Senate) these reports led to questions about government policy on coping with the impact of climate change. The answer came in the form of the publication of the government's policy programme: Adaptation Space and Climate or ARK (Adaptatie Ruimte en Klimaat).[1] An inventory of climate change impact, made by the national Environment and Nature Planning Agency, was the first step (MNP, 2005). Then the policy strategy was developed, starting with some scientific 'quick scans' called 'route planner' in which the main challenges were highlighted.

Not surprisingly, the policy started as a challenge to spatial development, because the most obvious issue that the country had to deal with was the problem of water safety. A paradigm shift towards the implementation of new water management, in which the mutual adaptation of space and water is the key, had only started (Wiering and Immink, 2006; Wolsink, 2006). The reinvented relation between space and water was easily connected to even larger problems that would emerge in the near future as a result of sea level rise and changing patterns of precipitation and river stream-flows.

In accordance with the Dutch political culture of consensus building, the new paradigm on water and space was also established in a covenant between different tiers of government. The National Governance Agreement on Water (NBW, 2003) was signed in 2003 by organizations of Dutch municipalities, the national organization of provinces, the organization of water boards, and the relevant ministries of the national government. In a similar way, a Climate Agreement between the national government and all other policy tiers was signed in November 2007. The general objective of the adaptation policy is making the country 'climate-proof' ('klimaatbestendig'). The meaning of that concept, however, is rather vague and the first steps in the development of the policy were to investigate the problems that the country will be confronted with in the next decades.

In September 2008 an advisory commission installed by the government – the 'Deltacommissie' – published a report on the protection of the country against flooding in the coming century. The report recommends a range of engineering projects which will cost between 1 to 1.5 billion euros each year (from 2010 to 2100). The surprising starting point in this report is 130cm sea level rise by the year 2100, which is far more than the estimates made by the IPCC in 2007.

Inventory of significant effects and required policies

The objective of climate proofing is to establish ecological, technical, economic and social systems that have the capacity to maintain functioning in a normal way in the face of substantial climate changes. The traditional water management paradigm (Wolsink, 2006) is based on resistance to extreme conditions, whereas the new idea that is accepted is that more flexibility is needed and the focus should be on resilience rather than resistance (Nelson et al, 2007). Furthermore, as resistance can be expressed in terms of the probability of calamities, 'climate proof' and resilience policies should be based on the impact that those extreme events would have on the normal functioning of the society. This implies that the concept of risk becomes a guiding principle for adaptation policy, which was already a commonly accepted key principle in climate change policy (Lorenzoni et al, 2005). The social acceptability of risk can only be established in the context of decision making, that is in a trade-off with other costs and benefits

and in relation to alternatives. In that context the perceptions and values of the people involved in that decision become major factors in the assessment and management of risk. Risks are not simply small or large: there are also essentially different types of risk that all require different policy approaches when decisions are taken (Klinke and Renn, 2002).

A fundamental aspect of risk is uncertainty, and as a result adaptation policy accepts substantial uncertainty. Nevertheless the Dutch government also decided that a part of the uncertainty could be taken away by doing research. The project 'Routeplanner' was set up to make 'quick-scans' of existing knowledge about emerging problems and identify the gap in knowledge (Veraart et al, 2006). The identified issues and policy measures have also been ranked in terms of rough assessments of costs and benefits (Van Ierland et al, 2007). The scans all stressed the enormous uncertainty in the knowledge about climate change and its consequences. Such consequences are dependant on socio-economic developments. Scenarios of the major planning agencies of the Dutch government (Janssen et al, 2006) have been used to analyse the impact of different paths of development on climate change impacts and the costs and benefits of adaptation measures.

Notwithstanding the uncertainties, there is no doubt that the most significant impact in the Netherlands of climate change concerns issues of water management. These effects and ideas about how to adapt the country to them, in particular through integrated spatial development and water management, are elaborated more extensively in the next sections. However, the inventory also includes impact in other domains that are sometimes connected to the problems that will arise in the water-space domain.

Related to spatial development and water issues is the adaptation of policy on nature protection. The significant impact that has been identified is affected by expected changes in precipitation patterns. Summers will most likely become dryer, and conditions in natural areas will become wetter during winters. As large parts of the Netherlands' nature reserves are wetlands, an even stronger impact can be expected from the changes in the ground water level: higher in winter and lower in summer. Those levels are very important for different reasons. First of all, in line with the old water control paradigm, as mentioned above, almost anywhere in the Netherlands ground water levels are under the full control of the water boards and are adapted to sector demands. In some cases this is important to natural areas because the character of the area may be influenced by different patterns of flows. In the majority of the cases, however, ground water levels are controlled to serve agricultural demands and for several decades ground water levels have been frequently lowered as requested by farmers to allow larger and heavier machines to operate on their land. As a result, soil dehydration has happened which has affected natural areas. Given that these conditions will be exacerbated by climate change, the regime of ground water control must be adapted and that change will have consequences for the agricultural sector. On the other hand, some climate conditions are becoming more favourable for farming, such as the length of the crop-growing season.

The impact of climate change on transport is only in some minor ways related to water management. The most significant impact will probably be the increase in number of days when low stream flows in the main rivers will constrain the carrying capacity of ships. The amount of freight transported on ships on those rivers is fairly large and the transport capacity will be seriously affected during dry summers. In a similar way the capacity of rivers and surface water available for the cooling of power generators in summers will be affected. This is compounded by simultaneously growing demand for electricity in summers which may partly be caused by the emergence of more intense 'heat islands' in cities.

A major climate change impact on housing will be on the choice of new building sites: planning and investments in new housing districts and industrial areas will have to change in response to the need for sustainable water management. The layout of building may also have to change to address increasing heat stress in inner cities. Other fields of climate change

impact that have connections to water management and spatial development include recreation and public health. In case of the latter, increased threats of diseases and stress related to water have been identified. However, the strongest effect on public health may occur as a result of more frequent flooding which may cause both physical and mental problems, as a result of stress.

In one of the ARK 'quick-scans' an assessment of options for adaptation by means of a multi-criteria analysis was made (Van Ierland et al, 2007). Out of the 96 options that were identified, the most important ones concerned water management and its relation with spatial development. The first four are:

- more space for water;
- spatial development steered by the concept of risk;
- risk management as baseline strategy;
- new institutional arrangements for water management and spatial development.

However, in the same study an assessment of the complexity of implementation demonstrated that important institutional changes were required for the implementation of most of these measures.

Bringing water management and spatial planning together

New policies

From the 1970s onwards, initiatives were taken to improve the coordination between water management and other policy sectors. Nevertheless integrated water management mainly resulted in innovations *within* the water management sector rather than between that and other policy areas such as spatial planning. For example, flood protection policies ended in stalemate. Raising and improving dikes increasingly encountered protest from environmentalists and inhabitants which considered these 'improvements' detrimental to the spatial quality of river areas (Wiering and Immink, 2006). Combined with a decline in political priority as a result of the economic recession in the early 1980s, for a long period no action was taken with regard to flood protection along the main rivers in the Netherlands (Wolsink, 2003).

This radically changed when in 1993 and 1995 flooding and near flooding occurred along the main rivers. This was a major shock to the public and political perspective. The comforting idea that everything was under control appeared to be an illusion. In its first reaction Rijkswaterstaat acted according to tradition. A new Delta Plan for the major rivers was drafted: 700 kilometres of dike reinforcements and emergency dikes were planned. To speed up implementation specific legislation was established, which made it possible to bypass existing planning and environmental legislation and to sideline opposition (Wiering and Driessen, 2001).

At the same time new alarming predictions of climate change were made. The fact that the existing situation did not provide the safety levels that were desired combined with the fact that future developments would make things even worse, opened a window of opportunity for policy change (Wolsink, 2006; Wiering and Immink, 2006).

The initiative was taken to establish a paradigmatic shift in water management through policy changes in three areas.[2] (At this stage the proposed paradigmatic shift has not been completed. Presently different paradigms exist side by side.) First of all, the approach to the management of the main rivers shifted towards an approach in which widening of the river – 'room for the river' – is the core element, combined with projects aiming at restoring the river's ecological integrity. Second, a new general philosophy for water management was developed by a special commission (CW21, 2000). This water policy for the 21st century is applicable on every spatial scale – from neighbourhood to country level – and to every physical environment – the city and the countryside. A third influential policy area was European water policy – the European Water Framework Directive (Grimeaud, 2004).

Parallel to the new Delta Plan for the main rivers, which reflected the traditional engineering approach, a new approach for the main rivers

was developed. Already in 1996 the Ministry of Transport, Public Works and Water Management and the Ministry of Housing, Spatial Planning and the Environment had issued a policy directive 'Room for the River'. It stated that in the short term all proposed developments in floodplain areas had to be abandoned and in the future sustainable protection against flooding should be based on a variety of physical measures that created more space for water. This directive developed into a national planning policy[3] that was officially established in 2007. This includes a wide variety of spatial measures. Calamity polders should be used to store water in the event of extreme high water levels. Different measures for widening the river are proposed. Where widening is impossible, bypasses are created. Altogether around 40 projects are planned along the main Dutch rivers.

The policy not only implies a substantive shift in policy, but also a procedural shift in dealing with river management (Van den Brink and Meijerink, 2006). The process architecture for this policy gave an important role to regional (provincial) governments which prepared a regional proposal in cooperation with regional and local stakeholders. However, the Ministry of Transport, Public Works and Water Management appeared to be reluctant to leave the old centralized culture behind. For a long time it assumed ultimate financial responsibility and discouraged creative solutions. Only later the possibilities for co-financing and public–private financing were considered a realistic option (Van den Brink and Meijerink, 2005). Furthermore, despite the ministry's attempts to include bottom–up input in policy development, different projects were initiated in a way that evoked enormous local resistance (Witsen, 2005).

The second strand of policy that emerged was first articulated in the report 'Water policy for the 21st century' (CW21, 2000). This was a report by a committee installed by the Ministry of Transport and Public Works and the Union of Water Boards with the explicit task of exploring the consequences of climate change for water management. It recommended 'dealing differently with water', and its conclusions were largely accepted by the government as the point of departure for future policy. This report considered the relationship between water management and spatial planning as key to successful policy. In other words, it 'spatializes' water management. First of all, it introduced a principle that has a strong spatial dimension. This argued that the starting point for spatial development and water management should be 'retaining–storing–draining', which means that the process of drainage of precipitation and surface water to the large rivers and ultimately the sea should be slowed down. In rainy seasons this would help to prevent flooding and water nuisance. In dry seasons water shortage should be avoided this way.

In implementing this principle spatial planning plays a crucial role. Firstly, the maintenance of the water system sets the pre-conditions for spatial planning. The underlying assumption is that in the past the adaptation of the water system to spatial development has had a negative impact on the natural capacity of water systems to handle large fluctuations in water and that this will need to change (De Vries, 2006). In terms of the new policy, the water system should structure spatial decision making. An important policy instrument is compulsory water assessment for spatial decisions and a wide range of statutory plans (Voogd, 2006). In essence this water assessment test is a protocol to ensure the timely involvement of the water authority with a stake in the plan and the careful consideration of the impact of the plan on the water system. Its purpose is to avoid negative impacts and, where negative impact cannot be avoided, mitigation or – as a last resort – compensatory measures should be taken. The water assessment test aims to lead water authorities (which in most cases are the regional water board, but in some cases are Rijkswaterstaat or the province) to play a more proactive role in the earlier stages of developing spatial plans. They should not only accept or dismiss plans but also contribute to creative solutions. One of the instruments that water boards have developed is the 'water opportunity map'. (Voogd, 2006). These, usually GIS-based, maps indicate suitability of a given land use from a water system perspective.

The second aspect of the relationship between spatial planning and water manage-ment, which was addressed in the report, concerns making space for water. This can be crudely summarized as meaning that, in order to prepare for changing patterns of precipitation and higher sea and river levels, more territory should be permanently or temporarily reserved for water. Based on a moderate scenario for climate change it is recommended that by 2030 an additional $1700km^2$ of land will need to be allocated to water. This is two and a half times as much land as is estimated to meet the demand for housing and economic development (De Vries, 2006).

The third strand of policy that changes the relationship between water management and spatial planning is the EU water framework directive and, more recently, the EU floods directive. By introducing a cross-border river basin approach these directives introduce a new area based approach in the Netherlands. The river basin approach is not familiar in the Netherlands. The water system has been manipulated to such an extent – even the flow of some rivers has been reversed – that natural watersheds have lost much of their importance. Furthermore, the main rivers are part of large cross-border river basins and, hence, the river basin approach requires cross-border cooperation, which is a complicated governance situation. Effectively dealing with the consequences of climate change, to which these EU directives attempt to contribute, requires an entirely different mindset for Dutch water management (Becker et al, 2007). Cross-border cooperation between different countries already exists for water quality improvement, for example, along the Rhine River. However, for water quantities such cooperative arrangements are new. The need to contribute to investments downstream on foreign soil will become a real possibility in the future.

Core arguments for spatialization of water management

The above shifts in governance have been given an important push by the debate on climate change and by a series of floods and near floods in the late 1990s. Nevertheless arguments for a change have been more diverse and many of them are older than the climate change discourse.

First of all, a spatial development process that largely ignores the risk of flooding is often considered not viable in the long run. In the Netherlands the economic heartland is predominantly situated in the area that is most prone to flooding. The westernmost urbanized part, the Randstad, lies below sea level in the delta of two large European rivers. These are partially land reclamations (polders) but most of the land that was originally at sea level has subsided as a result of peat oxidation caused by pumping out the water. This process of decline is continued by current ground water level control and is furthermore accelerated by geological processes.

From a risk approach, in which the likelihood of flooding must be multiplied by the impact of such an event, a vicious circle has developed. Flood defence measures decrease the likelihood of a flooding event and then an area subsequently becomes more attractive for development. This development leads to more disastrous consequences in the case of flooding. As a result of the increased potential impact a demand for further flood defences is created to maintain the same risk levels. This process has been referred to as 'control paradox' (Wiering and Immink, 2006) and in the Netherlands this process has contributed to the 'bath tub effect': higher dikes increase the difference between the water level and the land, which means that if the embankment breaks in the event of a flood, the district's water level will rise faster than it would without the embankment, thus seriously threatening the lives of those residents who cannot move quickly (Van de Ven, 1996). Another part of the safety paradox is the practice of assessing new developments and infrastructure projects individually. While each project individually does not increase risk levels significantly, their combined impact does lead to significantly riskier circumstances. In other words these developments increase the aggregated risk: a process referred to as the 'escalator effect' (Parker, 1995). Despite all the investments in the Netherlands in flood defence, the overall risk has

probably increased in the last decades (Smits et al, 2006).

Secondly, the insight gained over the last decade is that the norm-based approach that focuses on the likelihood of flooding, provides a false sense of safety. The idea that engineering can with a great amount of certainty determine the chance of flooding has been put in doubt. Not only do more uncertainties exist than the engineering approach suggests, the uncertainties are likely to increase as a result of climate change (Milly et al, 2002; Tol et al, 2003). This is obvious for changed river stream flow patterns, but also affects other parts of the water system. For example, in the summer of 2003 a peat dike collapsed near Amsterdam and the town of Wilnis was flooded. In this case, involving not a river but a waterway which was used for drainage of water pumped out of the polder ('boezem'), it was not too much water but a shortage of water that caused the problem. The peat dike dehydrated and crumbled. This example of system failure fell outside the risk model. Until then, flood risk was based on the estimation of water levels and the probability that these levels would go above the height of the dike.

Thirdly, while 'Room for the River' has been presented as the answer to the higher water levels that will result from climate change, it originates from changed views on river management. The effects of the traditional engineering approach – with the 1950s Delta Plan as its zenith – proved to not only have significant ecological but also maintenance disadvantages. For example, the maintenance costs of the Delta project are very high and much higher than estimated in the 1950s. Furthermore, in wide and shallow systems the energy of the tides is absorbed to a much higher degree than in waters with limited contact with the streambed and banks of the channel. In narrower and deep systems, floods develop more force and damage (Smits et al, 2006). In the new paradigm an essential starting point is that water systems that are based on natural characteristics are more resilient.

The above phenomena give, or should give, rise to different changes in spatial planning. Firstly, flood risk should guide spatial decision making more effectively. The Netherlands Institute for Spatial Research has shown that risks are spatially more differentiated than the present legal norms suggest. The differences in risk could be used to guide location decisions. For example, low-lying polders that would fill up fast after a breach of a flood barrier should be avoided. Furthermore, people need to be more aware of the differences in risk and take more responsibility themselves, instead of turning to the government to guarantee safety, through, for example, insurance systems (Pols et al, 2007).

Secondly, spatial planning should be more oriented towards the fact that the Netherlands is a water rich country. In the 20th century water was generally considered to be a threat, a nuisance, or at least a waste of space in spatial development. The first plans for redeveloping the Amsterdam Eastern Docklands in the early 1980s indicated that the water basins should be reclaimed. For various reasons this did not happen. Redeveloped docklands, in which the morphology of the harbour has been maintained, are particularly appreciated for their large spaces of open surface water. This example shows that opportunities to combine water and spatial development are often not used to advantage. In general, spatial strategies should be developed that are more compatible with the characteristics of the water-rich delta (Smits et al, 2006).

Conclusions: The emergence of a new paradigm in water management

Climate change has given an important push to rethinking the relationship between water management and spatial planning in the Netherlands. At the same time it must be stressed that the process of finding new ways to deal with water and spatial planning are underway. The main direction of 'spatializing' water management has been set out, but not everybody – especially within the water sector – is convinced that a significant change of policy is needed. Furthermore, implementation of spatialized

water management has just started. The challenges ahead are summarized below under two broad categories: firstly, allowing more local variation in dealing with water and, secondly, fundamentally changing the way in which policies are made.

From standardization to location specific solutions

The recent attempts to shift the paradigm of water management concerns a fundamental turn away from defensive water control. The new paradigm recognizes that full control is an illusion or at least cannot be sustained in the 21st century (Enserink, 2004; Wolsink, 2006). In the new paradigm of adaptive water management, the focus is primarily on the interface between water management and spatial planning. The urgency of the need to radically adapt the relationship between water management and spatial development has been reinforced by the rapidly developing policy for climate change adaptation.

The paradigm change in the Netherlands runs parallel to similar shifts elsewhere: for example, where technocratic infrastructures for water resources are challenged, and in new approaches to water quantity and quality in France and the UK (Brown and Damery, 2002; Howe and White, 2004; Pottier et al, 2005). Because of the Netherlands' record and advocacy of the engineering, full 'command-and-control' approach in water management, combined with its history of comprehensive spatial planning, the refocusing of the sector in the Netherlands on the integration of water and spatial management is particularly interesting. However, further governance innovations will be needed to complete the paradigm change and make it effective. We conclude this chapter with a tentative list of issues that need to be resolved.

Regarding the way water is handled, the accepted priorities now are retaining location-specific water by means of natural retention and storage. This should replace the current practice of rapid drainage that usually only moves problems from one place to another. It also moves away from the current habits of choosing solutions that include hard, inflexible infrastructures such as dikes, concrete embankments, channelled riverbeds and pumping systems. If the technocratic approach of defensive control has to be abandoned, this also means that the uniformity in norms and solutions should disappear. Standardization must be replaced by tailor-made designs that not only respect social and geographical conditions but also use the local characteristics creatively. This respect for local identity will lead to a large diversity in design.

Most of these ideas still need fundamental conceptual development. Obviously, the existing institutional framework does not fit this radically different kind of water management that is fundamentally entangled with spatial development. Equally important as the new approach to handling water and space, are new frameworks for governance.

Towards new ways of governance

As new, mutually adapted space and water management requires new types of decisions that do not belong to the current repertoire of knowledge and practice, much organizational and policy-oriented learning is needed. As in other countries, such as the UK, the full integration of spatial planning and water management therefore requires 'institutional change at all levels, including new forms of governance' (Howe and White, 2004, p422). Institutional capacity that fully applies all knowledge resources, relational resources and mobilization capacity should support such learning (Healey, 1997). Governance characteristics of the old paradigm must be replaced by new ones that serve the new ways of management (Wolsink, 2006).

First of all, because uniform and standardized measures should no longer be applied, significant parts of decision making must be decentralized and increasingly become the subject of multi-level governance. The use of local identity requires substantial and powerful input from local stakeholders, contribution of tacit and situation specific knowledge and values. Collaborative processes in which pluralism of

societal values are acknowledged should replace current technocratic and hierarchical procedures. Collaboration requires communication based on principles of reciprocity: communication serving mutual learning and trust (Dietz et al, 2003). Deliberative decision making is needed that applies open and participatory processes replacing the currently existing procedural focus and closed arenas. Nevertheless, projects are still planned very much top–down, facing huge institutional barriers or in fact continuing old practices more than introducing new solutions. New hierarchical instruments in the spatial planning system have in some cases reinforced top–down tendencies in decision making on river management where local stakeholders have been proponents of values that fit more to the new paradigm than the policies that the national and regional authorities tried to implement (Wolsink, 2006). Specific facilities that should serve the risk approach, such as calamity polders along the main rivers, have also been planned so centrally, that the emerging local conflicts have soon led to stalemate (Roth and Warner, 2007).

Institutional change is usually an uphill battle, and changes that would support the prevalence of local identity and building up resilience have hardly been discussed so far. For example, the new large scale adaptation projects, as introduced in 2008 by the Deltacommissie, seem to reinforce the old tradition of centralized and top–down decisions.

Second, the governance principle of consensus building is not easy to realize, because among the relevant stakeholders a clear clash of policy cultures related to technocratic and societal approaches exists (Wiering and Immink, 2006). While spatial planners have become very much oriented towards consensus building, water management is dominated by an engineering ethos.

Third, for organizations that are connected to the current control paradigm and that are full of people educated in that culture, it is hard to open the arena for other actors representing different types of knowledge. Obviously, this issue is not only Dutch, as in other western countries the observation is made that the 'full control' paradigm has created stagnation within the development of scientific knowledge. Science has become isolated from societal stakeholders, meaning that water management policy often uses outdated knowledge (Falkenmark, 2004). This is a serious barrier for creating innovative solutions that are needed to make the country climate proof.

Fourth, while official policy documents recognize the obstinacy of outdated water management practices, they tend to focus on public resistance as the key obstruction. The official image is that anybody with a stake in current practices may obstruct the implementation of change. It is questionable whether the new institutions that authorities are creating (rules, procedures, new bodies) will be sufficient to involve stakeholders constructively (Tol et al, 2003). Several projects have already revealed these problems: for example, the large ecological restoration of the Meuse – the 'Maaswerken' project. After strong dissatisfaction was expressed by the public and local stakeholders the plans had to be reconsidered. In their study on the Maaswerken, Van der Meulen et al (2006) concluded that overall the level of stakeholder involvement had been insufficient, even though it complied with pertinent environmental and planning legislation.

Finally, as the applicability of blueprints is low when anticipation to uncertain futures is at stake, strategic ways of planning have to emerge. The horizons of plans will necessarily become wider as a result of the ongoing process and uncertainties of climate change. Constantly changing water conditions and resilient spatial configurations require flexible governance, whereas current institutions and knowledge mainly serve the defensive status quo.

To conclude, while many initiatives have been taken to fundamentally shift basic premises in Dutch water management the actual institutional changes that have been achieved so far cannot yet be classified as 'fundamental' or 'paradigmatic' (Wiering and Arts, 2007). Again, the recent Delta programme (Deltacommissie, 2008) can serve as an example. This programme largely ignores uncertainty instead of introducing flexible adaptations that can be changed when faced with unforeseen developments. It simply takes the highest forecast sea level rise and the highest probability of low streamflows of the rivers in summer as a starting point. The programme is also, in line with the traditional paradigm, once again emphasizing the use of technology to resist change instead of promoting resilience. (The report of the Deltacommission (2008) mentions the term 'resilience' (or Dutch: 'veerkracht') only one time (in a reference from 1998)). The Delta programme is full of enhancement of existing infrastructure and building new infrastructure to deal with that threat: new and higher dikes and strengthening the coast by broadening the coastal strip with enormous amounts of sand (Deltacommissie, 2008, pp12–13).

Obviously, we are dealing with a practice – water management – that finds more strategic ways of planning as alien. Efforts to implement some of the new water-space concepts have already shown the urgent need for a change in planning approach. Dealing with uncertainty, as opposed to creating certainty in current practice, is a key challenge.

As in many other countries, the Netherlands has to face an uphill battle in adapting the country to climate change in practice. Although traditional risk management and the vulnerability approach should be integrated, the practice is often still dominated by the old paradigm with its 'bias in institutional culture towards technical concerns as the overriding criteria for management and action and ... inherent assumptions about public irrationality...' (Brown and Damery, 2002, p420). The process of adapting water management and spatial planning to new philosophies of sustainable spatial development in the age of climate change is work in progress.

Notes

1 Ark is Dutch for National Programme for the Adaptation of Space and Climate.
2 A fourth area of policy in which important shifts are to be expected is coastal zone management, which for practical reasons is left out of the discussion as it would require too much space to elaborate this line of policy. For an elaborated account, see Meijerink (2005).
3 Planologische Kernbeslissing – Spatial Planning Key Decision.

References

Becker, G., Aerts, J. and Huitema, D. (2007) 'Transboundary flood management in the Rhine basin: Challenges for improved cooperation', *Water Science and Technology*, vol 56, no 4, pp125–135

Brown, J. D. and Damery, S. L. (2002) 'Managing flood risk in the UK: Towards an integration of social and technical perspectives', *Transactions of the Institute of British Geographers*, vol 27, pp412–426

Bruijn, H. de, Ten Heuvelhof, E. (2004) 'Process arrangements for variety, retention and selection', *Knowledge, Technology and Policy*, vol 16, no 4, pp91–108

CW21 (2000) 'Waterbeleid voor de 21e eeuw' (Water Policy for the 21st century), Ministry of Transport and Public Works/Union of Waterboards, The Hague

Deltacommissie (2008) 'Samen werken met water Een land dat leeft, bouwt aan zijn toekomst', Bevindingen van de Deltacommissie, The Hague

De Vries, J. (2006) 'Climate change and spatial planning below sea-level: Water, water and more water', *Planning Theory & Practice*, vol 7, no 2, pp229–233

Dicke, W. (2001) *Bridges and Watersheds: A Narrative Analysis of Water Management in England, Wales and the Netherlands*, Aksant, Amsterdam

Dietz, T., Ostrom, E. and Stern, P. C. (2003) 'The struggle to govern the commons', *Science*, vol 302, pp1907–1912

Disco, C. (2002) 'Remaking "nature": The ecological turn in Dutch water management', *Science Technology and Human Values*, vol 27, no 2, pp206–235

Dolfing, B. (2000) 'Waterbeheer geregeld? Een historisch bestuurskundige analyse van de institutionele ontwikkeling van de hoogheemraadschappen van Delfland en Rijnland 1600–1800', PhD thesis, University Leiden

Enserink, B. (2004) 'Thinking the unthinkable: The end of the Dutch river dike system? Exploring a new safety concept for the river management', *Journal of Risk Research*, vol 7, nos 7–8, pp745–75

Falkenmark, M. (2004) 'Towards integrated catchment management: Opening the paradigm locks between hydrology, ecology and policy-making', *International Journal of Water Resources Development*, vol 20, pp275–281

Faludi A. (1987) *A Decision Centred View of Environmental Planning*, Pergamon Press, Oxford

Faludi, A. (2005) 'The Netherlands a country with a soft spot for planning', in B. Sanyal (ed) *Comparative Planning Cultures*, Routledge, New York, pp285–307

Grimeaud, D. (2004) 'The EU Water Framework Directive', *Review of European Community & International Environmental Law*, vol 13, pp27–40

Healey, P. (1998) 'Building institutional capacity through collaborative approaches to urban planning', *Environment and Planning A*, vol 30, pp1531–1546

Howe, J. and White, I. (2004) 'Like a fish out of water: The relationship between planning and flood risk management in the UK', *Planning, Practice & Research*, vol 19, pp415–425

IPCC (2001) 'Climate Change 2001: Impacts, adaptation, and vulnerability', contribution of Working Group II to the Third Assessment Report of the Intergovernmental Panel on Climate Change', www.ipcc-wg2.org, accessed 30 June 2008

Janssen, L., Okker, V. and Schuur, J. (2006) 'Centraal Planbureau, Natuur- en milieplanbureau, Ruimtelijk Planbureau', Welvaart en Leefomgeving, een scenariostudie voor Nederland in 2040, CPB/MNP/RPB, Den Haag/Bilthoven

Klinke, A. and Renn, O. (2002) 'A new approach to risk evaluation and management: Risk-based, precaution-based, and discourse-based strategies', *Risk Analysis*, vol 22, no 6, pp1071–1094

KNMI (2006) KNMI Klimaatscenario's 2006, Royal Netherlands Meteorological Institute, www.knmi.nl/klimaatscenarios/knmi06/index.html, accessed 30 June 2008

Lorenzoni, I., Pidgeon, N. F. and O'Connor, R. E. (2005) 'Dangerous climate change: The role for risk research', *Risk Analysis*, vol 25, no 6, pp1387–1398

Meijerink, S. (2005) 'Understanding policy stability and change: The interplay of advocacy coalitions and epistemic communities, windows of opportunity, and Dutch coastal flooding policy 1945–2003', *Journal of European Public Policy*, vol 12, no 6, pp1060–1077

Milly, P. C. D., Wetherald, R. T., Dunne, K. A. and Delworth, T. L. (2002) 'Increasing risk of great floods in a changing climate', *Nature*, vol 415, pp514–517

MNP (2005) 'Milieu- en Natuurplanbureau. Effecten van klimaatverandering in Nederland', MNP, Bilthoven

NBW (2003) 'Nationaal Bestuursakkoord Water' (National Governance Agreement Water) 3/7/2003, The Hague

Nelson, D. R., Adger, W. N. and Brown, K. (2007)

'Adaptation to environmental change: Contributions of a resilience framework', *Annual Review of Environment and Resources*, vol 32, pp295–419

Parker, D. J. (1995) 'Floodplain development policy in England and Wales', *Applied Geography*, vol 15, pp341–363

Pols, L., Kronberger, P., Pieterse, N. and Tennekes, J. (2007) 'Overstromingen als ruimtelijke opgave', NAI/RPB, Rotterdam/Den Haag

Pottier, N., Penning-Rowsell, E., Tunstall, S. and Hubert, G. (2005) 'Land use and flood protection: Contrasting approaches and outcomes in France and in England and Wales', *Applied Geography*, vol 25, pp1–27

Roth, D. and Warner, J. (2007) 'Flood risk, uncertainty and changing river protection in the Netherlands: The case of "calamity polders"', *Tijdschrift voor Economische en Sociale Geografie*, vol 98, no 4, pp519–525

Saeijs, H. L. F. (1991) 'Integrated water management: A new concept from treating of symptoms towards a controlled ecosystem management in the Dutch delta', *Landscape and Urban Planning*, vol 20, pp245–255

Smits, A, J. M., Nienhuis, P. H. and Saeijs, H. L. F. (2006) 'Changing estuaries, changing', *Hydrobiologica*, vol 565, pp339–355

Tol, R. S. J., Van der Grijp, N., Olsthoorn, A. A. and Van der Werf, P. E. (2003) 'Adapting to climate change: A case study on riverine flood risks in the Netherlands', *Risk Analysis*, vol 23, pp575–583

Van den Brink, M. and Meijerink, S. (2006) 'De spagaat van Verkeer en Waterstaat: Het project Ruimte voor de Rivier', *Stedebouw en Ruimtelijke Ordening*, vol 87, no 2, pp22–26

Van der Meulen, M. J., Rijnveld, M., Gerrits, L. M., Joziasse, J., van Heijst, M. W. I. M. and Gruijters, S. H. L. L. (2006) 'Handling sediments in Dutch river management: The planning stage of the Maaswerken river widening project', *Journal of Soils and Sediments*, vol 6, no 3, pp163–172

Van de Ven, G. (1996) 'The Netherlands and its Rivers', *Tijdschrift voor Economische en Sociale Geografie*, vol 87, pp364–370

Van Heezik, A. (2007) *Strijd om de rivieren: 200 jaar rivierenbeleid in Nederland of de opkomst en ondergang van het streven naar de normale rivier*, HNT Historische producties, Den Haag/Haarlem

Van Ierland, E., de Bruin, K., Dellink, R. B. and Ruijs A. (eds) (2007) 'A qualitative assessment of climate change adaptation options and some estimates of adaptation costs', Routeplanner, deelprojecten 3, 4 and 5, Wageningen University

Van Steen, P. J. M. and Pellenbarg, P. H. (2004) 'Water management challenges in the Netherlands', *Tijdschrift voor Economische en Sociale Geografie*, vol 95, no 5, pp590–598

Veraart, J. A., Opdam, P. F. M., Nijburg, C., Makaske, A., Luttik, J., Neuvel, J. M. M., Brinkman, S., Pater, F. de, Meerkerk, J., Leenaers, J., Graveland, J., Wolsink, M., Klijn, E. H. and Rietveld, P. (2006) Quickscan, Kennisaanbod- en leemten in klimaatbestendigheid (Quickscan, Knowledge availability and gaps on climate durability), Routeplanner 2, Wageningen University

Voogd, H. (2006) 'Combating flooding by planning: Some Dutch experiences', *disP*, vol 1, pp50–58

VROM (2004) *The National Spatial Strategy: Creating Space for Development*, summary part A, The Ministry of Economic Affairs, The Ministry of Agriculture, Nature and Food Quality, The Ministry of Transport, Public Works and Water Management, The Ministry of Housing, Spatial Planning and the Environment, The Hague

Wiering, M. A. and Arts, B. J. M. (2006) 'Discursive shifts in Dutch river management: "Deep" institutional change or adaptation strategy?', *Hydrobiologia*, vol 565, pp327–338

Wiering, M. A. and Driessen, P. P. J. (2001) 'Beyond the art of diking: Interactive policy on river management in the Netherlands', *Water Policy*, vol 3, pp283–296

Wiering, M. A. and Immink, I. (2006) 'When water management meets spatial planning: A policy arrangements perspective', *Environment and Planning C: Government and Policy*, vol 24, pp423–438

Witsen, P. P. (2005) 'Ieder voor zich in Rivierenland', *Blauwe Kamer*, vol 5, pp32–38

Wolsink, M. (2003) 'Reshaping the Dutch planning system: A learning process?', *Environment and Planning A*, vol 35, pp705–723

Wolsink, M. (2006) 'River basin approach and integrated water management: Governance pitfalls for the Dutch Space–Water–Adjustment Management Principle', *Geoforum*, vol 37, no 4, pp473–487

WRR (2006) 'Climate strategy: Between ambition and realism', Netherlands Scientific Council for Government Policy, Amsterdam University Press, Amsterdam, www.wrr.nl/content.jsp?objectid=4318, accessed 30 June 2008

16

Climate Change and Flood Risk Methodologies in the UK

Andrew Coleman

Introduction

In the United Kingdom, the likely effects of climate change will include increased flood risk as precipitation increases and becomes more intense and there are more extreme weather 'events', such as those that caused loss of life and disruption to much of England in the summer of 2007. The probable extent of the financial damage to the UK economy – and the implications for Government spending if the effects are to be reduced – was estimated by the Foresight 'Future Flooding' study (Department of Trade and Industry (DTI), 2004). It examined a range of impacts, including sea level rise and storminess, using the climate change projections and scenarios of potential social and economic changes in society generated by the UK Climate Impacts Programme (UKCIP). The project found that annual damage from flooding may rise in real terms from around £100 million at present to between £460 million (under the more 'community orientated' *Local Stewardship* scenario) and £2500 million (under the more 'consumerist' *World Markets* scenario) by 2080. The Foresight findings influenced the Government's long term strategy for flood risk and coastal erosion, 'Making Space for Water'

(Department for Environment, Food and Rural Affairs (Defra) 2005).

This chapter will compare the basic approaches to flood risk management in the spatial planning systems of England, Wales and Scotland and how the likely impacts of climate change are being incorporated. It will then describe in more detail the 'risk-based' approach and the roles of flood mapping, classifying the vulnerability of end users and flood risk appraisal and assessment at different spatial scales. Finally it will indicate potential future directions for flood planning policy in the UK.

The policy context

Climate change and new development, unless it is carefully planned and managed will increase flood risk. This has been recognized in the UK in specific national planning policy statements on the subject:

- Planning Policy Statement (PPS) 25 'Development and Flood Risk', in England (Department for Communities and Local Government (CLG), 2006).
- Technical Advice Note (TAN) 15 'Development and Flood Risk', in Wales

(Welsh Assembly Government (WAG), 2004).
- Scottish Planning Policy (SPP) 7 'Planning and Flooding' (Scottish Executive, 2004).

In England and Wales planning policy for flood risk was first introduced in 1992. Each subsequent revision of planning policy has strengthened and resulted in an increase in the complexity of planning policy over time, further complicated by increased national devolution. However, planning policy in one country has influenced the development of policy in others. In England and Wales, the Environment Agency has an important role in advising decision makers and developers on flood risk issues. In Scotland this role is performed by the Scottish Environmental Protection Agency (SEPA).

Flood risk: What is it?

Flooding is a natural process and takes many forms. In spatial planning, the forms of flooding that have received most attention are fluvial flooding (from rivers) and tidal flooding (from the sea or in tidal estuaries). However, other significant forms of flooding are rainfall running off the ground surface ('pluvial'), overwhelmed sewers and drains, groundwater and flooding from infrastructure such as reservoirs, canals and other artificial sources.

Flooding has a great positive influence on the natural landscape and biodiversity. Encouraging or allowing flooding in the right place and at the right time can help to meet many of the aims of sustainable development. However, the planning systems of England, Wales and Scotland are primarily concerned with influencing flooding when it poses a risk to property or people. Flood *risk* is a function of the *probability* (or likelihood – normally expressed in terms of '1 in X years' or in percentage terms) of flooding occurring and its *consequences*. If there are no or few adverse consequences of flooding – even if it is very probable – then there is little reason for the planning system to intervene to manage it. Planning has an important role to play in influencing both the probability of flooding

happening and the consequences if and when it does.

A major issue for flood management policy is the distribution of costs between the public sector (through spending on flood risk management or emergency flood relief), the insurance industry and individual householders, businesses and landowners. Stakeholders such as the insurance industry (represented by the Association of British Insurers) have influenced the development of flood risk planning policy by lobbying government and indicating that insurance for damages from flooding may be withdrawn if steps are not taken to reduce risk by building flood risk infrastructure to protect existing assets or by tougher planning policies that seek to limit new development in the floodplain.

Planning policy in England, Wales and Scotland separate flood risk (based largely on probability) into three categories, summarized in Table 16.1.

There are broad similarities between countries in defining risk zones. All three subdivide risk zones into three categories. The lowest category is the one where river and sea flooding is unlikely to be a planning constraint. In England and Scotland this is defined as less than a 1 in 1000 chance in any year. In England, the 'high risk' zone uses different probabilities for river and sea flooding – in effect sea flooding is regarded as twice as risky as river flooding. The rationale for this (probably partly based on the experience of the 1953 floods that killed just over 300 people on the east coast of England) is that coastal flooding can be sudden and more violent – often being accompanied by high winds and storm surges. There is a corresponding knock-on effect on the upper threshold for the English medium zone. In Wales, the 'medium' zone uses a unique (in the UK context) definition of land that has flooded in the past as evidenced by sedimentary deposits. The 'high risk' zone has a threshold (1 in 1000 chance in any given year) that is equivalent to the lower threshold of the English 'medium' zone. While this gives the immediate impression that flooding is regarded in Wales as a more serious constraint to development, it must be noted that probability classification is only part of the risk

Table 16.1 Definitions of flood risk zones and areas by country

Country	Flood risk definitions			Comments
	'High risk'	'Medium risk'	'Low risk'	
England	**Flood Zone 3a:** *River flooding:* 1:100 or greater risk *Sea flooding:* 1:200 risk **Flood Zone 3b:** Functional floodplain 1 in 20 or greater risk or as agreed between the local planning authority and EA.	**Flood Zone 2:** *River flooding:* 1:100 to 1:1000 *Sea flooding:* 1:200 to 1:1000	**Flood Zone 1:** *River and sea flooding:* less than 1:1000	Flood Zones do not take account of flood risk management infrastructure. Zones 2 and 3 shown on EA Flood Map. Zone 3b to be determined through Strategic Flood Risk Assessments
Wales	**Zone C:** *River and sea flooding:* 1:1000 or greater. **C1:** Developed areas of the floodplain and served by significant infrastructure, including flood defences. **C2:** Areas of the floodplain without significant flood defence infrastructure.	**Zone B:** *River and sea flooding:* Areas known to have flooded in the past evidenced by sedimentary deposits.	**Zone A:** *River and sea flooding:* Little or no risk	TAN 15 only uses the term 'high risk of flooding' in relation to Zone C. The terms 'medium' and 'low risk' of flooding are not used. Zones C1 and C2 incorporate consequences and 'residual' risk. Zone C derived from EA Flood Map. Zone B is taken from British Geological Survey drift data.
Scotland	**Area 3:** *River and sea flooding:* greater than 1:200 **Area 3(a):** Already built-up **Area 3(b):** Undeveloped and sparsely developed areas.	**Area 2:** *River and sea flooding:* 1:1000 to 1:200	**Area 1:** *River and sea flooding:* less than 1:1000	Note – terminology differs: 3. Medium to high risk area 2. Low to medium risk area 1. Little or no risk area Based on SEPA maps of probability. Areas 3(a) and (b) include some recognition of existing development and 'residual risks'.

Note: 'River' flooding = fluvial or watercourse; 'Sea' flooding = tidal / coastal.

Sources: Adapted from PPS25, TAN15 and SPP7.

equation and the sensitivity of 'receptors' (people, land and buildings), the policy approach to previously developed land and flood resilience and resistance are also relevant.

The Scottish definitions of 'medium risk' and 'high risk' zones do not differentiate between river and sea flooding – using the 1 in 200 annual probability threshold as the upper limit of the 'medium' and lower limit of the 'high' zones. Both the Welsh and Scottish 'high risk' zones are subdivided according to whether land is previously developed and protected by flood defences. This approach incorporates consideration of the 'consequences' of flooding into the zonation scheme. PPS25 introduced a new definition of high risk zones that ignores the presence of defences and whether land has been developed before. This approach is a 'purer' consideration of flood probability and allows a clearer demarcation of the consequences of flooding by taking into account the vulnerability of 'receptors' (people and buildings) and measures to reduce flood risk such as defences and flood resistant and resilient construction. It also arguably makes it clearer that it should not be automatically assumed that defended areas will continue to be defended and that owners of previously developed land may not enjoy an automatic expectation that it will receive planning permission for redevelopment. There is a real likelihood that, in the long term, climate change and restricted public funding for flood risk infrastructure may require a gradual retreat from the most vulnerable parts of the floodplain.

PPS25 contains strong policy encouragement to planners to address increased flood risk threats posed by climate change to existing development:

> ...where climate change is expected to increase flood risk so that some existing development may not be sustainable in the long-term, Local Planning Authorities should consider whether there are opportunities in the preparation of Local Development Documents to facilitate the relocation of development, including housing to more sustainable locations at less risk from flooding.
> (CLG, 2006, para 7)

Planning for flood risk

PPS25 states that a 'risk-based' approach should be used by planners. The model it uses is called 'source–pathway–receptor' similar to that used in PPS23 'Planning and Pollution Control' (CLG, 2004). The overall aim of the PPS is

> to ensure that flood risk is taken into account at all stages in the planning process to avoid inappropriate development in areas at risk of

flooding, and to direct development away from areas at highest risk. Where new development is, exceptionally, necessary in such areas, policy aims to make it safe without increasing flood risk elsewhere and where possible, reducing flood risk overall.

The PPS25 Practice Guide (CLG, 2008a) – published a year and a half after the PPS itself – details a 'Flood Risk Management Hierarchy' which divides the steps that planners should take into:

1 Assess: Appropriate flood risk assessment;
2 Avoid: Apply the sequential approach;
3 Substitute: Apply the sequential approach at site level;
4 Control: for example, sustainable drainage systems, flood defences;
5 Mitigate: for example, flood resilient and resistant construction.

The forward planning approach adopted by PPS25 requires regional and local planning bodies in England to:

- *appraise risk* – identifying areas at flood risk and carrying out assessments or appraisal of that risk to feed in to sustainability appraisals;
- *manage risk* – adopting planning policies that reduce and manage flood risk, including using the 'sequential approach' in the location of new development (see below);
- *reduce risk* – safeguarding land needed for current and future flood risk management from development, reducing flood risk to and from new development through location, layout and design, incorporating sustainable drainage systems and using opportunities offered by new development to reduce the causes and impacts of flooding;
- *a partnership approach* – working with other agencies involved in flood risk management and ensuring that spatial plans reflect flood management plans, such as shoreline management plans and catchment flood management plans and emergency planning.

Table 16.2 PPS25's approach to matching flood risk zones to vulnerability

Flood risk vulnerability classification	Essential infrastructure	Water compatible	Highly vulnerable	More vulnerable	Less vulnerable
Flood Zone					
Zone 1	✔	✔	✔	✔	✔
Zone 2	✔	✔	Exception test required	✔	✔
Zone 3a	Exception test required	✔	✘	Exception test required	✔
Zone 3b	Exception test required	✔	✘	✘	✘

Note: ✔ development is appropriate; ✘ development should not be permitted.

The 'sequential approach' is the central element of controlling flood risk in UK spatial planning. Its planning policy roots lie in Department of the Environment Circular 30/92 (DoE, 1992) which stated that '(the) Government therefore looks to local authorities to use their planning powers to guide development away from areas that may be affected by flooding, and to restrict development that would in itself increase the risk of flooding'. It is less explicit (but still present) in Scottish and Welsh planning policy, but is most clearly stated in PPS25. It can be summarized as steering new development to areas at the least probability of flooding. It is designed to be applied at regional, sub-regional, local authority and lower spatial scales – right down to individual sites and buildings.

The sequential approach used in PPS25 matches the vulnerability of appropriate end uses ('receptors') to flood risk. This is summarized in Table 16.2. This is based partly on Government research on the vulnerability of receptors (Defra / Environment Agency, 2002), but it also reflects the need to allow essential transport and power infrastructure and social infrastructure – as long as lower risk sites have been sought first, there is transparent justification for the developments, they will remain safe from flooding and not increase flood risk elsewhere.

As part of the sequential approach outlined above, PPS25 also contains a sequential test and exception test which are used in spatial plans and development control decisions. Similar to the sequential test that applies to retail uses in town centres, the PPS25 sequential test seeks to guide new development to the lowest probability flood zone. If there are no 'reasonably available' sites in that zone decision makers have to try to locate new development in flood zone 2 and only after reasonably available alternatives there have been exhausted, can flood zone 3 be considered.[1] PPS25 sets out the policy requirements and aims in each flood zone. In flood zones 2 and 3, some uses must pass an 'exception test' and others 'should not be permitted' – that is, there is a very strong policy presumption against such development.

In flood zone 3, the policy aims include trying to relocate existing vulnerable uses out of the floodplain. This is arguably one of the most challenging aspects of the PPS and will require planners to cooperate with landowners, developers, regeneration agencies, the public, the Environment Agency and other bodies with flood risk responsibilities. But the spatial planning system introduced in England in 2004 and the strengthening of the links between sustainable community strategies and local development frameworks (CLG, 2008b) should enable all parties around the negotiating table to seek opportunities to find land (even if it is 'greenfield' land) for development that make settlements and communities more resilient to climate change.

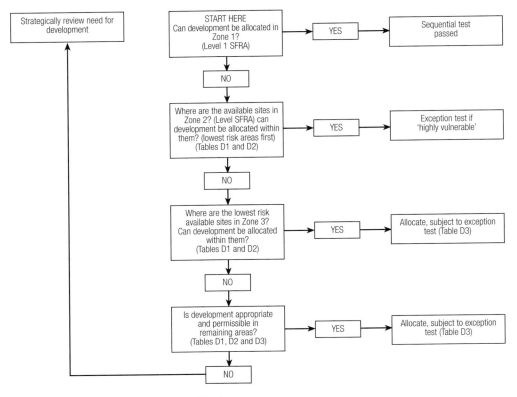

Note: 1 Other sources of flooding need to be considered in Flood Zone 1.

Source: CLG (2006a)

Figure 16.1 Application of the sequential test at the local development document level

Tools for appraising risk

This section describes the main tools used by planners to appraise and describe flood risk:

Flood maps

Maps are one of the most important tools in planning for flood risk. In each country the statutory environmental protection agency has responsibility for preparing maps of flood risk to inform the public, their own flood risk management activities and spatial planners. In England and Wales, the Environment Agency publishes a flood map on its website which shows the boundaries of flood zones 2 and 3 (see Table 16.1 above), not taking account of flood risk infrastructure that will reduce the probability and

consequence of flooding. This map is based on a number of sources including historical data and modelling. While it gives a good overview of flood probability and an overall idea of how the sequential test (in England) will apply, it does not provide the detailed information that most planners require. More detailed maps are provided to local planning authorities in CD format. The Environment Agency is in the process of mapping areas where surface water flooding is a problem which will help ensure that flood risk assessments and planning applications take this into account. Defra has consulted on improving surface water flooding and a system of surface water management plans is likely to be introduced in England – with responsibility lying with local authorities for producing them.

The policy 'zones' used in Wales are defined in the Welsh Assembly Government's

Development Advice Map and are based on the best available information considered sufficient to determine when flood risk issues need to be taken into account in planning future development. Three development advice zones are described on the maps, to which are attributed different planning actions. The maps are based on the Environment Agency's extreme flood outlines (zone C) and the British Geological Survey (BGS) drift data (zone B).

In Scotland, SEPA have produced a map using a generalized procedure for estimating flood frequency and a national digital elevation model (DEM). The flood map shows the areas of Scotland estimated to have a 1 in 200 or greater chance of being flooded in any given year. It does not recognize areas where the risk is reduced by flood prevention or alleviation measures. The maps will be reviewed regularly to take into account additional hydrological data and changes in the DEM, so accounting for climate change.

Flood risk assessment

All three countries use flood risk assessment as the main tool for assessing flood risk in the spatial planning process. Flood risk assessment is used here as a generic term – in Wales the term 'flood consequences assessment' is used and in England the terms 'regional flood risk appraisal' and 'strategic flood risk assessment' are used for regional and local plans respectively. The main difference between the countries is that in England, PPS25 requires it to be applied at all stages in the planning system – to regional spatial strategies, local development documents (or other sub-regional development proposals) and planning applications, whereas in Wales and Scotland it is only applied to planning applications. In Scotland, a 'comprehensive drainage assessment' may also be required to be submitted with planning applications for large scale proposals, in areas where drainage is already constrained or otherwise problematic, or if there would be off-site effects.

The purpose of flood risk assessment in the spatial planning system can be summarized as

being to measure existing flood risk and estimate future flood risk, taking account of new development and other factors such as climate change. Flood risk assessments for spatial plans should also be used to inform strategic environmental assessment or sustainability appraisal and to apply the sequential approach to allocations and decision making. In England, the different spatial levels at which flood risk assessments are required in spatial planning are closely related to those required for flood risk management activities, as set out in catchment flood management plans (CFMP). Flood risk assessments prepared for spatial planning should inform and be informed by those carried out for flood risk management (see Figure 16.2).

Responsibility for preparing flood risk assessments for spatial plans lies with the plan makers, whereas for planning applications, the developer prepares an assessment that is submitted as part of the supporting information accompanying the application.

The main factors affecting the scope and content of flood risk assessments are:

* the spatial level to which it applies;
* the sources of flooding at that spatial level;
* the development it is assessing;
* the 'lifetime' of the spatial plan or development being assessed.

In England, PPS25 and its Practice Guide (CLG, 2008a) provide guidance on the scope, content and criteria for flood risk assessments. In flood zones 2 and 3, all but the smallest planning applications require a flood risk assessment to be submitted with the application. In flood zone 1, large applications (greater than one hectare) require an assessment – in order to check that surface water runoff is controlled. The basic requirements for an assessment of flood risk at all spatial levels in England is set out in Annex E of PPS25, as summarized in Figure 16.3.

In Wales, similar guidance is given in TAN15 for flood consequence assessments. In Scotland, a separate drainage assessment, or drainage impact assessment, may also be required when applying for planning permission. This may include existing drainage systems and problems, infiltration,

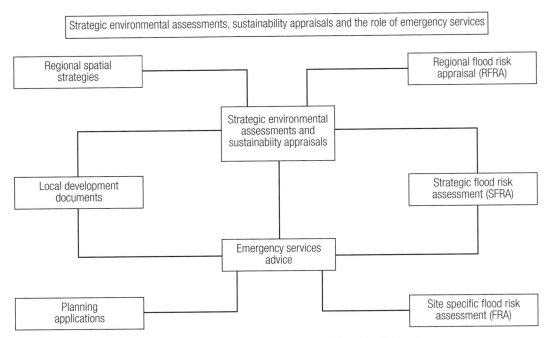

Figure 16.2 Relationship between flood risk assessments and spatial planning (England)

groundwater, surface water flow, foul and storm water disposal, SuDS (sustainable drainage systems) and drainage related flooding issues. Flooding from drains should be included in a flood risk assessment in England if it is identified as a probable source of flooding in pre-application discussion with the local authority and the Environment Agency.

Climate change is likely to increase flood risk over the lifetime of a development and therefore has to be specifically included in flood risk assessments. The PPS25 Practice Guide specifies the minimum lifetime of residential development to be 100 years unless special circumstances (such as time-limited permissions) dictate otherwise. For other developments, the developer should justify the lifetime chosen. Climate change is taken into account in strategic flood risk assessments for local plans using sea level allowances based on UKCIP projections. These are used as inputs to models to predict changes in the boundaries of the flood zones in 100 or 60 years time. Land use allocations are made on the basis of which *current* flood zone they fall into but related policies are required to

ensure that the resulting development is resilient to increased flood risk due to climate change. For example, this may mean that more vulnerable uses have to be placed on higher parts of the site or on upper storeys or that floor levels have to be raised.

PPS25 introduced a new requirement for regional planning bodies to prepare regional flood risk appraisals (RFRA) to inform their regional spatial strategies (RSS) on flood risk issues. These are intended to be 'broad brush' appraisals of flood risk that use much pre-existing data, such as Environment Agency flood maps, catchment flood management plans and shoreline management plans. RFRAs highlight flooding issues that local planning authorities should address through their strategic flood risk assessments (SFRAs). They also inform the sustainability appraisal (incorporating strategic environmental assessment) of the RSS. Because many of the regional strategies were well advanced when PPS25 was published, most RFRAs have been retro-fitted to proposed spatial patterns of growth (despite PPG25 (Planning Policy Guidance 25) containing

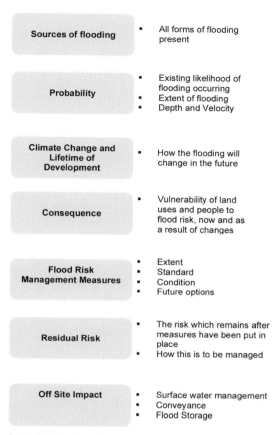

Sources of flooding	▪ All forms of flooding present
Probability	▪ Existing likelihood of flooding occurring ▪ Extent of flooding ▪ Depth and Velocity
Climate Change and Lifetime of Development	▪ How the flooding will change in the future
Consequence	▪ Vulnerability of land uses and people to flood risk, now and as a result of changes
Flood Risk Management Measures	▪ Extent ▪ Standard ▪ Condition ▪ Future options
Residual Risk	▪ The risk which remains after measures have been put in place ▪ How this is to be managed
Off Site Impact	▪ Surface water management ▪ Conveyance ▪ Flood Storage

Source: CLG (2006, Annex E)

Figure 16.3 Recommended content of flood risk assessments

similar policy requirement for flood risk to be considered at a regional scale) (Office of the Deputy Prime Minister (ODPM), 2001). There have been few examples of RSSs being changed to reflect strategic flood risks. One notable exception is changes proposed in the Panel Report on the East Midlands Regional Plan that recommends not allocating additional housing growth until a coastal strategy is in place – recognizing the vulnerability of the area to coastal flooding (East Midlands Regional Assembly, 2007). In future, if flood risk is to be given greater weight at a regional or sub-regional level, regional planning bodies should reduce housing allocations in local authority areas with relatively higher proportions of land in the floodplain, rather than rely on the

completion of SFRAs and policies that seek flood resilience and resistance measures to be imposed.

Modelling

As noted above, many flood risk assessments include an element of modelling of existing and future flood risk. The PPS25 Practice Guide contains general advice on computer modelling to complete a flood risk assessment. As a statutory consultee for spatial plans and planning applications in England and Wales, the Environment Agency advises regional planning bodies, local planning authorities and developers on the content of flood risk assessments and provides comments on the resulting documents. For planning applications, this will include whether the proposed development is likely to be 'safe' taking into account climate change over its lifetime. The Environment Agency has developed internal guidance on safety issues taking into account the vulnerability of receptor populations and the depth and speed of onset of flooding. The PPS25 Practice Guide (paras 4.27–4.61) outlines the 'safety' issues that developers and local planning authorities should take into account when making planning decisions – including safe access and egress. TAN15 similarly contains indicative guidance on 'tolerable' conditions (in terms of depth, velocity, rate of rise and inundation) for different types of development.

Catchment flood management plans (CFMP) are prepared using modelling outputs. A Modelling and Decision Support Framework (MDSF) provides the following functions for CFMP:

- assessment of flood extent and depth;
- calculation of economic damages due to flooding;
- calculation of social impacts due to flooding including population in flood risk area and their social vulnerability;
- presentation of results for a range of cases to assist the user in the selection of the preferred policy;

- procedure for estimating uncertainty in the results for each case;
- framework for comparing flood damages and social impacts as an aid to policy evaluation;
- archiving of cases.

Policy implementation

A large proportion of the objections that the Environment Agency makes to planning applications in English flood zones is because of the lack of an adequate flood risk assessment (62 per cent of all objections made on flood risk grounds in 2006/07) (Environment Agency, 2007). As this report only covered the first four months of PPS25's existence, it is too early to gauge whether PPS25 has made a significant difference to this figure. However this has been one of the most disappointing aspects of local planning authorities' and developers' responses to climate change.

Likewise, it is too early to come to a firm conclusion on the difference that PPS25 has made to the assessment of flood risk at regional and local authority levels. Many RSSs and local development documents (LDDs) were already in preparation when PPS25 was introduced and it is debatable that flood risk had been given much weight in shaping land use allocations, despite advice in PPG25, the consultation draft of PPS25 and PPS1 (ODPM, 2005) making it clear that it should be. Certainly, for many RSS reviews that were 'in the pipeline' when PPS25 emerged, the regional flood risk assessments had little influence on the broad locations of growth.

A Defra project to assess the coverage, adequacy and use of SFRAs and to make recommendations for 'good practice' is due to be published in 2009 (Defra, 2008a). The main objectives of the research are:

1 Establish an evidence base on the quality and effectiveness of SFRAs. This includes assessing the current status of SFRAs across local authorities, who are completing them, methods and models used and their direct relationships to other planning plans and strategies (including local development documents and sustainability appraisal).

2 Use a number of case studies to consider the relative influence of SFRAs upon planning decisions and policies. These case studies will be selected to cover a range of geographical, development and flood risk situations. The case studies will also be selected to enable a full assessment of technical and modelling methods currently used in the development of SFRAs within England.

3 Provide recommendations for the future development of SFRAs which can be fully integrated into the new PPS25 practice guide. This will include consideration of the potential barriers to the preparation of effective SFRAs and potential future mechanisms for the most effective use of existing data, models and information relevant to flood risk and drainage activity.

Future directions

A number of factors indicate that flood risk policy will need to be updated in the next few years. Revised UKCIP climate change projections in 2009 may result in a change to the allowances contained in PPS25 for sea level rise and other climate change variables affecting flooding. These should be taken into account when carrying out modelling for flood risk assessments. They should also affect decision makers' allocations of land for future development – especially within and at the edges of existing flood zones, the boundaries of which are likely to change.

New duties concerning surface water flooding in England are expected, as a result of the Pitt Review's recommendations following the 2007 floods and the government's 'Future Water' strategy (Cabinet Office, 2008; Defra, 2008b). Surface water management plans may be mandatory in areas where surface water flooding is likely and spatial planning will be expected to help deliver surface water flood risk reduction. In England, sustainable drainage system maintenance responsibilities may come to rest with local authorities (rather than a number of bodies) and the automatic right to connect new surface water drains to the existing sewer system may be amended to a conditional right.

In evidence to the Pitt Review, the Government revealed its intention to develop a full evaluation strategy for PPS25. This will seek to measure the effectiveness of PPS25 and the new call-in powers by drawing together data from a range of sources that monitor PPS25, including:

- data on planning applications approved against Environment Agency advice;
- feedback from regional government offices on how PPS25 is being reflected in regional and local plans;
- feedback from stakeholders;
- Defra research into the coverage and adequacy of SFRAs.

If this evaluation leads to a review of PPS25, then this could be expected to influence planning policy in Wales and Scotland. Likewise, changes in the policy in the devolved administrations could also influence any revision of PPS25.

Environment Agency maps of likely coastal erosion impacts in England will inevitably force planners to review their perceptions of the sustainability of existing and projected future development patterns – especially on low lying and 'soft' coastlines in the east and south-east of England. They are also likely to result in a revision of outdated English planning policy on the coast.

In all three countries, new flooding legislation is likely to establish new responsibilities for the environmental protection agencies and local authorities to manage flood risks. In England, the Floods and Water Bill will propose a new strategic overview role for the Environment Agency for all forms of flood risk management. Local authorities will take responsibility for surface water flooding in their own communities, supported by the Agency.

Conclusion

Flood risk policy in England, Wales and Scotland has evolved considerably since its emergence in the early 1990s. It has become stronger and more complex and now incorporates the implications of climate change into decision making. The basic methodology in use is one of appraising, managing and reducing risk. A number of tools, such as flood maps and flood risk assessments, are available to planners and their expert advisers to support policy aims. It is likely that climate change and other influencing factors will force further toughening of policy and the further development of tools such as new surface water flooding maps. Unless UK governments are to commit to very significant increases in spending on flood defences or to accept a higher exposure to risk, the long term precautionary direction should be to move new development away from areas at risk and introduce incentives for existing development to do the same.

Acknowledgements and Disclaimer

The views expressed in this chapter are not necessarily the official position of the Environment Agency. The author acknowledges the help of colleagues, particularly Geoff Gibbs.

Note

1 'reasonably available' is not defined in the PPS itself, but more guidance is in the PPS25 Practice Guide, e.g. 'suitable, developable and deliverable'.

References

Cabinet Office (2008) 'Learning lessons from the 2007 floods', www.cabinetoffice.gov.uk/thepittreview/final_report.aspx, accessed 11 September 2008

CLG (Department for Communities and Local Government) (2004) 'Planning Policy Statement 23: Planning and pollution control', www.communities.gov.uk/planningandbuilding/planning/planningpolicyguidance/planningpolicystatements/planningpolicystatements/pps25/, accessed 30 June 2008

CLG (2006) 'Planning Policy Statement 25: Development and flood risk',

www.communities.gov.uk/planningandbuilding/
planning/planningpolicyguidance/planning
policystatements/planningpolicystatements/
pps25/, accessed 30 June 2008

CLG (2007) 'Planning Policy Statement 1A: Planning
and climate change', Department for
Communities and Local Government, London

CLG (2008a) 'Planning Policy Statement 25:
Development and flood risk Practice Guide',
Department for Communities and Local
Government, London

CLG (2008b) 'Planning Policy Statement 12: Local
development frameworks', Department for
Communities and Local Government, London

Department of Environment (DoE) (1992) 'MAFF
Circular 30/92' (Welsh Office Circular 68/92),
London

Defra / Environment Agency (2002) 'Flood and
Coastal Defence R+D Programme FD2321:
Flood risks to people', Phase 2 Project Record,
www.defra.gov.uk/science/Project_Data/
DocumentLibrary/FD2321/FD2321_3438_
PR.pdf, accessed 7 September 2008

Defra (2005) 'Making space for water: Taking forward
a new Government strategy for flood and coastal
erosion risk management in England' – first
Government response to the autumn 2004
Making space for water consultation exercise,
www.defra.gov.uk/environ/fcd/policy/strategy/
firstresponse.pdf , accessed 11 September 2008

Defra (2008a) –'Quality and influence of strategic
flood risk assessments in the planning process',
Science and Research Projects FD2610 Land Use
Planning, http://randd.defra.gov.uk/Default.aspx?
Menu=Menu&Module=More&Location=None
&Complet ed=0&ProjectID=15572#
CentralDocuments, accessed 8 September 2008

Defra (2008b) *Future Water: The Government's Water
Strategy for England*, The Stationery Office,
London

DTI (2004) 'The Foresight Future Flooding Project',
Department for Trade and Industry,
www.foresight.gov.uk/OurWork/CompletedProje
cts/Flood/index.asp, accessed 7 September 2008

East Midlands Regional Assembly (2007) 'Report of
the examination in public of draft regional spatial
strategy for the East Midlands',
www.gos.gov.uk/497296/docs/229865/Panel_Re
port.pdf, accessed 11 September 2008

Environment Agency (2007) 'High level target 5
development and flood risk in England 2006/7',
www.environment-agency.gov.uk/aboutus/
512398/908812/1351053/1449570/?version=
1&lang=_e, accessed 7 September 2008

Office of the Deputy Prime Minister (2001)
'Planning Policy Guidance Note 25: Development
and flood risk', ODPM, London

Office of the Deputy Prime Minister (2005)
'Planning Policy Statement 1: Delivering sustainable development', ODPM, London

Scottish Executive (2003) 'National Flooding
Framework', www.scotland.gov.uk/Topics/
Environment/Water/Flooding/
national-framewok, accessed 8 September 2008

Scottish Executive (2004) 'Scottish Planning Policy
(SPP) 7: Planning and flooding',
www.scotland.gov.uk/Resource?Doc/47210/
0026394.pdf, accessed 1 May 2009

Welsh Assembly Government (2002) *Planning Policy
Wales*, WAG, Cardiff

Welsh Assembly Government (2004) 'Technical
Advice Note (TAN) 15: Development and flood
risk', www.scotland.gov.uk/Resource?Doc/
47210/0026394.pdf, accessed 1 May 2009

Part 3

Implementation, Governance and Engagement

Introduction to Part 3

Abid Mehmood

Discussions in this part resonate some of the arguments developed in earlier chapters around the issues of integrating mitigation and adaptation options and paradigm development in spatial planning. However, the approach here is more related to actions and implementation of innovative policy tools. It focuses on the importance of futures thinking in spatial planning for adaptive measures and discusses the significance of governance relations, with particular emphasis on leadership, the role of individuals in institutions, partnership for policy implementation, and public engagement in the implementation processes. Therefore, the chapters in this part can be divided into three broad themes:

- role of scenarios and modelling;
- policy implementation;
- governance and public engagement.

Role of scenarios and modelling

The first three chapters in this part specifically address the importance of forecasting tools, especially scenario building as tools for planning to identify measures for climate change at different spatial scales. In Chapter 17, Wilson stresses that a futures focus of spatial planning will help prepare for climate change consequences and reduce uncertainty. She examines the studies carried out to develop spatial scenarios in Europe at regional, national and supra-national

(EU) levels that have taken account of climate change both in urban and rural areas. A comparison of planning scenarios in the UK and the Netherlands reveals the use of different spatial and temporal horizons. Reasons can be attributed to the respective governance frameworks, national cultures and regional socio-economic conditions. Also, in both cases, the potential of integrated climate change scenarios remains underutilized. By identifying the advantages of and barriers to scenario building, Wilson argues in favour of scenarios as tools to enhance the adaptive capacity for tackling climate change.

Hall, in Chapter 18, promotes an integrated approach to forecasting the impacts of climate change at urban and regional levels. Advocating a quantitative impact assessment, he stresses the need to address both mitigation and adaptation in an integrated manner to manage the adaptive capacity of the natural and built environment. From the long-term analysis of flood-risks and coastal erosion in England and Wales, it is suggested that 'a range of plausible futures' can demonstrate the portfolio of options available to planners and policy-makers for mitigation and adaptation strategies. The technological improvements in visualisation and climate modelling also mean wider stakeholders' involvement in the decision-making processes.

An exploration of the functional role of green infrastructure is provided by Gill et al in Chapter 19. Based on the findings from 'Adaptation Strategies for Climate Change in Urban Environments (ASCCUE)' project, the chapter discusses how urban green infrastructure can help sustain the natural systems in the incidence of climate change. This is analysed by modelling the adaptation potential of urban green infrastructure within Greater Manchester. This model incorporates the potential effects of urban heat island as well as changes in hydrology and energy exchange in the urban environment. The quantified analysis under different emissions scenarios reflects the adaptive capacity of green infrastructure to moderate the natural processes. The authors stress that such infrastructure (trees, open spaces, private gardens, etc.) should be strategically planned to avail of optimum functionality because, unlike open green spaces, it requires regular maintenance (e.g. proper irrigation) in order to adapt to extreme events such as droughts.

Policy implementation

Langlais in Chapter 20 provides an account of variable responses from Swedish municipalities in terms of both adaptation and mitigation measures. By looking at the relationships between national and municipal policies, Langlais examines the adaptation barriers and acknowledges the role of 'visionary individuals' in some municipalities for initiating local actions under similar institutional and governance conditions. He analyses the effects of two subsequent national plans that encouraged the municipalities to actively engage in mitigation measures. These incentives gave major impetus to the development of networks of eco-municipalities and those engaged in reducing greenhouse gas emissions. The author argues for the role of planners in building knowledge and expertise at local levels to face the challenges of climate change.

Talking of the visionary individuals, Chapter 21 reflects on the personal experience of Allan Jones who has led the climate change policies and projects in Woking and London. Jones discusses the political, policy and planning challenges, strategies and measures during the course of developing low-carbon energy projects in the two places. The town of Woking led by

example through a range of energy and water efficiency measures which put it on the path to low energy consumption. In London, the introduction of Congestion Charge and Low Emission Zone resulted in the reduction of CO_2 emissions. Other measures such as the London Plan 2004, Mayor's Energy Strategy and London Development Agency's 'Green Alchemy' report have also been catalysts for adaptation and mitigation. Jones pinpoints the London Climate Change Agency's efforts to introduce low-carbon decentralized energy through a number of energy service companies through public–private partnerships across London. The chapter also describes the efforts of the C40 Climate Leadership Group as a global network of large cities for urban action to tackle the climate challenge.

Governance and public engagement

Although aspects of governance and public participation have been implicit in the earlier chapters of Part 3, the final two chapters of the book address these more explicitly within the UK planning and policy perspectives. A discussion over the development of climate change policy as a multilevel process in the UK is provided by Bulkeley in Chapter 22, in which she reflects on the impacts of climate change pressures on the governance structures. Maintaining that spatial planning is shaped both through vertical processes of governing (national–local) and horizontal networks (partnerships between state and non-state actors across different scales), Bulkeley examines how such processes of governing have facilitated or impeded the implications for achieving low carbon and more resilient futures. She explores the emergence of planning governance in relation to energy supply, energy demand, and adaptation as the critical climate change policy areas. The chapter argues that spatial planning is not merely a delivery mechanism but a mediating process to avail of 'productive accommodation'.

The final chapter of this book by Haggett focuses on the public engagement side of the climate change governance. Using the examples of renewable energies and, more specifically, wind farm development proposals in Europe she identifies three forms of engagement including engagement as: 'information provision', 'consul-tation' and 'deliberation'. From the analysis of these forms, Haggett examines the new planning legislation in the UK to discuss how planning measures at different scales can help or hinder public participation, interest, and engagement.

17

Use of Scenarios for Climate Change Adaptation in Spatial Planning

Elizabeth Wilson

Introduction

Many reports (such as the Stern Review of 2006 (Stern, 2007), and the EU Green Paper on 'Adapting to Climate Change in Europe' (Commission of the European Communities (CEC), 2007)) point to the important role of spatial planning in adapting to climate change. Spatial planning is one area of public policy intervention with an explicit focus on future horizons, and it has critically important outcomes in terms of activities, the built form and the natural environment. The built environment has a design life of perhaps up to 100 years, and the overall settlement pattern and urban form has even greater longevity. Nevertheless, even though the science and use of scenarios in climate change projections might suggest the need for a long-term view, UK planners and planning authorities have, until recently, been inhibited in taking a long-term perspective or in engaging with futures thinking. This chapter offers some reflections on the reasons for that, exploring some of the ethical and conceptual issues around futures thinking and futures-oriented action. The issue of climate change, however, is prompting new interest in the use of scenarios and futures thinking, and the chapter examines examples of regional spatial planning for adaptation in the Netherlands and the UK. It draws conclusions about the ways in which the understanding and experience of climate change is changing planning.

The Intergovernmental Panel on Climate Change (IPCC) has generated scenarios of climate change which, through the sequence of its reports on the science and implications for policy of climate change, have been extremely influential in prompting policy responses, especially in measures to mitigate climate change (Tomkins and Amundsen, 2008), for instance through the Kyoto Protocol and the European Union Emissions Trading Scheme (EU ETS). However, adaptation to climate change has been a relatively slow area of policy to develop. In the field of spatial planning, policies to reduce energy demand, conserve energy or promote the move to alternative fuels, have become integrated into adopted plans and implemented through a range of measures, especially at the city level (Bulkeley and Betsill, 2003; Lindseth, 2004; Davies, 2005; Betsill and Bulkeley, 2007). However, even if mitigation measures are adopted, it is expected that, because of lags in the climate system, some climate change is unavoidable, and we therefore need to be able to adapt to these changes. The IPCC concludes that 'There is high agreement and much evidence that with current climate change mitigation policies and related sustainable development practices, global greenhouse gas (GHG) emissions will continue to grow over the next few decades' (IPCC, 2007, p6). Moreover, there is substantial evidence that climate is already changing (European Environment Agency (EEA), 2005; CEC, 2007; IPCC, 2007; Jenkins et al, 2007).

But spatial planning, as with some other sectors, has only recently adopted policies for adaptation to these unavoidable impacts of climate change. The reasons for this lie in some generic issues of adaptation: while mitigation is a global issue, but one requiring action at international, national, local and individual levels, adaptation is particularly significant at local or regional levels. It may seem surprising that adaptation is given less priority at municipal level than mitigation, as the costs of inaction, and hence benefits of local action, would be felt locally. However, adaptation is unlikely to have a set of nationally agreed targets in the same way in which mitigation outcomes (such as targets for renewables) are cascaded or imposed down to regional or local governments (Wilson, 2008). Moreover, even across these scales, the distribution of costs and benefits is different, which raises issues of power and adaptive capacity: adaptation may require action by or on behalf of those often already most vulnerable in socio-economic terms and perhaps least able to adapt. While it is agreed now that mitigation and adaptation should be integrated (in order to avoid maladaptation, and to promote synergies and complementarity) (Swart and Raes, 2007; Howard, Chapter 2), there are difficulties in expressing the balance of costs and benefits: the latest IPCC report concludes that 'A wide array of adaptation options is available, but more extensive adaptation than is currently occurring is required to reduce vulnerability to climate

change. There are barriers, limits and costs, which are not fully understood' (IPCC, 2007, p14). In Europe, a study for the EEA also concludes that, while adaptation has a very important role in reducing the costs of climate change across Europe, there is currently little quantified information on these costs or the costs of adaptation itself (EEA, 2007). There may also be conceptual and political difficulties in persuading governments, institutions and businesses to take action on both mitigation and adaptation fronts, without undermining the case for their effectiveness (Wilson, 2006a).

Sustainability and the future

It is also possible that there are other, more fundamental, explanations for the late response of planning to climate change adaptation. It might be expected that the essence of sustainability thinking should include sustaining into the future, and that planning for sustainable development would emphasize the longer term. However, it is the contention of this chapter that the emphasis on the integrative and intra-generational aspects of sustainable development has obscured that of inter-generational justice and long-term horizons.

The concept of sustainability has generated an immense and not always enlightening literature, much of it around competing conceptions and definitions, springing from the early definition of the Brundtland Commission (World Commission on Environment and Development (WCED), 1987). Debates in the 1990s elaborated on the many meanings of sustainability (Jacobs, 1999; Davoudi et al, 2001), on the links between, and need to integrate, the economic, social or environmental elements (Kenny and Meadowcroft, 1999), and on the challenges to the fundamental ecological support systems in balancing or trading-off these elements. There has also been a developing field of work on environmental justice, on the distribution (within current generations) between those who have and those who have not (Dobson, 1999; Agyeman et al, 2001). But another vital dimension of sustainability has been relatively overlooked by those engaged with public policy (although not by political philosophers). This is the regard to the long-term future. The very meaning of sustainability assumes a view of the future as worthy of our regard and consideration. While Brundtland gives explicit attention to not compromising the ability of future generations to meet their needs, her definition has focused attention more on the needs of current generations for social and political equity. But 'sustainability obliges us to think about sustaining something into the future, and justice makes us think about distributing something across present and future' (Dobson, 1999, p5). While not all sustainability questions (for instance, those concerning the substitutability principle of natural versus man-made capital) directly raise issues of justice, and certainly not all theories of justice have a counterpart in environmental sustainability, the two are inextricably linked when we consider what it is that is to be sustained over time, and what rights or obligations are owed by one generation to another.

However, this position of separation of inter- and intra-generational issues is changing: during this century, the issue of climate change has risen up the agenda of policy making and public consciousness (IPCC, 2007), and raises evident and fundamental issues for distributive justice and both intra-generational and inter-generational justice (Page, 2006), especially in relation to adaptation (Paavola and Adger, 2006). Responding to the issue of climate change therefore requires a way of reconciling these two conceptions of sustainability.

Planning and the future

It might be expected that, of all the public policy areas, land use planning would have embraced these two elements of sustainability. Planning after all is an inherently future-oriented activity, and the products of decisions about land use – the buildings and infrastructure – are designed to remain for at least 20, and often 60–100 years (Graves and Phillipson, 2000; Shaw et al, 2007), and of course in many cases survive for much longer. Land use planning outcomes set the spatial patterns for the activities of future generations. But reflections on the purpose and focus of spatial planning have tended to focus more on other aspects of sustainability than on our obligations to future generations. In particular, they have – as with the more general discourse around sustainability – focused on questions of the balance, trade-off or compensation of the economic, social and environmental elements of sustainability (Davoudi et al, Chapter 1), especially in relation to the claim that land use planning (as a realm for resolving contested spaces) is familiar with these trade-offs (Owens and Cowell, 2002; Rydin, 2003). Given the difficulties with the Brundtland definition of sustainable development, Rydin (2003) argues that the tendency has been to adopt the three-factors (environmental, social and economic) model, whether as a set of overlapping dimensions or as a framework in which the (physical) environmental dimension underpins the others. In as much as a view is taken of the future, it remains a very cautious one, with planning horizons, at least in the UK, still conventionally limited to 15–20 years. Indeed, this limited horizon has perhaps been reinforced by the shift of policy attention to the 'spatial' as opposed to the traditionally narrow land use focus of UK planning, with a renewed emphasis on integration across broader policy areas such as biophysical and cultural domains (Harris and Hooper, 2004).

One area where there has been an explicit recognition of the future has been through the use of visioning exercises, often linked to community participative events through Local Agenda 21 processes (O'Riordan and Voisey, 1997; Ball, 2001; Shipley, 2002) or at the macro-scale to strategic visioning exercises associated with European trans-national spatial planning (Nadin, 2002). There have been calls for planning to revive its traditions of futures thinking and to do more to explore strategic issues explicitly under different scenarios as a way of assessing their potential outcomes and making better decisions (Albrechts, 2004; Friedmann, 2004). A number of factors may be changing planning's aversion to futures thinking. Marshall suggests that, in the late 1990s, the European move to a more strategic scale of planning, as well as the environmental agenda, has caused a rethink, with planners taking a more strategic perspective beyond both administrative boundaries and conventional planning time horizons (Marshall, 1997).

Use of future scenarios

Climate change is a further prompt for this rethinking. In the fields of climate change and strategic policy making, there has been a rapidly developing literature and policy interest in developing not predictions or visions of the future but storylines or scenarios for the future. Much of this work in the late 20th century stemmed from futures-thinking in military and strategic management fields, perhaps brought on by apocalyptic visions at the turn of the 20th century and the onset of the 21st, and now employed especially in speculative finance, macro-economic policy, business planning and climate change policy. Methods to reduce uncertainty about the future include the use of horizon-scanning, outlooks, scenarios and models of the future. Scenarios can be described as:

> *Coherent, internally consistent and plausible descriptions of future states of the world, used to inform future trends, potential decisions or consequences.*
>
> (UK Climate Impacts Programme (UKCIP), 2001, p4)

They can be used for a variety of purposes: the principal one is to envision a range of possible futures, and to allow for the exploration of these possibilities, without policy or resource commitment. They also provide a springboard for more creative thinking, allowing us to step outside the conventions of current thinking and structures, and can communicate relatively unstructured ideas. More instrumentally, they offer potential to build consensus amongst different interests or stakeholders, or to achieve policy buy-in. They can be particularly useful in testing the robustness of a policy, plan or strategy, and in this way can recognize and accommodate the inherent

uncertainties of not knowing the future which face decision makers (Department for Trade & Industry (DTI), 2002). Options are evaluated and preferences selected against criteria, including robustness with respect to a range of futures.

There are many examples of futures thinking around the globe, principally (if not explicitly) from a markedly western, liberal or neo-liberal perspective, such as that of the Global Scenario Group, convened in 1995 by the Stockholm Environment Institute, to elaborate the requirements for a transition to a more sustainable future (Raskin et al, 2002). In Europe, there have been a number of recent studies using scenarios to project possible land use futures, and integrating some element of climate change. For instance, the European Spatial Planning Observation Network (ESPON) study of spatial scenarios up to 2030 in the context of European enlargement was based on drivers of disparities in wealth at enlargement; external factors of globalization, energy price-rise, climate change and immigration; and internal factors of EU policies and population change (ESPON, 2006). The EEA has undertaken its own scenario study, PRELUDE (EEA, 2006), in which it explores the impacts of future development on European landscapes and biodiversity up to 2035. This also included an element of climate change. The EEA's Urban Sprawl in Europe (Ludlow, 2006) study for selected urban areas up to 2025 similarly included drivers for sprawl in the context of EU policies of the internal market, competitiveness, sustainable development and the Cohesion and Structural Funds, but did not include climate change as a major driver. For rural land use, the EU projects on assessment of Climate Change effects on land use and ecosystems (ACCELERATES) and Aquatic and Terrestrial Ecosystem Assessment and Monitoring (ATEAM) assess the vulnerability of agriculture and ecosystems under different climate change scenarios (Berry et al, 2006). While many of these scenario exercises have drawn out the policy implications of their findings, they were not developed specifically to guide policy. However, climate change scenarios have been explicitly developed for this purpose (see Hall, Chapter 18).

Climate change and socio-economic scenarios

Climate change is now recognized as one of the most serious challenges facing humankind. Given the immense complexities and hence uncertainties of modelling global climate, and especially the uncertainties of human behaviour in response to climate change (such as the adoption of carbon-reduction policies), the IPCC studies have adopted a set of scenarios for high, medium and low levels of emissions at periods representing broadly the 2030s, 2050s and 2080s (IPCC, 2007). The studies offer projections of future temperatures, precipitation and other climate factors, with judgements of likelihood on both average changes and the frequency and intensity of extreme events such as intense rainfall, prolonged drought, heatwaves or storminess. These climate scenarios are based on the Special Report on Emissions Scenarios (SRES), representing qualitative storylines which have been input into the IPCC's vulnerability, impact and adaptation assessments. But it is important to recognize two points.

Firstly, as Hulme and Dessai (2008) argue, while climate scenarios are useful in responding to climate change in pedagogic, motivational or practical ways, the scenarios themselves need to be recognized not just as the outcome of isolated modelling exercises, but as the outcome of a process of negotiation and social construction between different stakeholders representing the funders, policy communities, scientists and decision makers. They demonstrate that, in the UK, this has meant a multi-layered process in which 'values, scientific capability and institutional capacity interact in complex ways' (Hulme and Dessai, 2008, p67).

Secondly, future climates will be experienced in societies that are different from ours today. The climate change scenarios that are most commonly used are associated with a set of marker storylines, based on internally consistent assumptions about the extent of globalization, economic growth and societal values: Scenario A1F1 is based on a rapidly growing but integrated and connected world with increased consumption of fossil fuels; Scenario A2 assumes a more divided or heterogeneous world with regionally oriented but slower growing economies; Scenario B1 is based on an integrated and dynamic international relations and markets with more take-up of ecologically friendly technology; and Scenario B2 assumes a slower growing, more divided but more ecologically friendly world with emphasis on local sustainability. While climate models are highly complex, at least some aspects of the models are based on well-understood physical processes, whereas there is less understanding of the interactions or possible speed of change of socio-economic factors. The SRES scenarios recognize that levels of emissions are not really the principal driver likely to affect socio-economic futures, but are rather the result of other socio-economic drivers such as industrialization and economic development, patterns of mobility, attitudes to the future, and attitudes to individual responsibility or to other groups of society.

In the UK, the UKCIP commissioned a study of socio-economic futures, building on the earlier work of the UK Foresight Programme, which employed two socio-economic dimensions – social values and governance systems – to generate scenarios. The social values axis covers patterns of individual consumption and policy making, with consumerist, short-term values at one end, and concerns for greater equity and long-term sustainability at the other. The governance dimension covers the scale and structure of political authority, with globalization at one end representing authority focused upwards from nation states, contrasted with regionalization at the other end (Eames and Skea, 2002) (see Figure 17.1). The UKCIP climate change and socio-economic scenarios have been used to considerable effect in the Foresight study of future flooding (UK Foresight, 2004), which evaluated flood-risks and social, economic and environmental costs, and possible policy responses, under different scenarios. It recommended a portfolio of measures – catchment-wide storage, land use planning and realigning coastal defences – to cope with a range of possible futures, provided they are

Source: UKCIP (2001, p19)

Figure 17.1 Four socio-economic scenarios for the UK

Source: Janssen et al (2006, p46)

Figure 17.2 Four socio-economic scenarios for the Netherlands

implemented in ways sensitive to economic and social considerations.

In the Netherlands, a similar study (Janssen et al, 2006) designed four scenarios to the 2040s: the two key drivers were the extent of individualism of public action, and the extent of international cooperation or national action. The resulting scenarios are shown in Figure 17.2. Both countries clearly employ a similar reasoning to the SRES scenarios. It is therefore important – especially for public policy areas such as land use and spatial planning – to be cognisant of both possible future climates and of the socio-economic scenarios underlying them, but also of alternative socio-economic futures of which the built environment is a part.

Use of scenarios in spatial planning

Through the outreach and stakeholder work of national agencies such as UKCIP or the Netherlands Environmental Assessment Agency (PBL), both the UK and the Netherlands have experience of making use of these scenarios in the regional or metropolitan context. The intention of the UKCIP was to promote the use of

climate change and socio-economic scenarios to enlighten and inform sectors and domains, including spatial planning, where climate impacts are expected to have particular significance. At the regional scale, UKCIP coordinated the publication of regional impacts studies, and then supported the development of climate change partnerships to raise awareness amongst regional stakeholders, with varying degrees of success (West and Gawith, 2005; Hedger et al, 2006). Two examples will illustrate two different issues arising from climate change – the urban heat island, and multiple flood-risk – as well as showing the different approaches to the use of scenarios.

London, UK: Scenarios and the urban heat island

London has recognized the serious implications of climate change for the maintenance of its role as a world city and for the well-being of its citizens. This has been reflected in major studies and policy initiatives. 'London's Warming' (London Climate Change Partnership (LCCP), 2002) used the UKCIP02 climate scenarios (Hulme et al, 2002). However, as the regional climate model does not distinguish between urban surfaces and more rural areas (it treats all surfaces as though they were vegetated), the scenarios do not allow for the London Urban

Heat Island (UHI), a significant urban feature. London is likely to experience higher summer temperatures both from global warming, and from the intensification of the UHI, exacerbated by increasing densities, an increase in heat outputs from air conditioning units, and reduced evaporative cooling due to drier summers (see Gill et al, Chapter 19). For London, this is particularly significant as it already has a distinct, nocturnal heat island, with, at times in central London, a temperature difference of 9°C above rural areas (Mayor of London, 2006a). Under the low and high scenarios, maximum temperatures for August (when the UHI is most pronounced) might be 2–3°C and 5–6.5°C higher than at present.

Other significant impacts of climate change for London were identified as air quality, water resources, flood risk (fluvial, urban drainage and tidal flooding/sea-level rise), subsidence and erosion, and biodiversity. 'London's Warming' used a tailored version of two of the UKCIP socio-economic scenarios (the Global Markets and the Regional Sustainability scenarios) to explore the potential social impacts of climate change on London. It suggested that the draft London Plan at that time (2002) constituted a hybrid scenario of its own, with elements of both. The scenarios were therefore used for a broad evaluation of policies and commitments being entered into in the near term (i.e. within the draft Plan). So, for instance, it is expected that not only will the heat island effect increase in summer, but under the Global Markets scenario there would be stronger demand for air conditioning (a contributor to the UHI) and for development of green spaces, with consequences for the least well-off, and danger of increasing 'cool poverty'. On balance, the report concludes that the social impacts are likely to be more negative than positive.

This scenario-exploration has prompted further research (for instance, on the experience of other major urban areas in providing lessons for London on adaptation measures (LCCP, 2006), and on the UHI (Mayor of London, 2006a)), and further changes to the approved London Plan of 2004 (Mayor of London, 2004). The scenario work significantly altered the perception of the seriousness and urgency of addressing the impacts of climate change on London and the need to adapt. The Mayor's Further Alterations (Mayor of London, 2006b) to the adopted Plan were specifically drafted to take more account of climate change, with a new objective to make London an exemplary world city in mitigating and adapting to climate change, and with an extended plan-horizon up to 2026. The UHI report concluded that 'the way in which London develops over the coming decades will play a critical role in determining the characteristics of London's UHI under conditions of climate change' (p19) (see Table 17.1). Current directions for policy to address the UHI issue are therefore through area action plans, street and building design, and regional

Table 17.1 Adaptation actions for urban heat islands: The link between policy and urban climate scales

Physical scale	Policy scale	Urban climate scale
Individual building/street (façade and roof construction materials, design and orientation)	Building regulations and building control Urban design strategy Local Development Framework	1–10m Indoor climate and street canyon
Urban design (arrangement of buildings, roads, green space)	Urban design strategy Area Action Plan Local Development Framework	10–1000m Neighbourhood scale, sub-urban variations of climate
City Plan (arrangement of commercial, industrial, residential, recreational and greenspace)	Sub-regional spatial strategy	1–50km City/metropolitan sale, UHI form and intensity

Source: Mayor of London (2006a, p18).

policies: for instance, a UHI Action Area in central London, within which existing greenspace will be preserved, and new greenspaces inserted; at street and building scale, cool roofs, green roofs and walls, seasonal shading, cool materials; and at regional level, the promotion of the green grid infrastructure and work with the London Boroughs (Nickson, 2007).

Rotterdam, the Netherlands: Scenarios and the future water city

The Netherlands is a country particularly at risk from certain impacts of climate change: not just increased flood risk and sea-level rise, but also impacts on agriculture, water resources, biodiversity, urban areas' air quality, and urban heat islands. As 60 per cent of the Netherlands lies below sea level, including Amsterdam and Rotterdam, and 70–80 per cent of its national wealth is generated within this area, flood risk for urban and rural areas is already a significant issue (see de Vries and Wolsink, Chapter 15). Even under a medium emissions climate scenario, the impacts for Dutch economy and society would be serious, with projected sea-level rise of 60cm by 2100. The Netherlands government is committed to climate-proof the country, a task in which spatial planning has a key role (Kabat et al, 2005; Cramer, 2007). The government has sponsored a major research programme Climate Changes Spatial Planning (CCSP), and is adopting a national Spatial Planning and Adaptation Strategy (ARK), involving departments of state, provinces, municipalities and water boards. The intention is to adopt a methodology that consistently uses the climate change and socio-economic scenarios for developing and evaluating adaptation projects, and to aim for long-term robust adaptation strategies.

Such strategies might include changes to the national ecological structure, coastal protection (for instance, through widening the dune network seawards), flood protection through compartmentalizing dykes, as well as changes in other sectors such as revised flood insurance, or new agricultural practices. The Netherlands is keen to see adaptation not just as a response to risks, but also as a way of creating opportunities, especially for innovative design and solutions. For instance, the promotion of the Delta Metropole could be redesigned as a Hydrometropole, in which people live with and from the water (Kabat et al, 2005); or the country could be elevated, with development taking place on land raised by bringing in sand from the North Sea to a 5m height (Aerts, 2007). The CCSP includes modelling of land use changes under different scenarios for the 2040s, and detailed applied studies are underway in a number of 'hotspots', including areas planned for urban growth but at risk of flooding.

The climate scenarios are already being used to incorporate the impacts of climate change into national and regional spatial plans. The National Spatial Strategy (VROM) 'Nota Ruimte' (2006) retains the Space for Rivers policy (see de Vries and Wolsink, Chapter 15), and anticipates that climate change will require more space or water. At the metropolitan level, the new Strategic City Vision for Rotterdam 2030 (Tillie, 2007) has been published alongside a Water Plan for the City which provides space for water in accordance with the national plan. The Water Plan for Rotterdam 2030 (Gemeente Rotterdam et al, 2007) celebrates the city's relationship with water, but recognizes that it is threatened by water from rising sea levels, increasingly intense precipitation, ground water and river discharges. The City Plan aims principally to have a stronger, more diverse economy, and a more attractive city, and so the intention is to use the plan for water to help solve the city's other problems in providing safety, quality housing, good public spaces and a healthy economy, offering an integrated vision and a programme of solutions. The Plan includes provision, under the legislative framework of the Dutch Water Test, for accommodating extra water-storage by 2050 of 240,000m³ water in the city and its neighbourhoods. Up to 2035, water solutions for Rotterdam include areas of change and restructuring, of the central river city (by the existing river frontage) and the canal city to the south. The intention is to implement these through networks of water squares to provide temporary storage capacity, areas designed for living alongside water-courses, and the provision of green roofs to retain rainwater.

This illustrates the use of scenarios of the future to introduce seemingly radical solutions within the short–medium term spatial planning framework.

Barriers to use of scenarios

These two examples have illustrated some of the advantages to be gained from the use of scenarios, but also some of the barriers to their use. Scenarios within spatial planning are not without conflicts. Besides the reasons advanced at the beginning of this chapter for our reluctance to think about the future long-term in terms of intra-generational equity, other difficulties are that scenarios are deliberately exploratory or imaginary, whereas policy is more concerned with normative desired outcomes. Policy evaluation itself finds it difficult to step outside the conventional appraisal systems, as illustrated by the report of the scenario testing of the London Plan Alterations (Mayor of London, 2006b) with its reliance on the business-as-usual growth agenda in evaluating the scope for alternative actions. The assessment goes to considerable lengths to argue that, despite what might seem conflicts between the Mayor's growth strategy for London and adapting to climate change impacts (such as between urban intensification and flood-water absorption), growth is essential to pay for adaptation in areas such as the transport system, open spaces, the avoidance of 'cool poverty', and flood management.

Secondly, exploratory scenarios may be preoccupied by conflicting current political visions and world-views, which may prevent consideration of broader or more radical future alternatives. The more radical options (such as a significant shift eastwards of the Netherlands population and economic activity, or a major abandonment of the City of London) are difficult to contemplate politically. There may be seemingly rational objections: for instance, it could be argued that, as such choices are longer term and likely to be expensive, spatial planning should focus on the shorter term 10–40 years: the uncertainties of projections and location of impacts mean that it is worth postponing some decisions. There are also of course likely to be contested issues of resources and compensation in the context of politically difficult decisions, such as scenarios that envisage significant movements of development or populations as an adaptation response.

Thirdly, there are different national cultural attitudes amongst professional and elected members to risk, uncertainty and the future. For spatial planners in the UK, there may be reluctance to adopt what may be seen as utopian visions, possibly based on hard-bitten experience with over-ambitious redevelopment schemes, or a fear of being open to charges of social engineering. However, the Netherlands has a tradition of longer term horizon-scanning and planning. Perhaps because it is a nation vulnerable to physical changes, its planning horizons are longer term than those in the UK. The coastal defences and dikes are designed for a 1:10,000 year event, a much higher standard of flood-risk than that in the UK (where the Thames flood defences, for instance, are designed for a 1:1000 year event). Working with future scenarios is therefore an experience of longer standing. Moreover, the Netherlands has a tradition of national spatial planning (unlike the UK, or at least unlike England), and the current national spatial plan has a time horizon of 2030, longer than any English spatial plan.

More fundamentally, it can be argued that our ways of conceiving the future are inextricably linked up with our notions of permanence: those who adopt a conservationist ethic, assuming the longevity of the built or natural environment, can be contrasted with those who experience feelings of vulnerability, isolation and impermanence (Campbell, 2004). Campbell argues that this represents a challenge for professional planners in engaging stakeholders, who experience different collective or individual notions of time and space, in visioning processes. More pragmatically, we know the past is no guide to the future, and recognize the feebleness or inaccuracy of past attempts either to predict or to project the future. This may be because of a rational acknowledgement of the practical diffi-culties of forecasting the future (Skaburskis and Teitz, 2003), and a failure to distinguish scenarios from forecasts; or it may be because popular opinion disparages those who have been over-confident about futures thinking (Collard, 2003).

Fifthly, the endemic short-termism of politi-cal processes in institutional frameworks and electoral systems may mean that, for politicians, a mentality of 'within my period of office' may be inevitable, with targets to be achieved and measures delivered before the next electoral cycle. However, the powers available to the Mayor of London (greater than those of other local government in the UK) may mean that more difficult 'longer term' decisions can be taken which put development paths on a differ-ent trajectory. In Rotterdam, action has been the result of partnership working – the City of Rotterdam and the Water Boards, for instance – with greater consensus on medium-term horizons.

Finally, the two examples discussed in this chapter both relate to larger scale, strategic planning decision-making, and it may be that lower tiers of planning governance assume that longer term planning horizons are only appro-priate at larger spatial scales – either for reasons of difficulties of down-scaling at smaller scales, or because, the closer to the ground, the more problematic might seem radically different futures.

It may also be that, even scenarios specifically developed for policy making, need to be clear about who the decision makers are, and what the decision points are, or when they occur. It is also possible that the differences between the discourses of the climate change policy commu-nity as scenario builders, and the spatial planning community as users, are significant, representing different conceptions of policy-space and appro-priate actions (Wilson, 2006b).

Conclusions

Despite these barriers to greater use of scenarios in thinking about the future, and in particular in adapting to climate change, this chapter has shown that there are immense benefits to be gained. Scenarios of climate change enable us to see issues hitherto hidden, such as the future impact of the UHI in London, or they can be used much more explicitly than they have been to provide a systematic evaluation of spatial plans, or indeed as in Rotterdam to generate alternative policy paths. The adaptive capacity of both countries, but particularly the UK, can be enhanced by using scenarios to look beyond the lifetime of the plan, and actively responding to this in plan-making and development decision-taking. This approach can enable us to adapt creatively to climate change.

The advent of concern about the impacts of climate change, the need to develop adaptive capacity and to undertake adaptive actions, and the generation of plausible scenarios up to the end of the 21st century, has had the effect of making us think seriously about the longer term sustainable future and our obligations to future generations, and this is a development to be welcomed. Planning needs to broaden its preoc-cupation with space, and to take cognisance of time. Giving attention to the long term, and having some idea of what may obtain in 50 or 100 years time, is not a recipe for inaction now, and certainly does not mean that efforts to resolve issues can be put off into the long term. Regard to the future is a coherent and worth-while position for spatial planning.

References

Aerts, J. (2007) 'Climate adaptation: A spatial planning perspective in cross-sectoral approaches', presenta-tion to Climate Changes Spatial Planning Conference, 12–13 September, The Hague, www.klimaatvoorruimte.nl/pro3/general/start.asp?i=0&j=0&k=0&p=0, accessed 27 June 2008

Agyeman, J., Bullard, R. and Evans, B. (eds) (2001) *Just Sustainabilities: Development in an Unequal World*, Earthscan, London

Albrechts, L. (2004) 'Strategic (spatial) planning challenged or the challenge of strategic (spatial) planning, Comment 5 in Strategic spatial planning and the longer range', *Planning Theory & Practice*, vol 5, no 1, pp63–64

Ball, J. (2001) 'Environmental future state visioning: Towards a visual and integrative approach to information management for environmental planning', *Local Environment*, vol 6, no 3, pp351–366

Berry, P. M., Rounsevell, M. D. A., Harrison, P. A. and Audsley, E. (2006) 'Assessing the vulnerability of agricultural land use and species to climate change and the role of policy in facilitating adaptation', *Environmental Science & Policy*, vol 9, no 2, pp189–204

Betsill, M. and Bulkeley, H. (2007) 'Looking back and thinking ahead: A decade of cities and climate change research', *Local Environment*, vol 12, no 5, pp447–456

Bulkeley, H. and Betsill, M. (2003) *Cities and Climate Change: Urban Sustainability and Global Environmental Governance*, Routledge, London

Campbell, H. (2004) 'Time, permanence and planning: An exploration of cultural attitudes', *Planning Theory & Practice*, vol 4, no 4, pp461–483

CEC (2007) Green Paper 'Adapting to climate change in Europe: Options for EU action', COM (2007) 354 Final, Office of Official Publications of the European Communities, Luxembourg

Collard, J. (2003) 'Tomorrow's people', *The Times Magazine*, 22 December, pp22–27

Cramer, J. (2007) 'Adaptation to climate change in spatial planning', speech to Climate Changes Spatial Planning Conference, 12–13 September, The Hague, www.klimaatvoorruimte.nl/pro3/general/start.asp?i=0&j=0&k=0&p=0, accessed 27 June 2008

Davies, A. (2005) 'Local action for climate change: Trans-national networks and the Irish experience', *Local Environment*, vol 10, no 1, pp21–40

Davoudi, S., Layard, A. and Batty, S. (eds) (2001) *Planning for a Sustainable Future*, Spon, London

Department for Trade & Industry (DTI) (2002) 'Foresight Futures 2020: Revised scenarios and guidance', DTI, London

Dobson, A. (ed) (1999) *Fairness and Futurity: Essays in Environmental Sustainability and Social Justice*, Oxford University Press, Oxford

Eames, M. and Skea, J. (2002) 'The development and use of the UK environmental futures scenarios: Perspectives from cultural theory', *Greener Management International*, vol 37, pp53–70

EEA (European Environment Agency) (2005) 'Vulnerability and adaptation to climate change in Europe', EEA Technical Report No 7/2005, EEA, Copenhagen

EEA (2006) 'PRELUDE (Prospective environmental analysis of land use development in Europe) scenarios', www.eea.europa.eu/multimedia/interactive/prelude-scenarios/, accessed 27 June 2008

EEA (2007) 'Climate Change: The cost of inaction and the cost of adaptation', EEA Technical Report No 13/2007, Office for Official Publications of the European Communities, Luxembourg

European Spatial Planning Observation Network (ESPON) Monitoring Committee (2006) 'Spatial scenarios and orientations in relation to the ESDP and Cohesion Policy', ESPON Project 3.2., Final Report, www.espon.eu/mmp/online/website/content/projects/260/716/index_EN.html, accessed 27 June 2008

Friedmann, J. (2004) 'Hong Kong, Vancouver and beyond: Strategic spatial planning and the longer range', *Planning Theory & Practice*, vol 5, no 1, pp49–67

Gemeente Rotterdam, Waterschap Hollandse Delta, Hoogheemraadschap van Schieland en de Krimpenerwaard and Hoogheemraadschap van Delfland (2007) 'Waterplan Rotterdam: Werken aan water voor een aantrekklijke stad', Gemeente Rotterdam DsV, Rotterdam

Graves, H. and Phillipson, M. (2000) *Potential Implications of Climate Change in the Built Environment*, Foundation for the Built Environment, London

Harris, N. and Hooper, A. (2004) 'Rediscovering the "spatial" in public policy and planning: An examination of the spatial content of sectoral policy documents', *Planning Theory & Practice*, vol 5, no 2, pp147–169

Hedger, M. M., Connell, R. and Bramwell, P. (2006) 'Bridging the gap: Empowering decision-making for adaptation through the UK Climate Impacts Programme', *Climate Policy*, vol 6, no 2, pp201–215

Hulme, M. and Dessai, S. (2008) 'Negotiating future climates for public policy: A critical assessment of the development of climate scenarios for the UK', *Environmental Science & Policy*, vol 11, no 1, pp54–70

Hulme, M., Turnpenny, J. R., and Jenkins, G. J. (2002) 'Climate change scenarios for the United Kingdom: The UKCIP02 Scientific Report', Tyndall Centre, School of Environmental Sciences, University of East Anglia, Norwich

IPCC (2007) 'Fourth Assessment Report: Climate Change 2007: Synthesis Report, Summary for Policy-makers', approved at IPCC Plenary XXVII Valencia, www.ipcc.ch/pdf/assessment-report/ar4/syr/ar4_syr_spm.pdf, accessed 27 June 2008

Jacobs, M. (1999) 'Sustainable development as a contested concept', in A. Dobson (ed) *Fairness and Futurity: Essays in Environmental Sustainability and Social Justice*, Oxford University Press, Oxford

Janssen, L., Okker, R. and Schuur, J. (2006) 'Welfare, prosperity and quality of the living environment', MNP, CPB, RPB, The Hague

Jenkins, G., Perry, M. and Prior, J. (2007) 'The climate of the United Kingdom and recent trends, UKCIP 08', Met Office Hadley Centre, Exeter

Kabat, P., van Vierssen, W., Veraart, J., Vellinga, P. and Aerts, J. (2005) Climate-proofing the Netherlands, *Nature*, vol 438, pp283–284

Kenny, J. and Meadowcroft, J. (eds) (1999) *Planning Sustainability*, Routledge, London

Lindseth, G. (2005) 'The Cities for Climate Protection Campaign (CCCP) and the framing of local climate policy', *Local Environment*, vol 9, no 4, pp325–336

London Climate Change Partnership (LCCP) (2002) 'London's Warming: The impacts of climate change on London', Technical Report, LCCP, London

LCCP (2006) 'Adapting to climate change: Lessons for London', GLA, London

Ludlow, D. (2006) 'Urban sprawl in Europe: The ignored challenge', EEA Technical Report 10/2006, Office for Official Publications of the European Communities, Luxembourg

Marshall, T. (1997) 'Futures, foresights and forward looks: Reflections on the use of prospective thinking for transport and planning strategies', *Town Planning Review*, vol 68, no 1, pp31–53

Mayor of London (2004) 'The London Plan: Spatial Development Strategy for Greater London', GLA, London

Mayor of London (2006a) 'London's urban heat island: A summary for decision-makers', GLA, London

Mayor of London (2006b) 'Scenario testing for the further alterations to the London Plan', GLA, London

Nadin, V. (2002) 'Visions and visioning in European spatial planning', in A. Faludi (ed) *European Spatial Planning*, Lincoln Institute of Land Policy, Cambridge, MA, pp121–137

Nickson, A. (2007) 'Preparing London for inevitable climate change', presentation to Hot Places, Cool Spaces symposium, 25 October, Amsterdam

O'Riordan, T. and Voisey, H. (eds) (1997) *Sustainable Development in Western Europe: Coming to Terms with Agenda 21*, Frank Cass, London

Owens, S. and Cowell, R. (2002) *Land and Limits: Interpreting Sustainability in the Planning Process*, Routledge, London

Paavola, J. and Adger, N. (2006) 'Fair adaptation to climate change', *Ecological Economics*, vol 56, no 4, pp594–609

Page, E. A. (2006) *Climate Change, Justice and Future Generations*, Edward Elgar, Cheltenham

Raskin, P., Banuri, T., Gallopín, G., Gutman, P., Hammond, A., Kates, R. and Swat, R. (2002) *Great Transitions: The Promise and Lure of Times Ahead*, Stockholm Environment Institute, Boston, MA

Rydin, Y. (2003) *In Pursuit of Sustainable Development: Rethinking the Planning System*, RICS Foundation, London

Shaw, R., Colley, M. and Connell, R. (2007) *Climate Change Adaptation by Design: A Guide for Sustainable Communities*, TCPA, London

Shipley, R. (2002) 'Visioning in planning: Is it based on sound theory?' *Environment and Planning A*, vol 34, no 1, pp7–22

Skaburskis, A. and Teitz, M. (2003) 'Forecasts and outcomes', *Planning Theory & Practice*, vol 4, no 4, pp429–442

Stern, N. (2007) *The Economics of Climate Change: The Stern Review*, Cambridge University Press, Cambridge

Swart, R. and Raes, F. (2007) 'Making integration of adaptation and mitigation work: Mainstreaming into sustainable development policies?', *Climate Policy*, vol 7, pp288–303

Tillie, N. (2007) 'Climate change, city change', paper presented to Metrex (Network of European Metropolitan Regions and Areas) Conference on Climate Change, Hamburg, 28 November–1 December 2007, www.eurometrex.org/euco2/DOCS/Hamburg/14Rotterdam.pdf, accessed 27 June 2008

Tompkins, E. L. and Amundsen, H. (2008) 'Perspectives on the effectiveness of the UNFCC in advancing national action on climate change', *Environmental Science & Policy*, vol 11, no 1, pp1–13

UKCIP (2001) 'Socio-economic scenarios for Climate Change Impact Assessment: A guide to their use in the UK Climate Impacts Programme', UKCIP, Oxford

UK Foresight (2004) 'Future Flooding: Executive Summary', Office of Science and Technology, London

VROM (Ministry of Housing, Spatial Planning and Environment) (2006) 'Nota Ruimte' (National Spatial Policy Document), VROM, The Hague

West, C. and Gawith, M. (eds) (2005) 'Measuring progress: Preparing for climate change through UKCIP', UKCIP, Oxford

Wilson, E. (2006a) 'Adapting to climate change at the local level: The spatial planning response', *Local Environment*, vol 11, no 6, pp609–625

Wilson, E. (2006b) 'Developing UK spatial planning policy to respond to climate change', *Journal of Environmental Policy & Planning*, vol 8, no 1, pp9–25

Wilson, E. (2008) 'Multiple scales for environmental intervention: Spatial planning and the environment under New Labour', *Planning Practice and Research*, vol 24, no 1, pp119–138

World Commission on Environment and Development (1987) *Our Common Future* (the Brundtland Report), Oxford University Press, Oxford

18

Integrated Assessment to Support Regional and Local Decision Making

Jim Hall

The challenges of decision making for mitigation and adaptation

Responding to climate change by mitigating greenhouse gas (GHG) emissions and adapting to the impacts of climate change is placing new and complex demands upon decision makers. The discipline of planning has always been one of dealing with multiple objectives and constraints over extended timescales. Climate change adds greatly to this challenge for a number of reasons. First, is the urgent need to mitigate GHG emissions (Stern, 2007). In the context of the built environment, this implies rapid introduction of energy efficient technologies and, more significantly, transformation of the existing building stock (Intergovernmental Panel on Climate Change (IPCC), 2007; White, 2000). It implies phasing out of fossil-fuel based transport systems, upon which we are now more or less dependent.

The demands of adaptation to climate change vary from place to place. In the UK, the south-east of the country stands out, for a variety of geographical reasons, as being particularly vulnerable to water scarcity (Environment Agency, 2007), flooding (Evans et al, 2004) and excessive urban heat (London Climate Change Partnership, 2002). Responding to these challenges requires a portfolio of measures that may involve reversal of entrenched patterns of demand and development. The need to adapt to climate change may conflict with the demands of mitigation (Klein et al, 2003). Thus, for example, intolerable urban temperatures increase energy demand for air conditioning, which in turn also increases heat emissions into urban areas, exacerbating the problem. Action to tackle climate change needs to be set in the broader context of sustainability, including issues of resource use, human well-being and biodiversity (Najam et al, 2003).

These issues have been addressed from a number of different perspectives in different chapters of this book. Here our concern is primarily with problems of decision making and, in particular, how a new generation of tools for quantified analysis of change can be used to support decision makers in the difficult process of responding to climate change at the scale of cities and regions (Rotmans et al, 2000; Rotmans and van Asselt, 2000). We argue that climate change and sustainable development more broadly are presenting unprecedented challenges for decision makers. These include:

- complexity of interactions;
- constraints of existing patterns of development;
- dynamics of change;
- an evolving policy context;
- interactions at spatial scales;
- uncertainty;
- portfolios of responses.

Complexity of interactions   Different aspects of responding to climate change (energy, transport, the built environment) interact with one another in complex ways. Planners need to know which of these interactions are actually significant in the context of a given decision or set of decisions.

Constraints of existing patterns of development   Responding to climate change involves designing a *transition* (Rotmans et al, 2001) from the existing state of the built environment to one which is no longer dependent on fossil fuels and is well adapted to climate change. The main challenge lies in the legacy of existing development and in the capacity to transform it on the necessary timescales. This requires creativity, but analysis is also essential in order to rigorously assess priorities and the potential efficiency of responses.

Dynamics of change   Climate change is a stimulus to plan over extended timescales, so, for example, the prospect of sea level rise on a millennial scale (Lowe et al, 2006) greatly extends the usual timescale of planning decisions. On the other hand, mitigation demands urgent action. Some of the impacts of climate change exacerbate problems that even in the absence of climate change were in need of timely attention (e.g. water resources in the southeast of England and flooding in urban areas). Climate change therefore requires action on a range of timescales.

An evolving policy context   The political and policy framework surrounding climate change is also rapidly evolving and will continue to do so. This in part is determined by the rapid evolution of climate science. In the face of this evolving decision context, decision making becomes regarded as a process of adaptive management (Willows and Connell, 2003).

Interactions at spatial scales   The climate problem is also striking in the range of spatial scales under consideration. Mitigation of GHG emissions is intended to have an aggregate effect at a global scale, but derives from a host of actions at smaller scales, for example, related to national energy and transport policies. Increasingly cities are becoming major actors within the mitigation arena (Betsill and Bulkeley, 2006). Adaptation to climate change is rooted in understanding of regional climate and vulnerability of specific systems and populations.

Uncertainty   Different facets of the climate challenge bring a variety of different uncertainties for decision makers. The targets for mitigation are constantly evolving with improving global climate science and the faltering UN negotiations. Understanding of potential climate impacts is also uncertain, because of the uncertainties in climate science and in the prospects for effective mitigation. These climate-related uncertainties need to be considered alongside incomplete knowledge of potential for other long term changes within society, for example, associated with technology or demography. Recognition of the importance of all of these types of uncertainty has stimulated evolution in the treatment of uncertainty in policy analysis (Hall and Solomatine, 2008).

Portfolios of responses   There is no single response to the challenge of climate change. Mitigation and adaptation will require a host of measures, implemented over different timescales. These portfolios need to be designed in coherent ways so that different elements compensate for each other's deficiencies and make the most of the opportunities that may arise. The composition of response portfolios and the plan for future implementation needs to evolve through time as new information becomes available and circumstances change.

How are decision makers to deal with these various unavoidable aspects of the climate problem? It is unrealistic to suppose that the issues mentioned above can dependably be incorporated in intuitive or qualitative approaches to decision making. The complexity is simply too great for any individual to cope with. Responding to these challenges therefore requires approaches that can provide quantified understanding of the complexities and uncertainties associated with climate change, and harness the capacity of teams, by providing a common evidence-based platform for analysis.

Fortunately, a new generation of quantified integrated assessment methods is emerging that is highly innovative in a number of respects. A serious attempt is being made to represent in a fairly complete way the relevant processes of change within the natural, human and built environments. So, for example, in integrated coastal assessments (Dawson et al, in press) processes of shoreline erosion forced by climate change are analysed in a coupled system of models that also incorporates the behaviour and deterioration of coastal defence infrastructure and changing patterns of human development on the coast (see also Kizos et al, Chapter 8). Assessments are increasingly integrated; that is, the same assessment system is used to study multiple climate impacts and GHG emissions. This is necessary in order to understand conflicts and potential synergies.

These types of quantified assessment are now remarkable in the spatial scales that they span. On the one hand it is necessary to model systems at a city or regional scale in order to understand aggregated effects and to leverage synergies, for example, through changing patterns of transport use. On the other hand, many impacts or adaptations exist at the scale of individual buildings, for example, in relation to vulnerability to sewer flooding. Geographical Information System (GIS) based analysis can span these scales through the incorporation of very high resolution datasets. As well as exploiting the potential of digital datasets, computer-based systems are more or less essential when the numbers of potential decision options and sequences of implementation are large. High performance and distributed computing mean that it is feasible to analyse and optimize very large numbers of options under a wide range of future conditions.

Whilst this type of integrated assessment is more or less essential for decision makers to implement responses to climate change, its adoption does present a new set of challenges in terms of the institutional capacity to use these tools and enable participation of stakeholders in the decision making process. The capacity of computerized systems to deal with large numbers of options and futures shifts the challenge of assessment from one of analysis to communication. Analysis can provide the impression of dependability which is not warranted by the underlying methods and data. It is important therefore to provide transparency in the communication of uncertainty. Advanced visualization techniques are essential tools for communication and web-enabled tools can open analysis out to enable public participation.

Having set out some of the challenges of analysis of effective planning decisions in the context of climate change, this chapter describes some of the features of integrated assessment systems and how they can support decision making. Specifically it examines the overall framework for integrated assessment at the scales of interest to planners and provides some examples in the context of climate impacts and adaptation.

Combining adaptation and mitigation

Adaptation and mitigation have tended to be dealt with separately in government and intergovernmental climate policy. However, at a regional and urban scale, there is a clear motive for dealing with them in an integrated way as part of an overall policy of sustainable development. Notwithstanding this clear motivation, there are barriers and limits to the integration of adaptation and mitigation. The timescales of mitigation and adaptation can differ. The benefits of mitigation activities carried out today will only be evidenced in several decades given the long residence time of GHGs in the atmosphere, whereas many effects of adaptation measures should be apparent immediately or in the near future. Adaptation, on the other hand, helps to respond immediately to climate variability as well as climate change.

Mitigation has global benefits, whilst adaptation typically takes place on the scale of an impacted system, either regionally or most likely locally. For example, mitigation investments in renewable energy sources will lower atmospheric carbon dioxide (CO_2) concentrations, with the gain being reduced global climate change. Whereas, adaptation investments, for example, in improved sea defences will only benefit the settlements and activities directly protected by such defences.

Adaptation and mitigation also differ in the extent to which their costs and benefits can be determined, compared and aggregated. Mitigation options serve to reduce GHG emissions for global benefit so where in the world mitigation takes place is irrelevant. Expressed as CO_2 equivalents, reductions in emissions can be compared with other mitigation options and if implementation costs are known, the cost-effectiveness of these options can be determined and compared (Klein et al, 2003). However, for a full cost benefit assessment the cost of impacts needs to be understood.

Benefits of adaptation are difficult to express as a single measure. A value can be obtained by subtracting the costs of implementing the adaptation options from the benefits of adaptation (i.e. the difference between the potential impact of climate change on a system assuming no adaptation and the residual impacts assuming adaptation). However, assessing and comparing adaptation benefits is fraught with the difficulties related to the uncertainty about and differences between the impacts avoided (Klein et al, 2003). Mitigation reduces the amount of adaptation required and its cost, while adaptation does not reduce mitigation costs.

Finally, the actors involved in adaptation and mitigation differ. Mitigation primarily involves the energy and transportation sectors in industrialized countries and to an increasing extent the energy and forestry sectors in developing countries. Compared to adaptation, the number of sectoral actors involved in mitigation is limited. They are typically well organized, closely linked to national planning and policy making and used to taking medium to long-term investment decisions. This is not to say that energy efficiency in households and industry is not important on an individual level. In contrast, actors involved in adaptation represent a large variety of sectoral interests including agriculture, tourism and recreation, human health, water supply, coastal management, urban planning and nature conservation. Barriers to mitigation and adaptation are also likely to be different.

Notwithstanding these differences between adaptation and mitigation, there is a strong motivation to deal with them in an integrated way in order to reduce conflicts and maximize opportunities for synergies. The form of the built environment and associated transport systems is a foremost instance of where adaptation (primarily to heat, but also to aspects of urban flooding and water scarcity) and mitigation (of emissions from buildings and transport) need to be dealt with in an integrated way. The framework illustrated in Figure 18.1 provides the basis for this type of integrated analysis.

Frameworks for analysis of processes of change

The aim of quantified integrated assessment is to provide a platform for analysis of alternative policy and design options. There are two sets of considerations in this type of analysis: (i) the set of policy options (or combinations of options) under consideration and (ii) the set of possible future scenarios in which the options might be expected to perform. Important dimensions of possible future scenarios are climate change scenarios and socio-economic scenarios.

Distinguishing between policy options and future scenarios is fraught with difficulties. In simple terms we think of policy options being part of a decision space over which the decision maker we are seeking to inform has some authority. On the other hand, possible future scenarios are situations that might materialize independently from the control of the decision maker. For planners and decision makers at a city and regional scale, climate change is one instance of possible scenarios that are outside their control. Similarly, European or national policy can be considered to be outside the control of planning decision makers, even if they might seek to influence these decision making processes. Thus targets for reduction of GHG emissions that are set at a national scale are effectively constraints on decision makers rather than options they have at their disposal.

The status of emissions targets is of interest because it has some bearing on the approach to appraisal of options for mitigation and adaptation. In impacts and adaptation studies, a set of alternative adaptation options, including a baseline option, is analysed in a set of possible futures, in order to identify the option(s) that are most desirable according to some criteria, which may include conventional cost–benefit criteria (with appropriate allowance for uncertainty). The analysis looks forward to see how options will perform under a range of plausible futures. In contrast, the more customary approach to emissions reduction studies (Tyndall Centre, 2005) is to begin with a target and then analyse the policies that may achieve that target, under a range of conditions, for example, for economic growth or technological change. This so-called 'backcasting' approach is prevalent in mitigation studies (see Wheeler, Chapter 10) because, for the time being, there is no widely accepted social cost of CO_2 emissions (or other GHGs) and in the absence of the 'benefit' side of the equation in cost–benefit analysis, appraisal of options for mitigation cannot be fitted within an optimising framework. The increasing penetration of carbon trading is stimulating an optimization approach to decision making in some sectors, but it is hard to see how this could be extended to the diverse functions of urban areas that planners are

concerned with. For the time being therefore, we can expect rather different frameworks of appraisal for mitigation and adaptation. This adds to the potential confusion in an already complex decision arena.

One approach to managing complexity and ensuring coherence in mitigation and adaptation studies is to restrict the range of futures under consideration to a relatively small number of narrative storylines. The Special Report on Emissions Scenarios (SRES) scenarios are an instance of this (IPCC, 2000), as are the Foresight Futures (Office of Science and Technology, 2002; UKCIP, 2001) (Figure 18.1) which have been quite widely used and adapted in the UK. Each of the four scenarios describes a contrasting plausible (but not equally likely) way in which UK society might evolve in the coming decades. Considerable additional detail has been added to these scenarios to enable their use in practice (UKCIP, 2001). Thus qualitative scenarios are used to provide an underlying rationale for quantified analysis of a set of possible futures. There are of course several (perhaps many) possible quantified futures that correspond to a given narrative scenario. The inevitable linguist vagueness of a narrative scenario and its interpretation in practice can allow a range of possible futures to be accommodated within a given narrative (Lempert et al, 2003).

Backcasting exercises can be illustrative for the purpose of analysing the achievability of emissions targets. However, a forward-looking approach for analysing possible transitions from the present into a set of possible futures is generally more useful in analysis of planning and design decisions. Setting the possible futures against a background of narrative scenarios helps make those futures intelligible to stakeholders and ensure consistency between the different dimensions of the analysis.

The overall framework developed by the Tyndall Centre for Climate Change Research for analysis of long term change in urban areas is illustrated in Figure 18.2. In keeping with the approach described above, the analysis framework begins with long term climate and socio-economic scenarios. The climate scenarios encompass the range of possible global emissions

Source: Office of Science and Technology (2002, p4).

Figure 18.1 The Foresight Futures scenario grid

scenarios and a reasonable range of climate sensitivities, which together provide a wide range of possible climate futures to which the city in question might be subject. The socio-economic scenarios are derived from a combination of global, national and regional projections. We have analysed population (at European, national and regional scales), gross domestic product (GDP) and regional gross value added (GVA) projections in order to specify a range of possible futures for population and the economy. These naturally have a wide range of associated uncertainty. Trends in fertility and mortality are reasonably well known, but the effects of migration on population are more difficult to predict. In the economy, we cannot rule out fundamental shifts in the functioning of the economy, for example, associated with technological transitions (Köhler, 2003).

The timescale of analysis should be appropriate for the decisions it is intended to inform. Major planning or infrastructure investment decisions, for example, associated with flood defence or water resources, can have a legacy of a century and beyond. Because of the potential for major technological and behavioural changes, it is less plausible to analyse regional and city-scale emissions over such a long timescale. Within the Tyndall Centre, city-scale emissions from the energy and transport sectors are being analysed to 2050.

The dotted line in Figure 18.2 illustrates that socio-economic and climate scenarios are not unrelated. However, even at a national scale the influence of mitigation policy on global climate is small. Therefore, for the purpose of regional and urban analysis, these two sets of scenarios can be decoupled.

The aim of socio-economic scenarios is to provide the variables necessary for analysis of emissions and climate impacts. For emissions analysis this requires particular attention to the energy and transport sectors of the economy. For climate impacts analysis, we require spatial projections of land use and population change, which are key determinants of vulnerability. To provide this resolution requires more detailed analysis of the economy, at a sectoral level, and spatially explicit analysis of population and land use. These high resolution simulations of economy and land use generate scenarios that are conditional upon broader economic and demographic assumptions, so they should not be mistaken for projections. Yet they are necessary to explore in a quantified way the implications of broader scale scenarios of change.

High resolution scenarios of land use, population, economy and climate provide the basis for analysis of policy options, including land use policy, transport investment, emissions reduction policy and adaptation measure. Land use, building type and occupancy are key determinants of rates of change in urban areas, so are a unifying strand within the scenario analysis.

The narrative background to the scenarios provides the basis for exploration of dimensions

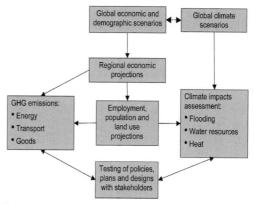

Figure 18.2 Overview of the integrated assessment methodology for GHG emissions and climate impacts analysis

of socio-economic futures that are not explicit in the economy–demography–land-use approach promoted here. Examples include the implications of lifestyle or technology changes that may have a significant bearing upon the long term policies under consideration.

Impacts assessment and adaptation studies

The framework established in Figure 18.2 provides the basis for analysis of climate impacts and options for adaptation, in that it provides climate variables and quantified spatial scenarios of changing vulnerability. However, to provide quantified projections of impacts requires climate variables at an appropriate resolution as well as physical understanding of the systems in question and how they may be adapted in future.

Analysis of climate impacts in the built environment and infrastructure systems begins with understanding of how those systems were designed for weather variability in a climate without a long term trend of change. The sophistication of treatment of weather variability differs significantly from sector to sector. For example, the water resources sector has a long tradition of dealing in probabilistic terms with rainfall and river flows, including sequences and spells of dry weather on a seasonal timescale. Analysis of sewer capacity has been based upon a design condition (typically the rainfall intensity that is exceeded with an annual probability of 0.033) applied uniformly over the study area. A similar approach to identifying a 'design condition' is adopted in designing buildings for wind loading and thermal comfort. In the latter case a typical 'design year' of temperature conditions is used as the basis for testing designs. The ease with which these approaches can be adapted to deal with future changes varies. Furthermore, the type of information available from Regional Climate Model (RCM) outputs may not be directly suitable for impacts studies. So, for example, water resources studies require long time-series of rainfall, which account for orographic differences across the catchment that

are not resolved in RCM topography. Urban drainage design requires rainfall intensities at sub-hourly durations. Stochastic weather generators (Burton et al, 2008) are therefore increasingly used to provide time series outputs of climate variables at the temporal and spatial resolutions necessary for impacts studies.

A further consideration is the resolution of heat in urban areas. The land surface schemes in RCMs are now incorporating the effect of urban areas and anthropogenic heat emissions but climate models are still some distance from resolving the interactions between the atmosphere and the built environment at the scale of streets and buildings. Yet reconfiguration of urban areas to limit the severity and impacts of urban heat is one of the foremost challenges to urban planners. Approaches to resolving this problem may be based upon nesting urban atmospheric models within an RCM or upon more empirical approaches to understanding the relationship between urban temperatures and building type.

Analysis of climate impacts and adaptation requires representation of infrastructure systems at a broad scale. This is because climate impacts are typically reflected in the aggregate performance of systems as well as at the level of individual installations. Moreover, critical performance is often associated with climatic extremes. A probabilistic treatment of system reliability is therefore required. An example of this type of broad scale systems analysis is provided by the Tyndall Centre's Regional Coastal Simulator, which is examining potential for coastal erosion and flooding over 50 kilometres of the north Norfolk coast in the UK. This comprises a mixture of soft cliffs and beaches adjoining low-lying coastal lowlands prone to flooding. Understanding the full implications of climate change on coastal areas requires integrated assessment of large lengths of interacting coastline (often termed sub-cells) over many decades. The study considered possible changes in coastal drivers such as sea-level rise, wave height and direction, morphological change analysis and its human implications in terms of erosion and flood risk. Forty-two scenarios combining sea-level rise, changes in wave climate and different coastal management options have been explored. This included links between erosion risk on the cliffed coast and flood risk on the neighbouring low-lying coast (Dawson et al, in press).

Figure 18.3 shows the coastal evolution under different climate change scenarios given a natural coast with no engineering control. While changes are greater with a larger rate of sea-level rise, the response is not a simple uniform retreat, and parts of the coast accrete more with the higher sea-level rise. If the effect of existing cliff defences are considered, the response becomes more complex, and the beaches at Happisburgh and further south may erode, significantly raising the flood risk under higher sea-level rise scenarios. Hence, the analysis shows an important trade-off between erosion and flood defence which planners must address as they consider how to respond to climate change.

A broader scale example of impacts analysis is provided by the UK Foresight Future Flooding project (Evans et al, 2004; Hall et al, 2003b). The

Foresight project looked 30 to 100 years into the future to examine the effects of climate and socio-economic change on flood risk. The Foresight Futures socio-economic scenario framework illustrated in Figure 18.1 was used as the basis for a quantified analysis of flood risk in England and Wales. A simplified national-scale flood risk analysis method (Hall et al, 2003a) was adapted to analyse the effects of changing precipitation and sea level rise. National-scale risk assessment is by no means straightforward, because of the need to assemble national datasets and then carry out and verify very large numbers of calculations. Increasingly, however, national-scale datasets are becoming available. Aerial and satellite remote sensing technologies are providing new topographic and land use data. In 2002 the Environment Agency, the organization responsible for the operation of flood defences in England and Wales, introduced a National Flood and Coastal Defence Database (NFCDD), which for the first time provides in a digital database an inventory of flood defence structures and their

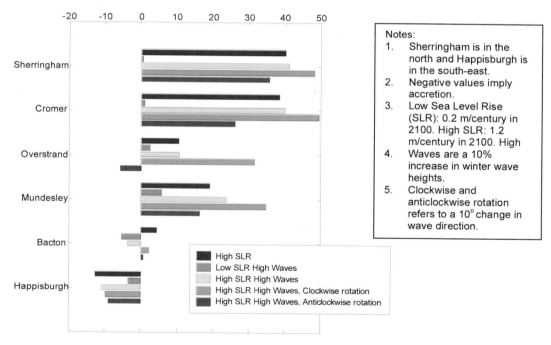

Source: Hall et al (2006, p26).

Figure 18.3 Erosion on the cliffed coast of north Norfolk over 100 years under varied climate change scenarios and assuming no engineering intervention

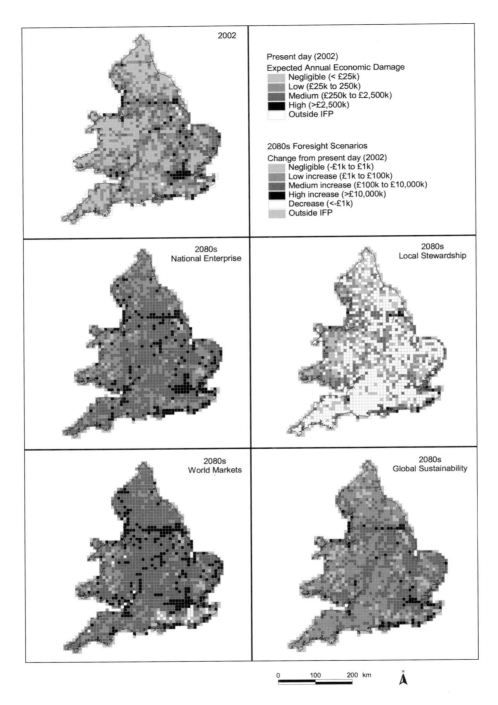

Figure 18.4 Typical outputs from the Foresight scenario analysis of changes in flood risk in England and Wales

overall condition. Together, these new datasets now enable flood risk assessments to be carried out that incorporate probabilistic analysis of flood defence structures and systems. Once the necessary datasets are held in a GIS they can then be manipulated in order to explore the impact of future flood management policy and scenarios of climate change.

A sample of the results is presented on a 10km grid in Figure 18.4. In all scenarios other than the low growth, environmentally/socially conscious Local Stewardship scenario, annual economic flood damage is expected to increase considerably over the next century assuming the current flood defence policies are continued in future. A roughly 20-fold increase by the 2080s is predicted in the World Markets scenario, which is attributable to a combination of much increased economic vulnerability (higher flood-plain occupancy, increased value of household/industrial contents, increasing infra-structure vulnerability) together with increasing flood frequency. Increasing risk is predicted to be concentrated in broadly the same areas as where it is currently highest. Coastal flooding makes an increasing contribution to total flood risk, increasing from 26 per cent in 2002 to 46 per cent in the 2080s. The largest increases are observed where both housing pressure is greatest and the standard of defence is most susceptible to climate change. This critical combination is most clearly seen in the World Markets and National Enterprise scenarios around the coastal strip of south-east England, East Anglia and south and north Wales. The area between Lancashire and the Humber, and the Thames Valley also sees a significant increase in exposure to economic loss. The drier climate in the south-east is reflected in reduced economic damage.

The Foresight flooding scenarios were used to understand the risks associated with future changes and where they are located. This formed the basis of a comprehensive qualitative and quantitative analysis of portfolios of options for responding to future changes.

Engaging with stakeholders

The framework for quantified scenario analysis that we have described provides the basis for building collective understanding of long term future changes. These are complex issues that are conceptualized in widely different ways. A common framework helps to understand long term impacts and interactions. In the contested situations that characterize many long term planning decisions, a common modelling frame-work can help to develop shared understanding of the problem and rapidly focus attention on the substantive areas of differences in objectives.

However, to achieve this aim of developing shared understanding requires intelligible and accessible analyses. There are compromises to be struck in definition of the space of possible options and scenarios, which on the one hand has to be sufficiently broad to allow analysis of all salient possibilities, but, on the other, should not be overwhelming. The complexity of analysis can be progressively introduced to participants, so that at the 'entry level' a rather constrained set of options is exposed, which can then be expanded in response to increasingly sophisti-cated queries.

Graphical user interfaces and advanced visualization provide essential tools for accelerat-ing stakeholder comprehension. The Tyndall Centre's Regional Coastal Simulator has been enabled with a combination of Virtual Reality and GIS tools (Figure 18.5). The proliferation of service-based software architectures opens opportunities for versions of this type of long term scenario analysis being made available on the web. This carries risks, as users of such web-based tools might test them in unexpected and inappropriate ways, developing incorrect under-standing of the problem. But, if well designed, freely available web-based tools or games can both educate and empower large groups of stakeholders.

Source: Brown et al (2007, p118).

Figure 18.5 Visualization of coastal simulations

Conclusions

The need for integrated assessment of climate change at a regional and city scale is only now beginning to be recognized and there are no complete demonstrations of the frameworks and concepts we have introduced in this chapter. This inhibits understanding of their potential, but does not undermine the need for integrated analysis of long term changes and appraisal of policy and design options in the context of those changes. Fortunately, the main components of this type of analysis do exist, and in this chapter we have provided examples of climate downscaling and broad scale high resolution analysis of long term change in cities and on coasts. This is being coupled with analysis of mitigation, in order to understand the synergies and conflicts between adaptation and mitigation.

In order to assess the sustainability of alternative policies and designs we need to be able to provide evidence of potential performance in the context of a range of possible futures. Because of the inherent uncertainties it is unrealistic to suppose that we can identify optimal solutions, but we can seek options that perform reasonably well under a range of plausible futures. We expect conditions and objectives to change in future, so we should be seeking options that are adaptable to those future changes.

We expect that the future will see more widespread use of quantified integrated assessment of climate change at a regional and city scale. This will involve sophisticated analysis of porfolios of options by technical decision makers but will also extend to engage non-technical stakeholders, including the general public, through the use of advanced visualization techniques and web-enabled tools.

References

Betsill, M. and Bulkeley, H. (2006) 'Cities and the multilevel governance of global climate change,' *Global Governance*, vol 12, no 2, pp141–159

Brown, I., Jude, S. R., Koukoulas, S., Nicholls, R., Dickson, M. E. and Walkden, M. J. A. (2007) 'Using virtual reality to simulate future coastal erosion: A participative decision tool?', in A. Lovett and A. K. London (eds) *Environmental Decision Making and GIS*, Taylor and Francis, London

Burton, A., Kilsby, C. G., Fowler, H. J., Cowpertwait, P. S. P. and O'Connell, P. E. (2008) 'RainSim: A spatial-temporal stochastic rainfall modelling system', *Environmental Modelling and Software*, vol 23, pp1356–1369

Dawson, R. J., Dickson, M. E., Nicholls, R. J., Hall, J. W., Walkden, M. J. A., Stansby, P., Mokrech, M., Richards, J., Zhou, J., Milligan, J., Jordan, A., Pearson, S., Rees, J., Bates, P., Koukoulas, S. and Watkinson, A. (in press) 'Integrated analysis of risks of coastal flooding and cliff erosion under scenarios of long term change', *Climatic Change*

Environment Agency (2007) 'Water for the Future: Managing water resources in the South East of England', discussion document, Environment Agency, Bristol

Evans, E., Ashley, R., Hall, J., Penning-Rowsell, E., Saul, A., Sayers, P., Thorne, C. and Watkinson, A. (2004) *Foresight, Future Flooding, Scientific Summary: Volume I, Future risks and their drivers*, Office of Science and Technology, London

Hall, J. W. and Solomatine, D. (2008) 'A framework for uncertainty analysis in flood risk management decisions', *Journal of River Basin Management*, vol 6, no 2, pp85–98

Hall, J. W., Dawson, R. J., Manning, L., Walkden, M., Dickson, S. and Sayers, P. (2006) 'Managing changing risks to infrastructure systems', *Civil Engineering*, vol 159, no 2, pp21–27

Hall, J. W., Dawson, R. J., Sayers, P. B., Rosu, C., Chatterton, J. B., and Deakin, R. (2003a) 'A methodology for national-scale flood risk assessment', *Water & Maritime Engineering*, vol 156, no 3, pp235–247

Hall, J. W., Evans, E. P, Penning-Rowsell, E. C., Sayers, P. B, Thorne, C. R. and Saul, A. J. (2003b) 'Quantified scenarios analysis of drivers and impacts of changing flood risk in England and Wales: 2030–2100', *Global Environmental Change Part B: Environmental Hazards*, vol 5, issues 3–4, pp51–65

IPCC (2000) *Special Report on Emissions Scenario*, Intergovernmental Panel on Climate Change, Geneva

IPCC (2007) 'Climate Change 2007: Mitigation of Climate Change, Summary for Policymakers', contribution of Working Group III to the Fourth Assessment Report of the Intergovernmental Panel on Climate Change, IPCC, Geneva

Klein, R. J. T., Schipper, L. E., Dessai, S. (2003) 'Integrating mitigation and adaptation into climate and development policy: Three research questions', Working Paper 40, Tyndall Centre for Climate Change Research, Norwich

Köhler, J. (2003) 'Long run technical change in an energy–environment–economy (E3) model for an IA system: A model of kondratiev waves', *Integrated Assessment*, vol 4, no 2, pp126–133

Lempert, R. J., Popper, S. W. and Bankes, S. C. (2003) *Shaping the Next One Hundred Years: New Methods for Quantitative, Long-Term Policy Analysis*, RAND, Santa Monica, CA

London Climate Change Partnership (2002) 'London Climate Change Partnership: A climate change impacts in London evaluation study', Final Report

Lowe, J. A., Gregory, J. M., Ridley, J., Huybrechts, P., Nicholls, R. J. and Collins, M. (2006) 'The role of sea-level rise and the Greenland ice sheet in dangerous climate change: Implications for the stabilisation of climate', in H. J. Schellnhuber, W. Cramer, N. Nakicenovic, T. Wigley and G. Yohe (eds) *Avoiding Dangerous Climate Change*, Cambridge University Press, Cambridge, pp29–36

Najam, A., Rahman, A. A., Huq, S. and Sokona, Y. (2003) 'Integrating sustainable development into the Fourth Assessment Report of the Intergovernmental Panel on Climate Change', *Climate Policy*, vol 3, no S1, ppS9–S17

Office of Science and Technology (2002) *Foresight Futures 2020: Revised Scenarios and Guidance*, Department of Trade and Industry, London

Rotmans, J. and van Asselt, M. B. A. (2000) 'Towards an integrated approach for sustainable city planning', *Journal of Multi-Criteria Decision Analysis*, vol 9, nos 1–3, pp110–124

Rotmans, J., Kemp, R. and van Asselt, M. (2001) 'More evolution than revolution: Transition management in public policy', *Foresight*, vol 3, no 1, pp15–31

Rotmans, J., van Asselt, M. and Vellinga, P. (2000) 'An integrated planning tool for sustainable cities', *Environmental Impact Assessment Review*, vol 20, no 3, pp265–276

Stern, N. (2007) *The Economics of Climate Change: The Stern Review*, Cambridge University Press, Cambridge

Thorne, C. R., Evans, E. P. and Penning-Rowsell, E. C. (eds) (2007) *Future Flooding and Coastal Erosion Risks*, Thomas Telford Ltd, London

Tyndall Centre (2005) 'Decarbonising the UK: Energy for a climate conscious future', Tyndall Centre for Climate Change Research, Norwich

UK Climate Impacts Programme (2001) 'Socio-economic scenarios for climate change impact assessment: A guide to their use in the UK Climate Impacts Programme', UKCIP, Oxford

White, R. M. (2000) 'Climate systems engineering', *Environmental and Engineering Challenges: A Technical Symposium on Earth Systems Engineering*, National Academy of Engineering, National Academy Press Washington DC

Willows, R. I. and Connell, R. K. (2003) 'Climate adaptation: Risk, uncertainty and decision-making', UKCIP Technical Report, UKCIP, Oxford

19

Planning for Green Infrastructure:
Adapting to Climate Change

Susannah Gill, John Handley, Roland Ennos and Paul Nolan

Introduction

In its 2007 report on the state of the UK urban environment, the Royal Commission on Environmental Pollution (RCEP) made a distinction between 'cumulative' and 'systemic' urban issues (RCEP, 2007). Cumulative issues can arise in any human settlement but may be exacerbated in towns and cities; an example of such an issue is building energy use. Systemic issues, on the other hand, arise because of the unique social, economic and environmental characteristics of urban settlements.

Urban areas have distinctive biophysical features in comparison with surrounding rural areas (Bridgman et al, 1995). This is partly due to the altered surface cover (Whitford et al, 2001). For example, energy exchanges are modified to create an urban heat island where air temperatures may be several degrees warmer than in the countryside (Graves et al, 2001; Wilby, 2003). The magnitude of the urban heat island effect varies in time and space as a result of meteorological, locational and urban characteristics (Oke, 1987). The urban heat island effect has consequent impacts on human health, energy consumption

and biodiversity. Hydrological processes are also altered in urban areas such that there is an increase in the rate and volume of surface water runoff (Whitford et al, 2001; Mansell, 2003). This can cause problems such as localized flooding from overwhelmed drains. Urban green infrastructure can help to restore natural processes (Hough, 2004), for example, by moderating temperature extremes through evaporative cooling and shading, and by providing permeable surfaces where rainwater can be intercepted, stored, and infiltrated into the ground (Whitford et al, 2001).

The RCEP report concludes that a new approach to urban governance is needed which places 'the natural environment ... at the heart of urban design and management' (RCEP, 2007, p83). It argues that whilst cumulative issues may be more amenable to national policy intervention, systemic issues require significant local powers in terms of planning and design (RCEP, 2007). This echoes an earlier report on the state of cities in Britain which called for a comprehensive approach to planning, urban design and management with a view to realizing the potential amenity value of the public realm (Urban

Task Force, 1999). The Urban Task Force argued that cities and towns should be designed 'as networks that link together residential areas to public open spaces and natural green corridors' with benefits for both people and wildlife (1999, p58). It follows that landscape planning and management should be based around multi-functional green networks or 'green infrastructure' (Handley et al, 2007) which can help urban areas to respond more flexibly to a changing set of environmental challenges, including climate change (RCEP, 2007).

In a changing climate the functionality of urban green infrastructure becomes increasingly important. The UK Climate Impacts Programme 2002 (UKCIP02) climate change scenarios suggest average annual temperature increases of 1–5°C by the 2080s, with greater increases in summer than in winter. The seasonality of precipitation changes, with winters up to 30 per cent wetter and summers up to 50 per cent drier by the 2080s (Hulme et al, 2002). Precipitation intensity increases, especially in winter, and the number of very hot days increases, particularly in summer and autumn (Hulme et al, 2002). There is likely to be significant urban warming over and above that expected for rural areas (Wilby, 2003; Wilby and Perry, 2006). Climate change in urban areas will be felt by both people and the built infrastructure. For example, it is estimated that the European summer heat wave in 2003 claimed 35,000 lives (Larsen, 2003). A recent report by the Department of Health suggests that although an improved tolerance to heat will reduce the impact of hotter summers, the increased frequency and intensity of heatwaves are a major concern (Donaldson and Keatinge, 2008). In contrast, between 1971 and 2003 annual cold-related mortality in the UK fell by more than 33 per cent (Donaldson and Keatinge, 2008). Incidents of flooding can also result in both physical and psychological illnesses (e.g. Shackley et al, 2001; Baxter et al, 2002; Reacher et al, 2004) and buildings are vulnerable to flooding depending on their location (Graves and Phillipson, 2000).

Urban green infrastructure, through the provision of cooler microclimates and reduction of surface water runoff, offers potential to help adapt cities for climate change. In this chapter, we present findings from a recent research project into 'Adaptation Strategies for Climate Change in Urban Environments (ASCCUE)'. One important aspect of this research modelled the adaptation potential of urban green infrastructure in relation to surface temperature and surface water runoff.

Characterization of the case study site

Greater Manchester was selected as the case study site. It is a large conurbation covering 1300km^2 with a population of 2.5 million. It was characterized here by Urban Morphology Type (UMT) mapping and a surface cover analysis, both based on aerial photograph interpretation (1997 images; resolution 0.25m) (Cities Revealed, 2008; Gill et al, 2008). Twenty-nine distinctive UMTs were grouped according to 13 primary categories (Table 19.1). The UMTs serve as spatial units linking physical features, human activities and natural processes. Their close affinity to urban land use planning categories and compatibility with the UK National Land Use Database classification (version 4.4; NLUD, 2003) enhances the transfer of ecological information into the planning system (Pauleit and Duhme, 2000). Within high, medium and low density residential UMTs the mean number of address points per hectare (a proxy for dwellings per hectare, using Ordnance Survey's MasterMap Address Layer data) is 47.3, 26.8 and 14.8, respectively. Whilst the UMT categories provide an initial indication of where patches and corridors of green may be expected they do not reveal the extent of green cover within the built matrix. Thus, a surface cover analysis, based on randomly placed points (e.g. Akbari et al, 2003), was undertaken for each UMT category using nine surface cover types: building, other impervious, tree, shrub, mown grass, rough grass, cultivated, water and bare soil/gravel.

Approximately 40 per cent of Greater Manchester is farmland, with the remaining 60 per cent representing the 'urbanized' area.

Table 19.1 Primary and detailed UMT categories

Primary UMT category	Detailed UMT categories
Farmland	Improved farmland; Unimproved farmland
Woodland	Woodland
Minerals	Mineral workings and quarries
Recreation and leisure	Formal recreation; Formal open space; Informal open space; Allotments
Transport	Major roads; Airports; Rail; River and canal
Utilities and infrastructure	Energy production and distribution; Water storage and treatment; Refuse disposal; Cemeteries and crematoria
Residential	High density residential; Medium density residential; Low density residential
Community services	Schools; Hospitals
Retail	Retail; Town centre
Industry and business	Manufacturing; Offices; Distribution and storage
Previously developed land	Disused and derelict land
Defence	Defence
Unused land	Remnant countryside

Residential areas account for almost half of the urbanized area and can be viewed as the 'matrix', representing the dominant landscape category in the urban mosaic (Forman and Godron, 1986). Recreation and leisure is the next major land use, covering 12 per cent of the urbanized area. These units are 'patches' within the built matrix (Forman and Godron, 1986). Industry and business, and previously developed land each cover 9 per cent of the urbanized area, whilst woodland accounts for 5 per cent.

Seventy-two per cent of Greater Manchester, or 59 per cent of the urbanized area, is evapotranspiring (i.e. vegetated and water) surfaces (Figure 19.1). This ranges from an average of 20 per cent evapotranspiring surfaces in town centres to 98 per cent in woodlands. Trees cover 12 per cent of Greater Manchester and 16 per cent of the urbanized area. Whilst woodland has 70 per cent tree cover, all other UMTs have below 30 per cent with town centres having 5 per cent cover. Importantly, residential areas account for almost half of urbanized Greater Manchester and 40 per cent of its evapotranspiring surfaces. Such surfaces largely occur within private gardens and as street trees. In high density residential areas built surfaces (i.e. building and other impervious surfaces) cover two thirds of the area, compared to half in medium density

and a third in low density areas. Tree cover is 26 per cent, 13 per cent and 7 per cent in low, medium and high density areas, respectively.

Quantifying the environmental functions

The UMTs, with their distinctive surface covers, formed one of the inputs into energy exchange and surface runoff models (Whitford et al, 2001). The models were run for the baseline 1961–1990 climate, as well as for the UKCIP02 Low and High emissions scenarios for the 2020s, 2050s and 2080s (Hulme et al, 2002). Results presented here are for the 1961–1990 baseline and the 2080s Low and High emissions scenarios. Temperature and precipitation inputs were calculated using daily time series output from a weather generator for Ringway (Manchester Airport) (Watts et al, 2004a; BETWIXT, 2005; Kilsby et al, 2007).

Model runs were completed for the UMT categories with their current form (i.e. using proportional surface covers from the urban characterization) as well as for a series of 'development scenarios' in which the green cover was altered. The 'development scenarios' were intended both to help understand the effects of

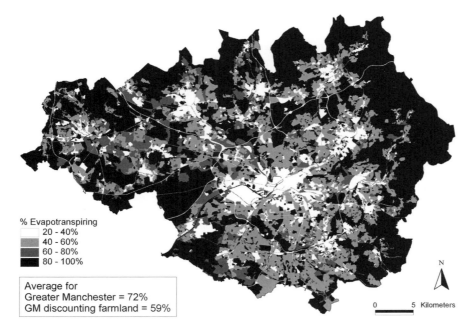

% Evapotranspiring
- 20 - 40%
- 40 - 60%
- 60 - 80%
- 80 - 100%

Average for
Greater Manchester = 72%
GM discounting farmland = 59%

N

0   5 Kilometers

Figure 19.1 Percentage of evapotranspiring (i.e. vegetated and water) surfaces in Greater Manchester (assuming the average from the photo interpretation of sample points within UMT categories)

current development trends (e.g. Duckworth, 2005; Pauleit et al, 2005) as well as to explore the potential of greening in adapting to climate change. They included residential and town centres plus or minus 10 per cent green or tree cover; greening roofs in selected UMTs; high density residential development on previously developed land; increasing tree cover by 10–60 per cent on previously developed land; residential development on improved farmland; and permeable paving in selected UMTs. Not all 'development scenario' results are presented here, but can be found in Gill (2006).

In addition, for the energy exchange model, model runs were completed where grass was excluded from the evapotranspiring proportion. This was intended to give some indication of the impact of a drought, when the water supply to vegetation is limited and hence there is reduced evaporative cooling. Grass may be the first type of vegetation in which this happens due to its shallow rooting depth.

Maximum surface temperatures

The energy exchange model has maximum surface temperature as its output and is based upon an energy balance equation (Tso et al, 1990; Tso et al, 1991; Whitford et al, 2001). The warming of the urban environment in summer is an important issue because of its implications for human comfort and well being (e.g. Eliasson, 2000; Svensson and Eliasson, 2002). Whilst air temperature provides a simple estimator of human thermal comfort, it is less reliable outdoors owing to the variability of other factors such as humidity, radiation, wind and precipitation (Brown and Gillespie, 1995). In practice, the mean radiant temperature, which in essence is a measure of the combined effect of surface temperatures within a space, is a significant factor in determining human comfort, especially on hot days with little wind (Matzarakis et al, 1999). As well as requiring input of the proportional area covered by built and evapotranspiring surfaces, the model also requires a building mass per unit of land, and various meteorological parameters

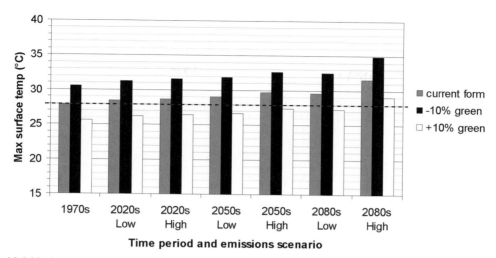

Figure 19.2 Maximum surface temperature for the 98th percentile summer day in high density residential areas, with current form and when 10 per cent green cover is added or removed. Dashed line shows the temperature for the 1961–1990 current form case

including air temperature. We modelled the 98th percentile average daily summer temperature (i.e. the average temperature occurring on approximately two days per summer).

Maximum surface temperature is dependent on the proportion of green cover, which becomes increasingly important with climate change. Currently the maximum surface temperature of woodlands (the least built up UMT) is 18.4°C, 12.8°C cooler than town centres (the most built up UMT) at 31.2°C. However, by 2080s Low, the maximum surface temperatures are 19.9°C in woodlands and 33.2°C in town centres and by the 2080s High, 21.6°C and 35.5°C. Thus by the 2080s, maximum surface temperatures may increase by 1.5–3.2°C in woodlands and 2–4.3°C in town centres, depending on the emissions scenario. The temperature difference between these UMTs increases to 13.9°C by the 2080s High. In high density residential areas (with an evaporating cover of 31 per cent) maximum surface temperatures could increase by 1.7–3.7°C by the 2080s depending on the emissions scenario; in low density areas (with an evaporating cover of 66 per cent) the increase is 1.4–3.1°C. The temperature difference between these residential UMTs increases from 6.2°C in 1961–1990 to 6.5–6.8°C by the 2080s, depending on the emissions scenario. This suggests that more highly built-up areas will warm more strongly than well vegetated ones.

Adding 10 per cent green cover to highly built-up areas, such as the town centre and high density residential UMTs, would cool them significantly and could keep maximum surface temperatures at or below the 1961–1990 baseline temperatures up to, but not including, the 2080s High (Figure 19.2). In high density residential areas maximum surface temperatures in 1961–1990 with current form are 27.9°C. Adding 10 per cent green cover decreases maximum surface temperatures by 2.2°C in 1961–1990, and 2.4–2.5°C by the 2080s Low and High, respectively. Thus, maximum surface temperatures decrease by 0.7°C by the 2080s Low and increase by 1.2°C by the 2080s High, in comparison to the 1961–1990 current form case. This is compared to temperature increases of 1.7–3.7°C by the 2080s Low and High if no change was made to surface cover. On the other hand, if 10 per cent green cover is removed maximum surface temperatures by the 2080s High would be 7–8.2°C warmer in high density residential and town centres, respectively, compared to the 1961–1990 current form case.

Surface runoff

The surface runoff model uses the curve number approach of the US Soil Conservation Service (USDA Natural Resources Conservation Service, 1986; Whitford et al, 2001). Surface cover is required as an input along with precipitation, antecedent moisture conditions and hydrologic soil type. We modelled the 99th percentile winter daily precipitation event (i.e. a precipitation event that occurs approximately once per winter). The 99th percentile daily winter precipitation is 18mm for 1961–1990, 25mm for the 2080s Low and 28mm for the 2080s High. Results presented here are for normal antecedent moisture conditions.

In general, the more built-up a UMT is, the more surface runoff there is. Additionally, soil type is very important with surface sealing having a more significant impact on runoff where soils have higher infiltration rates, such as sandy soils. The proportion of rainfall converted to runoff is lower on such soils than on slower infiltrating soils, such as clays. On a sandy soil, 32 per cent and 74 per cent of an 18mm precipitation event will be converted to runoff in low density residential and town centres, respectively; on a clay soil this changes to 76 per cent and 90 per cent. By the 2080s there will be increased precipitation, an increased volume of surface runoff, and a larger percentage of the precipitation contributes to surface runoff. The total runoff over Greater Manchester for an 18mm rainfall event is 13.8 million m^3; the 28mm rainfall event, which has 55.6 per cent more rain than the 18mm event, would produce 82.2 per cent more runoff at 25.2 million m^3.

The 'development scenarios' suggested that, whilst increasing green cover reduces the volume of surface water runoff, this would not be sufficient to cope with the increased precipitation anticipated with climate change. For example, increasing tree cover by 10 per cent in the residential UMTs would reduce runoff from these areas from a 28mm precipitation event by 5.7 per cent. However, this does not keep the future runoff at or below the runoff levels for the baseline 1961–1990 current form case. In fact, by the 2080s High, runoff from high density residential areas will still be approximately 65 per cent higher than the 1961–1990 current form case even when tree cover is added. However, whilst runoff from such extreme winter events does increase by the 2080s High compared to 1961–1990, if 10 per cent tree cover is added by this time there will be 14 per cent less runoff from residential areas than if it remains with current tree cover.

The occurrence of drought

Adding green cover in built-up areas has significant potential in moderating surface temperatures with climate change. This cooling effect is the result of evapotranspiration by vegetation and is dependent on a supply of water. In the absence of irrigation, water may be reduced during droughts, which may be more frequent and intense with climate change (e.g. Watts et al, 2004a; Watts et al, 2004b; BETWIXT, 2005).

We undertook a simplified water balance for the Greater Manchester UMTs under current and future climate scenarios, to demonstrate when grass may be water stressed and evapotranspiration restricted. This combined available soil water (data from the National Soil Resource Institute), incoming precipitation (from UKCIP02 5km climate change scenarios) and outgoing evapotranspiration (from the daily weather generator for Ringway). The method was developed to include actual evapotranspiration alongside potential evapotranspiration (Rowell, 1994; Allen et al, 1998). The findings suggest a significant increase in the duration of droughts with climate change (Figure 19.3). In 1961–1990, most of Greater Manchester experienced no or little water stress. By the 2080s Low only a few UMT units on the northern and eastern fringes of the conurbation where rainfall is highest will experience no water stress; other areas will have up to $4\frac{1}{4}$ months of stress. By the 2080s High all the UMT units will experience some water stress, with the number of months varying from $2\frac{1}{2}$–$5\frac{1}{4}$ in the average year.

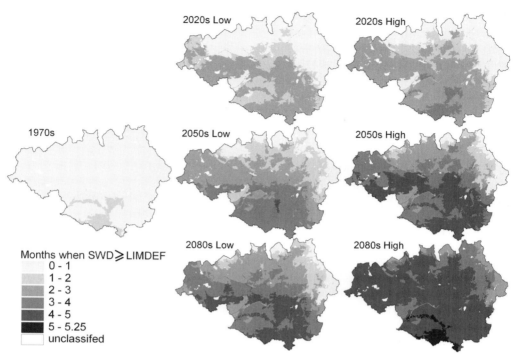

Figure 19.3 Months per year when grass experiences water stress (i.e. the soil water deficit (SWD) is greater than or equal to the limiting deficit (LIMDEF))

A further energy exchange model run, for the case when grass dries out and stops evapotranspiring, showed that the biggest change in maximum surface temperature would be found in UMTs where grass forms a large proportion of the evapotranspiring surfaces. For example, in schools which often have large playing fields, maximum surface temperature would increase by 13.8°C in 1961–1990 and 14.7–15.6°C by the 2080s Low and High. Rivers and canals became the coolest UMT, with maximum surface temperatures of 19.8°C in 1961–1990 and 21.2–22.9°C by the 2080s Low and High. This modelling work does not include the effect of shading by trees on surface temperatures; however, pilot measurements suggested that the shade provided by mature trees can keep surfaces cooler by as much as 15.6°C. This shading effect of trees would become even more crucial during droughts, when the evaporative cooling effect of vegetation may be reduced.

Climate adaptation via the green infrastructure

These results show that urban green infrastructure can offer significant potential to moderate the increase in summer temperatures expected with climate change, through evaporative cooling and shading. Adding 10 per cent green cover in high density residential areas and town centres would keep maximum surface temperatures at or below 1961–1990 baseline levels up to, but not including, the 2080s High. Green infrastructure is most effective at moderating surface runoff, through the interception, storage and infiltration of rainwater, on sandier soils. However, whilst runoff from extreme winter rainfall events does increase by the 2080s High compared to 1961–1990, if 10 per cent tree cover is added by this time there will be 14 per cent less runoff from residential areas than if it remains with current tree cover. Still, the total volume of runoff increases regardless of changes to surface cover. There is significant potential, therefore, to capture, store and use this water for irrigating green infrastructure during droughts, to ensure that its cooling functionality is maintained when it is most needed. This research has thus demonstrated that the benefits of green infrastructure go well beyond consideration of amenity and that opportunities will have to be taken to ensure an adequate water supply to vegetation in times of drought. Unless provision for irrigation is made there will be conflict as green infrastructure will require irrigating at the same time as water supplies are low and restrictions may be placed on its use. Currently during periods of water shortages urban vegetation is often the first target of a 'drought order'.

Studies of land cover change have found that vegetated and pervious surfaces are actually decreasing. For example, between 1975 and 2000 there was a 5 per cent decrease in vegetated surfaces in residential areas in Merseyside (Pauleit et al, 2005). In Keighley, pervious surface cover within house curtilages decreased by 15–21 per cent, depending on residential density, between 1971 and 2002. This change was linked to the creation of hard standings as a result of increased car ownership (Duckworth, 2005). The results presented in this chapter suggest that such changes will reduce our capacity to adapt to climate change.

In order to adapt to climate change, it is crucial that this functional role of urban green infrastructure is well understood. The idea of the functionality of green infrastructure is reflected in the Millennium Ecosystem Assessment (MEA, 2005), which has attracted attention from the UK Department for Environment, Food and Rural Affairs (Defra, 2007). This report classified ecosystem services into: supporting services (such as soil formation, nutrient cycling and photosynthesis), provisioning services (such as food, water, timber and fibre), regulating services (that affect climate, floods, disease, wastes and water quality), and cultural services (that provide recreational, spiritual and aesthetic benefits). In this chapter we have been concerned with the local climate and water regulating services provided by urban green infrastructure and their importance for climate change adaptation.

The increasing attention given to green infrastructure planning in the UK (e.g. URBED, 2004; Kambites and Owen, 2006; North West Green Infrastructure Think Tank, 2007) is welcome as the focus on functionality offers a vehicle through which to plan for climate change adaptation (Gill et al, 2007; Shaw et al, 2007). Indeed, green infrastructure has been defined as 'an interconnected network of green space that conserves natural ecosystem values and functions and provides associated benefits to human populations' (Benedict and McMahon, 2002, p12). Importantly, green infrastructure incorporates all green and blue elements, not just those in public ownership. In addition to providing climate adaptation, green infrastructure offers a range of other environmental, social and economic benefits in urban areas (e.g. Givoni, 1991; URBED, 2004). This makes the use of green infrastructure an attractive climate adaptation strategy. Moreover, green infrastructure may also help in reducing greenhouse gas (GHG) emissions, or in mitigating climate change. For example, vegetation can reduce solar heat gain in buildings and thus reduce the demand for mechanical cooling through air conditioning, which emits GHGs and further intensifies the urban heat island through waste heat (e.g. Niachou et al, 2001; Onmura et al, 2001; Papadakis et al, 2001).

Faced with the challenge of climate change and recognizing the multifunctional benefits of green infrastructure, strategies, policies, plans and programmes should seek to:

- protect critical environmental capital where green infrastructure assets have a demonstrable level of climate functionality. This includes town centre parks, flood plains, and areas where the soil has a high infiltration capacity.
- ensure that there is no net loss of green infrastructure cover.
- undertake creative greening to enhance the green infrastructure where urban form is largely established. This could include street tree planting (e.g. Red Rose Forest, 2006; The Mersey Forest, 2007), green roofs, and building façades. Particular attention should be given to the public realm in town centres

to ensure a sufficient range and quality for human comfort (Walsh et al, 2007), and to new planting in locations where a low green infrastructure cover combines with socio-economic deprivation and/or human vulnerability. Where green spaces currently exist they should be enhanced so that they provide multiple climate-related functions. This could include the incorporation of water features, increasing tree cover, and the provision of sustainable drainage systems (SuDS) such as swales, infiltration, detention and retention ponds (CIRIA, 2000; Mansell, 2003).

- take opportunities to improve levels of green infrastructure provision during urban restructuring and new developments. This is vital given the long life time of buildings from 20 to over 100 years (Graves and Phillipson, 2000).
- employ innovative measures to secure an alternative water supply to sustain the functionality of greens infrastructure during times of drought, such as through rainwater harvesting, the re-use of greywater, making use of water in rising aquifers under cities where present, and floodwater storage.

Implementation through strategies, policies, plans and programmes

The planning system has an important role to play in adapting places to climate change (see Davoudi et al, Chapter 1). In the UK, at a national level this has been reflected in the supplement to Planning Policy Statement 1 on 'Planning for Climate Change' (DCLG, 2007c). This supplement recognizes the importance of green infrastructure for urban cooling in particular, but also the role of natural drainage systems. Regional Spatial Strategies will be key in setting out policies at a regional level. For example, the draft Regional Spatial Strategy for the North West of England has tackling climate change as an urgent regional priority as well as a green infrastructure policy which states that

plans, strategies, proposals and schemes should: identify, promote and deliver multi-purpose networks of greenspace, particularly where there is currently limited access to natural greenspace or where connectivity between these places is poor; and integrate green infrastructure provision within existing and new development, particularly within major development and regeneration schemes.
(NWRA, 2006)

Local authorities will need to reflect Regional Spatial Strategies when producing their Local Development Frameworks. Ideally there would be a strong green infrastructure policy within the Core Strategy, which reflects its climate change adaptation functions. Green infrastructure could potentially also be incorporated at this level as a Supplementary Planning Document. Kambites and Owen (2006, p492) have emphasized the need to embed green infrastructure planning in statutory planning systems as a 'normal part of the preparation and review of development plans' in order to ensure widespread and sustained commitment.

It is important, however, that other strategies, plans, policies and programmes outside of the statutory planning system take into account climate change adaptation via the green infrastructure and that these are integrated as far as possible with the planning system. Adapting to climate change is one of a number of the Government's new performance indicators for local authorities and local authority partnerships, which could help to strengthen partnership working on this issue through Local Strategic Partnerships and Local Area Agreements (DCLG, 2007a; 2007b). In North West England green infrastructure planning is taking place at the sub-regional and city-regional levels, in part as a result of the focus at this level within the Regional Economic Strategy (NWDA, 2006a). Plans at this level can take into account the fact that neither green infrastructure nor climate change respect local administrative boundaries. Similarly, the North West Climate Change Action Plan includes an action to 'undertake scoping studies to assess future regional risks, opportunities and priorities for the potential for green infrastructure, including regional parks, to adapt and mitigate for climate change impacts and commence implementation of findings' (NWDA, 2006b). This action will explore the climate adaptation functions of green infrastructure considered in this chapter alongside others including upstream flood storage (e.g. Thomas and Nisbet, 2007), soil stabilization (e.g. Nisbet et al, 2004), storm protection in coastal areas, as a habitat and corridor for wildlife, and as a recreation facility to reduce visitor pressure to areas which have a higher vulnerability to climate change (e.g. McEvoy et al, 2006). It will also consider the climate mitigation functions of green infrastructure such as carbon sequestration and storage, woodfuel to replace fossil fuels, timber to replace materials with higher embedded energy, and reducing the need to travel by car by creating green routes around urban areas.

Conclusion

Green infrastructure offers significant potential in adapting our towns and cities for climate change. It provides vital ecosystem services, which will become even more critical with climate change. This green infrastructure resource, which includes street trees, private gardens and city parks must be strategically planned, designed and managed to ensure that its climate-related functionality is maximized.

It is essential that strategies, policies, plans and programmes at all levels recognize the functional importance of green infrastructure in adapting to climate change and seek to maximize this role. In particular, they should seek to protect critical environmental capital, ensure no net loss of green cover, promote creative greening, enhance the functionality of existing green spaces, improve levels of green infrastructure provision during restructuring and new developments, and ensure that measures are taken to sustainably irrigate green infrastructure elements so that they continue to provide evaporative cooling during droughts.

The creative use of green infrastructure is one of the most promising opportunities for climate change adaptation. This opportunity needs to be recognized in the statutory planning process at all scales from the Regional Spatial Strategies, through Local Development Frameworks to development control within urban neighbourhoods.

Acknowledgement

This research was undertaken as part of the 'Adaptation Strategies for Climate Change in Urban Environments (ASCCUE)' project, one of a consortium of projects under the 'Building Knowledge for a Changing Climate (BKCC)' umbrella, funded by the UK Climate Impacts Programme (UKCIP) and the Engineering and Physical Sciences Research Council (EPSRC) (GR/S19233/01).

References

Akbari, H., Rose, L. S. and Taha, H. (2003) 'Analyzing the land cover of an urban environment using high-resolution orthophotos', *Landscape and Urban Planning*, vol 63, no 1, pp1–14

Allen, R. G., Pereira, L. S., Raes, D. and Smith, M. (1998) 'Crop evapotranspiration: Guidelines for computing crop water requirements', FAO Irrigation and Drainage Paper 56, UN Food and Agriculture Organization, Rome, Italy

Baxter, P. J., Moller, I., Spencer, T., Spence, R. J. and Tapsell, S. (2002) 'Flooding and climate change', in P. Baxter, A. Haines, M. Hulme, R. S. Kovats, R. Maynard, D. J. Rogers and P. Wilkinson (eds) *Health Effects of Climate Change in the UK*, Department of Health, London

Benedict, M. A. and McMahon, E. T. (2002) 'Green infrastructure: Smart conservation for the 21st century', *Renewable Resources Journal*, vol 20, no 3, pp12–17

BETWIXT (2005) 'Built EnviromenT: Weather scenarios for investigation of Impacts and eXTremes. Daily time-series output and figures from the CRU weather generator', www.cru.uea.ac.uk/cru/projects/betwixt/cruwg_daily/, accessed 30 June 2008

Bridgman, H., Warner, R. and Dodson, J. (1995) *Urban Biophysical Environments*, Oxford University Press. Oxford

Brown, R. D. and Gillespie, T. J. (1995) *Microclimate Landscape Design: Creating Thermal Comfort and Energy Efficiency*, John Wiley & Sons, Chichester

CIRIA (2000) *Sustainable Drainage Systems: Design Manual for England and Wales*, Construction Industry Research and Information Association, London

Cities Revealed (2008) 'Cities revealed', www.cities-revealed.com/, accessed 30 June 2008

DCLG (2007a) 'The new Performance Framework for Local Authorities and Local Authority Partnerships: Single set of national indicators', Department for Communities and Local Government, London

DCLG (2007b) 'National indicators for Local Authorities and Local Authority Partnerships: Handbook of definitions', Draft for Consultation, Department for Communities and Local Government, London

DCLG (2007c) 'Planning Policy Statement: Planning and climate change, Supplement to Planning Policy Statement 1', Department for Communities and Local Government, London

Defra (2007) 'Ecosystem services: Living within environmental limits', Department for Environment, Farming and Rural Affairs, www.ecosystemservices.org.uk, accessed 30 June 2008

Donaldson, G. and Keatinge, W. (2008) 'Direct effects of rising temperatures on mortality in the UK', in S. Kovats (ed) *Health Effects of Climate Change in the UK*, Department of Health, London, pp81–90

Duckworth, C. (2005) 'Assessment of urban creep rates for house types in Keighley and the capacity for future urban creep', unpublished MA thesis, University of Manchester

Eliasson, I. (2000) 'The use of climate knowledge in urban planning', *Landscape and Urban Planning*, vol 48, nos 1–2, pp31–44

Forman, R. T. T. and Godron, M. (1986) *Landscape Ecology*, John Wiley & Sons, New York

Gill, S. E. (2006) *Climate Change and Urban Greenspace*, unpublished PhD thesis, University of Manchester, www.greeninfrastructurenw.org.uk, accessed 30 June 2008

Gill, S. E., Handley, J. F., Ennos, A. R. and Pauleit, S. (2007) 'Adapting cities to climate change: The role of the green infrastructure', *Built Environment*, vol 33, no 1, pp115–133

Gill, S. E., Handley, J. F., Ennos, A. R., Pauleit, S., Theuray, N. and Lindley, S. (2008) 'Characterising the urban environment of UK cities and towns: A template for landscape planning', *Landscape and Urban Planning*, vol 87, pp210–222

Givoni, B. (1991) 'Impact of planted areas on urban environmental quality: A review', *Atmospheric Environment*, vol 25B, no 3, pp289–299

Graves, H., Watkins, R., Westbury, P. and Littlefair, P. (2001) *Cooling Buildings in London: Overcoming the Heat Island*, BRE & DETR, London

Graves, H. M. and Phillipson, M. C. (2000) *Potential Implications of Climate Change in the Built Environment*, BRE, Centre for Environmental Engineering, East Kilbride

Handley, J., Pauleit, S. and Gill, S. (2007) 'Landscape, sustainability and the city', in J. F. Benson and M. Roe (eds) *Landscape and Sustainability*, second edition, Routledge, London, pp167–195

Hough, M. (2004) *Cities and Natural Process*, Routledge, London

Hulme, M., Jenkins, G., Lu, X., Turnpenny, J., Mitchell, T., Jones, R., Lowe, J., Murphy, J., Hassell, D., Boorman, P., McDonald, R. and Hill, S. (2002) *Climate Change Scenarios for the United Kingdom: The UKCIP02 Scientific Report*, Tyndall Centre for Climate Change Research, School of Environmental Sciences, University of East Anglia, Norwich

Kambites, C. and Owen, S. (2006) 'Renewed prospects for green infrastructure planning in the UK', *Planning Practice and Research*, vol 21, no 4, pp483–496

Kilsby, C. G., Jones, P. D., Burton, A., Ford, A. C., Fowler, H. J., Harpham, C., James, P., Smith, A. and Wilby, R. L. (2007) 'A daily weather generator for use in climate change studies', *Environmental Modelling and Software*, vol 22, no 4, pp1705–1719

Larsen, J. (2003) 'Record heat wave in Europe takes 35,000 lives', *Earth Policy Institute*, 9 October 2003, www.earth-policy.org/Updates/Update29.htm, accessed 30 June 2008

McEvoy, D., Handley, J. F., Cavan, G., Aylen, J., Lindley, S., McMorrow, J. and Glynn, S. (2006) *Climate Change and the Visitor Economy: The Challenges and Opportunities for England's Northwest*, Sustainability Northwest and UK Climate Impacts Programme, Manchester and Oxford

Mansell, M. G. (2003) *Rural and Urban Hydrology*, Thomas Telford, London

Matzarakis, A., Mayer, H. and Iziomon, M. (1999) 'Applications of a universal thermal index: Physiological equivalent temperature', *International Journal of Biometeorology*, vol 43, pp76–84

MEA (2005) *Ecosystems and Human Well-Being*, synthesis, Island Press, London

Niachou, A., Papakonstantinou, K., Santamouris, M.,

Tsangrassoulis, A. and Mihalakakou, G. (2001) 'Analysis of the green roof thermal properties and investigation of its energy performance', *Energy and Buildings*, vol 33, no 7, pp719–729

Nisbet, T., Broadmeadow, S. and Orr, H. (2004) *A Guide to Using Woodland for Sediment Control*, Forest Research, Farnham, Surrey

NLUD (2003) 'Land use and land cover classification – version 4.4', National Land Use Database, www.nlud.org.uk, accessed 30 June 2008

North West Green Infrastructure Think Tank (2007) 'North West Green Infrastructure Guide', www.greeninfrastructurenw.org.uk, accessed 30 June 2008

NWDA (2006a) 'Northwest Regional Economic Strategy', North West Regional Development Agency, Warrington

NWDA (2006b) 'Rising to the challenge: A climate change action plan for England's Northwest 2007–09', Northwest Regional Development Agency, Warrington

NWRA (2006) 'The North West Plan: Submitted draft Regional Spatial Strategy for the North West of England', North West Regional Assembly, Wigan

Oke, T.R. (1987) *Boundary Layer Climates*, Routledge, London

Onmura, S., Matsumoto, M. and Hokoi, S. (2001) 'Study on evaporative cooling effect of roof lawn gardens', *Energy and Buildings*, vol 33, no 7, pp653–666

Papadakis, G., Tsamis, P. and Kyritsis, S. (2001) 'An experimental investigation of the effect of shading with plants for solar control of buildings', *Energy and Buildings*, vol 33, no 8, pp831–836

Pauleit, S. and Duhme, F. (2000) 'Assessing the environmental performance of land cover types for urban planning', *Landscape and Urban Planning*, vol 52, no 1, pp1–20

Pauleit, S., Ennos, R. and Golding, Y. (2005) 'Modelling the environmental impacts of urban land use and land cover change: A study in Merseyside, UK', *Landscape and Urban Planning*, vol 71, nos 2–4, pp295–310

Reacher, M., McKenzie, K., Lane, C., Nichols, T., Kedge, I., Iversen, A., Hepple, P., Walter, T., Laxton, C. and Simpson, J. (2004) 'Health impacts of flooding in Lewes: A comparison of reported gastrointestinal and other illness and mental health in flooded and non-flooded households', *Communicable Disease and Public Health*, vol 7, no 1, pp1–8

Red Rose Forest (2006) 'Green Streets', www.redroseforest.co.uk/forestpro/greenstreets.html, accessed 30 June 2008

Rowell, D. L. (1994) *Soil Science: Methods and Applications*, Longman Scientific and Technical, Harlow

Royal Commission on Environmental Pollution (RCEP) (2007) *The Urban Environment*, TSO, Norwich

Shackley, S., Kersey, J., Wilby, R. and Fleming, P. (2001) *Changing by Degrees: The Potential Impacts of Climate Change in the East Midlands*, Ashgate, Aldershot

Shaw, R., Colley, M. and Connell, R. (2007) *Climate Change Adaptation by Design: A guide for sustainable communities*, Town and Country Planning Association, London

Svensson, M. K. and Eliasson, I. (2002) 'Diurnal air temperatures in built-up areas in relation to urban planning', *Landscape and Urban Planning*, vol 61, no 1, pp37–54

The Mersey Forest (2007) 'Green Streets Merseyside: February 2007–December 2008', www.merseyforest.org.uk/files/GreenStreetsPROJECTBUS-plan 8%20summary.doc, accessed 30 June 2008

Thomas, H. and Nisbet, T. R. (2007) 'An assessment of the impact of floodplain woodland on flood flows', *Water and Environment Journal*, no 21, pp114–126

Tso, C. P., Chan, B. K. and Hashim, M. A. (1990) 'An improvement to the basic energy balance model for urban thermal environment analysis', *Energy and Buildings*, vol 14, no 2, pp143–152

Tso, C. P., Chan, B. K. and Hashim, M. A. (1991) 'Analytical solutions to the near-neutral atmospheric surface energy balance with and without heat storage for urban climatological studies', *Journal of Applied Meteorology*, vol 30, no 4, pp413–424

Urban Task Force (1999) 'Towards an urban renaissance', final report of the Urban Task Force chaired by Lord Rogers of Riverside, Department of the Environment, Transport and the Regions, E & FN Spon, London

URBED (2004) *Biodiversity by Design: A guide for sustainable communities*, Town and Country Planning Association, London

USDA Natural Resources Conservation Service (1986) 'Urban hydrology for small watersheds', United States Department of Agriculture, Washington DC

Walsh, C. L., Hall, J. W., Street, R. B., Blanksby, J., Cassar, M., Ekins, P., Glendinning, S., Goodess, C.

M., Handley, J., Noland, R. and Watson, S. J. (2007) 'Building knowledge for a changing climate: Collaborative research to understand and adapt to the impacts of climate change on infrastructure, the built environment and utilities', Newcastle University, Newcastle upon Tyne

Watts, M., Goodess, C. M. and Jones, P. D. (2004a) 'The CRU Daily Weather Generator', Climatic Research Unit, University of East Anglia, Norwich

Watts, M., Goodess, C. M. and Jones, P. D. (2004b) 'Validation of the CRU Daily Weather Generator', Climatic Research Unit, University of East Anglia, Norwich

Whitford, V., Ennos, A. R. and Handley, J. F. (2001) 'City form and natural process: Indicators for the ecological performance of urban areas and their application to Merseyside, UK', *Landscape and Urban Planning*, vol 57, no 2, pp91–103

Wilby, R. L. (2003) 'Past and projected trends in London's urban heat island', *Weather*, vol 58, no 7, pp251–260

Wilby, R. L. and Perry, G. L. W. (2006) 'Climate change, biodiversity and the urban environment: A critical review based on London, UK', *Progress in Physical Geography*, vol 30, no 1, pp73–98

20

A Climate of Planning:
Swedish Municipal Responses to Climate Change

Richard Langlais

It is important to know why the climate change issue moves some communities, more than others, to respond *actively* and *concretely* to its implications, and whether this is different from how they are responding to more general calls for sustainable development. What are the conditions that succeed in making a shift to *acting* on climate change? This is especially interesting since there are widespread laments that this is *not* happening. When active, concrete response emerges, is it because key individuals have made the difference? To what extent have other stakeholders been critical to their success? Are particular local conditions necessary? In the absence of direct threats, is the response due to informed global altruism, or crass economic calculation, or even both? In our work with these questions in Sweden, we want to know if the increasing focus on climate change also marks a significant turning point in the work to achieve sustainable development.

Action on climate change gains relevance from, but is not explained by, its relation to sustainable development. To what extent is that relation vital? Part of the answer lies in comparing the basic premises and messages, in asking

how similar they really are, and in analysing how various actors have interpreted them. We feel that the answers to these questions will reveal not only if this is a turning point, but whether its implications are sustainable in the long term. By studying the responses that are being made at local, municipal levels, we seek to discover if there is something different about climate change, why that is so, and how that is being transformed into effective social change.

In Sweden, mitigation, rather than adaptation, has so far been the dominant response to the issue of climate change by the country's 290 municipalities (Langlais et al, 2007). It is possible that its particular focus of sustainable development work implies that the issue of climate change, more generally, is qualitatively different from the other focuses. It is intriguing to consider that the essential differences between an approach dominated by an attitude of mitigation and another by adaptation are connected to the current transition from a nation state to a market state world (Royal Dutch/Shell Group, 2005; Bobbitt, 2008). As yet, it is not entirely clear what the long-term consequences of such a difference will imply.

If we take sustainable development as a context for climate change response, we see that by now it has been discussed on numerous policy levels. The Brundtland Commission stabilized it as the slogan of a global consensus in the 1980s, and it has been moving in a top–down manner ever since, from global to international, to national and regional and on 'down' to local levels. Now that climate change is becoming a pressing issue at those levels, with varying degrees of conceptual blending, it is possible that this marks a 'tectonic shift' in the overall direction and quality of response, where the first local-level actions are generating feedbacks (now bottom–up) to the national and international levels. At the same time, the relations of power and the space for change are being rearranged. The local appears in some cases to be superseding the national, and the relative weights of different jurisdictions are in flux. We need to know to what degree this description is correct, since it indicates the manner, directions and speed with which planners, stakeholders and decision makers will be expected to (re)act.

Swedish municipalities and the origin of climate change response

From a Swedish perspective, climate change work at the municipal level began in a tangible, albeit modest, way as a part of the initial moves to implement Local Agenda 21, in 1996. Although the municipalities pledged to act as leading examples of sustainable development, the element of climate-change-related work in the Local Agenda 21 documents was limited. A key feature of that campaign was the determination to strive to include various municipal actors and the general public in local environmental work. It was only during the latter part of the 1990s, stimulated by the adoption of the Kyoto Protocol in 1997, and its focus on the issue of greenhouse gas (GHG) emissions, that climate change emerged as a more explicit part of municipal policies. A number of municipalities then adopted a more active response to the climate issue by incorporating related objectives in their Local Agenda 21 documents. When the national government's environmental protection agency (Swedish: Naturvårdsverket) offered grants through its Local Investment Programmes (LIPs), and its Local Climate Investment Programme (KLIMP), which had a tendency to encourage commitments more towards all forms of energy-related issues, municipalities became more active in their work with, above all, decreasing their emissions of GHGs. A penchant for mitigation with regard to climate change was becoming entrenched as the primary approach on climate change (Rylander, 2005).

An overwhelming majority of the mitigation measures undertaken, therefore, have been within the energy sector, where improvement in energy efficiency, renewable district heating and providing advice on energy issues are the most common actions. The LIP and KLIMP programmes have been increasing the municipalities' incentives to work more actively on climate-change-related issues. Measures in the traffic sector, on the other hand, have been rarer, while those that have been adopted have been limited to such items as expanded bicycle paths and improvements in public transportation. The success of the LIP and KLIMP investment subsidies indicates that there is at least a modicum of support for the work with climate change issues among decision makers at the national level (Rylander, 2005), even if most of the measures that have been implemented are of a more reactive character.

In the first years of this decade there was much uncertainty about where responsibility for dealing with climate change issues rested. There were calls for studies of the manner in which the government and the Riksdag, the Swedish parliament, were dealing with the challenge; it was felt that ambiguity prevailed. The issue of unclear division of responsibility was a concern for officials in the municipalities, counties and other authorities, who could imagine that even though they should have certain responsibilities in relation to climate matters, they were nevertheless unsure of exactly how those were defined. Almost none of the important actors had analysed how they would adapt to climate change. A majority, on the other hand, had undertaken analyses of the effects of a changing climate on their own localities. Concrete adaptation measures that were identified as being in response to climate change, however, were extremely rare, with only half-a-dozen or so exceptions among the total of 290 municipalities. A consensus was beginning to form around the need for more concrete action (Rummukainen et al, 2005).

For the purposes of our studies, and in order to achieve any distinguishable contrast between those municipalities that stand out as proactive and the more passive mainstream, it has been necessary to create a 'normal' background against which they can be more easily discerned. The next section sketches that background by describing the obligations that all Swedish municipalities have, and share, as well as the opportunities and forms of collaboration that are at their disposal for responding to climate change. The objective is not only to clarify the state's role regarding both the current national environment objectives and the local investment subsidies available from the state, but also to indicate the various possibilities that the municipalities have for participating in different networks concerned with climate change issues.

Municipal obligations and possibilities: The Swedish Environmental Protection Agency and Sweden's 16 environmental objectives

At the national level, the Ministry of the Environment is responsible for the environment and climate change interface. A number of different authorities are represented within the Ministry, where The Swedish Environmental Protection Agency (Naturvårdsverket) is a central administrative authority in the environmental arena. It works to promote sustainable development, based on the environmental objectives set by parliament. The agency's main objective is to set central standards and to coordinate and evaluate environmental work throughout the country. Above all, this includes both the duty to inform and to ensure that environmental laws are followed, while also assisting the government and the parliament on policy, proposing legislation where needed. Furthermore, the agency functions by serving as a guide to other central, regional and local authorities with respect to environmental issues. Another important part of the work carried out by the agency is to clarify how laws should be interpreted, through the creation of regulations, manuals, and so on, and by holding general councils on the topic.

The Swedish Environmental Protection Agency interacts with various public authorities, private companies and sector organizations. It has also set up and participated in a number of different networks, both in Sweden and at the EU level. In Sweden, both the counties and the municipalities are important collaborative partners, since they are responsible for environmental issues on the regional and local levels. A central focal point in this respect is the Swedish Riksdag's list of 16 environmental objectives (the first 15 were adopted in April 1999, the 16th in November 2005.). The Swedish Environmental Protection Agency is not, however, the only body responsible for all environmental objectives. Other authorities and agencies also have a role to play, for example, The National Board of Housing, Building and Planning, The National Chemicals Inspectorate and The Swedish Radiation Protection Authority are responsible for the environmental objectives that concern their various areas of responsibility.

In relation to the environmental objectives, municipal obligations remain poorly focused. A general overview shows that municipal responsibilities with respect to climate change are rather diffuse, with the municipalities not actually obliged to do anything in particular. The municipalities are, however, responsible for the maintenance of a 'good habitat' at the local level and, according to parliament, have overall responsibility for local implementation of national environmental and public health objectives. The municipalities are also considered to have an important role to play in pursuing a dialogue with their citizens with regard to how the environmental objectives are to be achieved.

The outlines of that work vary, although the introduction of environmental management systems and routine coordination with the local Agenda 21 work commonly occur.

The Agenda 21 processes in Sweden's municipalities have provided a good basis from which to develop the work with the environmental objectives already outlined. In addition, there are the so-called Environmental Quality Standards, initially introduced when the Environmental Code was implemented, on 1 January 1999. Environmental Quality Standards are a novel governing tool in the context of the Swedish environmental legislative system. It is not always clear what applies when implementing the regulations. It is important to emphasize that for certain issues that have been pursued by the introduction of the Environmental Quality Standards there are no simple and straightforward answers. In some cases a legal precedent remains to be set.

Currently, no specific Environmental Quality Standard has been set for the climate change issue, further emphasizing the current and somewhat inadequate situation with regard to municipal obligations. According to the first environmental objective, municipal obligations seem to focus primarily on encouraging change and the provision of a supportive environment. It is worth noting here, however, that the municipalities are required by law (Sweden, 1977) to establish an up-to-date energy plan, for the supply, distribution and use of energy. The plan is to be established by the municipal council.

Municipal obligations and possibilities: State investment subsidies, LIP and KLIMP

Another part of understanding the background for the active progress of climate change mitigation measures in Sweden are the two programmes of state investment subsidies, the LIP and the KLIMP.

Both parliament and the government have been active in encouraging the environmental work already being done in Sweden's municipal-

ities by offering subsidies. LIP and KLIMP were designed to both stimulate and support the municipalities in their environmental work. The LIP was a more general environmental subsidy, while KLIMP has the more focused aim of instigating climate-change-related measures. Both subsidies, however, have required municipal co-financing. A closer look at each of them follows.

LIP – Local Investment Programme

The LIP came into force once parliament passed regulation (Sweden, 1998) stipulating that state subsidies to local investment programmes will increase society's ecological sustainability. A secondary aim was to stimulate employment. LIP's rationale was summed up in four intentions; it was to:

- start from the municipality's unique situation by setting priorities for identifying local environmental problems;
- be based on a bottom–up approach and each municipality's on-going Agenda 21 programme;
- be linked to each municipality's overall environmental work;
- ensure that the 'global' end result of its programmatic focus will be magnified beyond the sum of its individual parts. (Sweden, 1998).

During the period 1998–2002, some 6.2 billion SEK (Swedish krona) were released by the state through The Swedish Environmental Protection Agency, and 211 investment programmes were implemented in 164 municipalities and in 2 municipal associations. Within the framework of those 211 investment programmes, some 1814 measures were undertaken. In total, approximately 27.3 billion SEK was spent on various LIP projects, while approximately 20.7 billion SEK of that total was used for specific environmental investments.

The municipalities that applied for these environmental subsidies could do so individually or in collaboration with other actors. The recipient municipality was also responsible for meeting the requirements of the regulations. There were, however, no specific requirements with respect to technological systems or solutions, since the main objective here was to decrease the environmental impact.

KLIMP – Local Climate Investment Programme

Following the discontinuation of LIP subsidies, KLIMP emerged in their place. The main objective of the KLIMP programme is to foster a reduction in GHG emission levels and, in connection with each specific KLIMP project, conduct information campaigns about its goals.

Since 2003, municipalities, counties, firms and other local actors have been able to apply for KLIMP subsidies from The Swedish Environmental Protection Agency. Between 'start up' in 2003, and May 2006, a total of some 1.1 billion SEK had been disbursed to various programmes. The total investment volume within the framework of KLIMP has, however, been 4.7 billion SEK since KLIMP-financing, as was the case with the LIP, requires co-financing from the receiving actor. Each KLIMP programme lasts for four years, after which the results are reported to The Swedish Environmental Protection Agency. The agency then makes a final assessment, which serves as the basis for calculating the final amount to be allocated to the KLIMP project; this assessment is in turn based on how the results correspond to the stipulated objectives of the project. As such, the final reporting of projects from the first four-year period took place in 2007. The parliament decided that during 2007 and 2008, KLIMP funding would be increased by some 395 million SEK per year (Swedish Environmental Protection Agency, 2007).

Among the KLIMP projects that attracted funding during the period between 2003 and 2006, two dominant areas emerged; these were energy production and distribution, and transport – mainly road traffic – which together were allocated more than 50 per cent of the total available funding.

Networks for municipal cooperation: The Swedish Eco Municipalities and the Swedish Network of Municipalities on Climate Change

There are a number of Swedish municipal cooperation initiatives that have taken the form of network structures, where members collaborate in order to raise the level of awareness about the climate change issue. Their intention is to support each other in climate-related work and to share knowledge and experience. The two largest networks are The Swedish Eco Municipalities (SEKOM) and The Swedish Network of Municipalities on Climate Change (Klimat Kommuner, or KK).

The Swedish Eco Municipalities (SEKOM)

The concept of eco-municipality was originally launched in the Nordic countries by the Finnish municipality of Suomussalmi, in 1980. In 1983, the concept was introduced via the Swedish municipality of Övertorneå, which was on the border with Finland, in the far north of the country. In that early stage, those municipalities were the only eco-municipalities in the Nordic countries (Sveriges Ekokommuner, 2008).

The work undertaken by Övertorneå in developing the eco-municipality concept eventually inspired many other Swedish municipalities. A network of eco-municipalities was formed in which 15 other municipalities participated. It was intended as a response to the need for support in their ecological sustainable planning that the municipalities were experiencing. In 1995, when the network was formed, a decision was also made to create a common secretariat for servicing the municipalities within the network. It is a non-profit association and has subsequently grown to include 68 municipalities. The member municipalities are represented by both a civil servant and a politician, something that is supposed to ensure the creation of better conditions and an increase in the importance of environmental and climate change issues within the municipalities.

The objective of the network is to help all of Sweden's municipalities to become sustainable. An important element of their interaction is that politicians and civil servants in the municipalities are able to exchange experience and knowledge concerning the climate change issue. This is enabled by, among other things, SEKOM's intranet and regular meetings.

During 2006, an environmental communication project was carried out between The Swedish Eco-Municipalities and the Swedish Network of Municipalities on Climate Change (see below). The project was financed to a certain extent by the Environmental Objectives Council through The Swedish Environmental Protection Agency, with the central objective being to increase the competence of responsible officials. Furthermore, an important part of the project was to communicate environmental issues on the local level in order to make it possible for municipalities and other organizations to implement the information campaigns that contribute to the achievement of Sweden's environmental objectives.

The Swedish Network of Municipalities on Climate Change

The KK is a network of municipalities working actively with the climate change issue and whose overall aim is to reduce GHG emissions. The network was founded in 2003, at the initiative of the municipality of Lund, which also serves as the host municipality. KK is primarily financed by The Swedish Environmental Protection Agency, and its members today include: Malmö, Lund, Luleå, Kristianstad, Växjö, Götene, Helsingborg, Hässleholm, Falköping, Lidköping, Mölndal, Olofström, Stockholm, Säffle, Södertälje, Uppsala, Åmål and Östersund. KK supports municipalities that want to take on the climate change issue. It is an instigator of the national climate work through its focus on the important possibilities, obstacles and driving forces that impact the results of climate change work. The networks also help to distribute infor-

mation and experience in working locally on climate issues, with a view to increasing knowledge levels in respect of the complex problems associated with it. KK also develops its international cooperation in order to gain contact with similar networks in other countries. In that way, KK hopes to set an example for all Swedish municipalities to follow, since they want each community to find another that they can compare themselves to (Klimatkommunerna, 2008).

Membership in KK is free, but there are certain requirements involved in becoming a member. The municipalities are required to establish environmental goals at the political level. These goals require the municipality to work for continuous reductions in greenhouse gas emissions, to set goals for the reduction of emissions, to create an action plan and to implement measures in order to decrease emissions and continuously inform the network about their work-in-progress, the obstacles and possibilities facing them. The overall objective of the network is that, by the end of 2008, at least half of Sweden's municipalities will have participated in the network's activities and that the network itself will have grown to approximately 25 members.

In the autumn of 2006, the network made it possible for small municipalities with a population of up to 15,000 to attend the Climate Coaching project, which was designed to aid small municipalities in responding to climate change. The project, with 23 participating municipalities, was run by KK itself, with the aim of creating the basis for long-term climate strategies for smaller municipalities. During the life of the project which began in January 2007, each participating municipality has received 'custom-made' help from a 'climate coach', based on an individual needs analysis. The objective is also to initiate work processes in order to create a climate strategy, update the municipality's energy plan and renew its programme of environmental objectives. Participating municipalities receive a contribution of some 50,000 SEK, which can be used for funding measures within the project. In return, each participating municipality signs a declaration of intent authorized by the municipal executive board. The declaration of intent contains, among other things, a promise that the municipal council will prescribe within two years, a proposal for their own climate strategy. The municipality also agrees to contribute, in the form of local research resources or measures, an amount rising to at least 50,000 SEK. The municipality also receives support from a 'mentor' municipality that is already a member of KK, and thus has the experience and 'critical mass' to contribute with suggestions and strategies for the local climate work.

In concluding this discussion of Sweden's climate-change-related networks of municipalities, there are some reflections that are worth mentioning. Although the history of the networks helps to explain part of the reason for the dominance of energy-related mitigation practices in the modes of municipal response, that explanation is limited by the fact that only about one-third of the municipalities are members of, or otherwise participate in, those networks. Other networks, such as the Aalborg Declaration, as well as ASPIRE (Achieving Energy Sustainability in Peripheral Regions of Europe) and Energy in Minds, are gaining more Swedish municipal membership. Those networks are EU-based and focused mainly on energy conservation.

Another observation about the Swedish networks is that membership in KK or SEKOM has not in itself appeared to guarantee that a municipality will be on the 'cutting edge' in their approach to climate change mitigation, even if that might be what some of their representatives would have liked. In fact, in our review of the municipalities' efforts, there was a striking contrast between the image created by a municipality's membership in either or both of the networks and the level of actual engagement that could be found.

Concluding discussion

The way from a habit of mitigation to a creative approach to adaptation is not so obvious, even in a small country with a decades-long tradition of continual transformation of its energy system,

notwithstanding that its activity has had other motivations than climate change response, per se. That is not to say that mainstreaming mitigation has been perfectly straightforward, either, while even more profound transitions must be forthcoming if new emissions targets are to be met, sooner rather than later. There is currently a great deal of uncertainty among the municipalities, for example, regarding their obligations with respect to what they see as vague national legal guidelines on climate change response. In spite of that, several of the state-sponsored measures that have been implemented with a view to stimulating municipal engagement in this area have been considered to be more successful. Such measures include the LIP and the KLIMP, as described above. Among smaller, less resource-rich communities, those that appear to have come furthest in their climate change work have for the most part received substantial assistance from those programmes. This has included their reliance on the advice of 'consultants', or mentors, from the 'Klimatkommunerna' network, who have helped smaller municipalities to plan their general work in a more climate-conscious manner. The verdict is still out, however, on the deeper reasons for those municipalities' success; it is a type of 'chicken or egg' dilemma. In other words, one can ask whether a municipality's application for assistance has been approved because the municipality has shown deep engagement in climate change issues, or, conversely, whether it has been the availability of grants that has generated their engagement.

The growing general acceptance of the need for mitigation measures has not necessarily led to an ever-increasing acceptance of other climate change responses; our work has revealed that in some municipalities this has been decreasing. In a small number of municipalities (9 of 290), the level of sustainable-development- and climate-change-related activity appears to have decreased. This has only been the negative extreme in an otherwise general change in the organization of sustainable development activities in the communities, in turn a result of the reduction of Agenda 21 projects in most municipalities. Many of the personnel who previously had responsibility for following through on the municipality's Agenda 21 activities have since left those positions. This is evidence, although of a negative kind, that the role of key individuals is essential in at least some cases for achieving a shift from merely discussing action, to concretely implementing it.

In the resulting policy vacuum, this has led to the further consequence that sustainable development and climate change activities have been 're-invented', or 'resurrected', in the municipalities' profiles, primarily through revision of already existing energy plans, environmental objectives, and so on. One of our respondents referred to this process as a kind of 'decisional archaeology' (in Swedish, 'beslutsarkeologi'). The work that had already been proceeding under other names and guises has now found a renewed drive and motivation in the discourse of climate change. While it is not yet a 'turning point', it does convey that the drive for sustainable development is at least entering the turn.

In our surveys of municipal response to climate change, we observe a broad diversity of initiatives emerging, even if much of what is already going on is old work under a new name. It is proving difficult to break out of the current paradigm, which has relegated much of the earlier diversity in sustainable development campaigns to being mostly about various sorts of energy-sector-related activities. Moving into a thorough and penetrating discussion about behavioural and lifestyle issues, for example, is only halting and shallow. The impetus seems present, however, which leads us nevertheless to expect a very different picture to emerge in the next few years.

When concrete actions do occur, to what extent have other stakeholders been critical to their success? The initiatives we have been observing occasionally highlight the increasing role of private corporations in joint projects with the municipalities, although still mostly within the biogas and ethanol manufacturing sector. This reflects the eagerness to envision the challenge of climate change as a stimulus to innovation and entrepreneurship. It builds on a widespread opinion in Sweden, both at the national and municipal levels, which holds that much of Sweden's current well-off situation is due to the development and maintenance of a growing, robust form of the 'knowledge economy'. This in turn is expected to apply as the basis for future growth.

The municipalities' most common concrete measures include the replacement of their vehicle fleets with so-called 'eco-cars', the expansion of district heating systems, the development of new energy plans, holding eco-driving courses for municipal employees and engaging energy consultation assistance for firms and private homeowners. Staging showings of Al Gore's film, *An Inconvenient Truth*, was often pointed out as being part of several municipalities' broader information campaigns.

Another kind of pattern is becoming evident among suburban and commuter-dominated municipalities, mostly in the greater Stockholm and Malmö regions. Such municipalities often display a lower level of engagement in climate change issues than what can be observed in most other municipalities. One reason is that they are instead preoccupied with, and often overwhelmed by, the problems associated with the heavy transit traffic that passes through their areas. That traffic is often impervious to the municipalities' efforts, since it is dominated by expressways, which function according to logics that rarely have anything to do with the municipalities they transect. As a consequence, the affected municipalities claim that since they don't even have sufficient resources for counteracting the problems created by local and through traffic, then mitigation actions with regard to those problems associated with climate change are viewed as beyond reach. Several municipalities were found to be exceptions to that situation (for example, Ekerö, Kungsbacka, Kungälv and Mölndal) and, upon further enquiry, the explanation was that they appeared to be operating from value bases that were more associated with public transit. This finds expression in the construction and planning of new building projects that are being designed to be both in proximity to, and creating synergies and harmonizing with, public transit facilities. In other municipalities, their location near large waterways and other bodies of water also seemed to encourage an enhanced level of climate change mitigation awareness, as did the presence of institutions of higher education, such as universities and other advanced academic and research facilities, although not necessarily of the more concrete kind.

In conclusion, Sweden's municipalities engage in a variety of approaches in responding to climate change. What are the underlying motives? Is the explanation always to be found in arguments for energy-sector-driven economic rationality? Often, yes, but a sizeable proportion (about one-fifth) of the municipalities are responding to locally perceived threats of increasing natural calamities, such as flooding, landslides and drought. Some municipalities see climate change issues permeate their overall work while most do no more than the law obliges them to. This was in sharp contrast to a handful of communities that do seem to be acting on the basis of an essentially altruistic concern for the state of the world in general, with only the vaguest concern for any direct gain locally, apart from moral satisfaction. Many municipalities, on the other hand, claim that they have environmental goals, specific climate work plans and programmes on climate change currently either 'in the pipeline' or coming up for decision; notwithstanding those claims, however, it was seen on closer observation that few of them had much of a concrete nature that they could point to as currently being underway.

In the most inactive, or even passive, municipalities, there is a palpable atmosphere of there being a 'lack of political will' available for addressing climate change issues. Once again, the need for clearer goals was often cited as an explanation for that dearth of motivation. The national environmental quality objectives are a favourite target for criticism, and often described by our respondents as being too abstract and not specifically applicable to their municipality. As expressed in Swedish, '*Mest skrik och ganska lite ull*' ('A lot of shouting and pretty little wool') was an illustrative comment that reflected the kind of sentiment often heard to describe the vague and ambiguous national environmental goals.

There is a paradox in the way that many small municipalities approach their climate change work. Some of the smallest municipalities claim that they are simply unable to play an active and concrete role in response to climate-change-related issues, because they are 'too small and thus lack the necessary resources.' At the same time, others claim that they are perfectly placed to carry out such work, precisely *because* they are so small, since they are not hindered by an overbearing level of bureaucracy. Here, again, the role of key, visionary individuals appears to be playing a strong role. Overall, there is an oft-expressed interest in learning more about how other municipalities were able to initiate and carry out concrete measures; the level of interest is high.

Finally, a composite derived from a number of studies indicates that the engagement of Swedish municipalities in responding to the challenge of climate change has increased dramatically in the period 2004–07 (Rummukainen et al, 2005; Rylander, 2005; Winblad, 2005; Schmidt-Thomé, 2006; Langlais et al, 2007; Sweden, 2007). Our work seeks evidence of concrete action and explanations for why some municipalities are much more active than others. This difference is striking, given that the institutional and governance conditions are similar. Although adaptation issues dominate in some communities, the focus of most is on mitigation. The role of planners in mitigation activities is often described as frustrating, since at the same time as it is of central importance, with high political priority, there is a lack of knowledge and experience in incorporating the broad generalities of climate change challenges into the detailed protocols and routines of everyday practice.

References

Bobbitt, P. (2008) *Terror and Consent: The Wars for the Twenty-First Century*, Alfred A. Knopf, New York
Klimatkommunerna [The Climate Municipalities] (2008) 'Klimatanpassning i skånska kommuner' [Climate adaptation in municipalities in Skåne], www.klimatkommunerna.infomacms.com/?page= start, accessed 15 August 2008
Langlais, R., Francke, P., Nilsson, J. and Ernborg, F. (2007) 'Turning point on climate change?

Emergent municipal response in Sweden: Pilot study', Nordregio Working Paper 2007:3, Nordregio, Stockholm

Royal Dutch/Shell Group (2005) *Shell Global Scenarios to 2005 – The Future Business Environment: Trends, Trade-Offs and Choices*, Peter G. Peterson Institute for International Economics, Washington DC

Rummukainen, M., Bergström, S., Persson, G. and Ressner, E. (2005) 'Anpassning till klimatförändringar' [Adaptation to climate change], SMHI Series RMK 106, Swedish Meteorological and Hydrological Institute, Norrköping, Sweden

Rylander, Y. (2005) 'Kommunernas klimatarbete – Klimatindex för kommuner 2005' [The municipalities' climate work – Climate Index for Municipalities 2005], Svenska Naturskyddsföreningen [Swedish Society for Nature Conservation], Stockholm

Schmidt-Thomé, P. (2006) *Integration of Natural Hazards, Risk and Climate Change into Spatial Planning Practices*, Geological Survey of Finland, Espoo, Finland

Sveriges Ekokommuner [Sweden's eco-municipalities] (2008) 'Information about SEkom', www.sekom.nu/, accessed 15 August 2008

Sweden, Miljödepartementet, Klimat- och sårbarhet-sutredningen [Ministry of the Environment, The Commission on Climate and Vulnerability] (2007) 'Sverige inför klimatförändringarna – hot och möjligheter' [Sweden and climate change – threats and opportunities], SOU 2007:60, Statens offentliga utredningar (SOU), Stockholm

Sweden, Riksdagen (1977) 'Lag om kommunal energiplanering' [Municipal Energy Planning Law], Svensk författningssamling 1977:439, Riksdagstryck, Riksdagen [Parliament], Stockholm

Sweden, Riksdagen (1998) 'Förordning om statliga bidrag till lokala investeringsprogram som ökar den ekologiska hållbarheten i samhället' [Ordinance on state assistance to local investment programmes that increase society's ecological sustainability], Svensk författningssamling 1998:23, Riksdagstryck, Riksdagen [Parliament], Stockholm

Swedish Environmental Protection Agency (2007) 'Vad är klimp?' [What is KLIMP?], www.naturvardsverket.se/sv/ Lagar-och-andra-styrmedel/ Ekonomiska-styrmedel/Investeringsprogram/ Klimatinvesteringsprogram-Klimp/ Vad-ar-Klimp/, accessed 31 August 2007

Winblad, S. (2005) 'Women in climate-policy-related decision-making: National Study for Sweden', personal communication, Climate for Change: Gender Equality & Climate Policy Project, Malmö, Sweden

21

Moving Cities Towards a Sustainable Low Carbon Energy Future:
Learning from Woking and London

Allan Jones, MBE

Introduction

Climate change is the greatest threat to life on our planet. For the first time in history more than 50 per cent of the world's population is living in cities. Cities are responsible for emitting 80 per cent of the world's greenhouse gas (GHG) emissions.[1] Cities are, therefore, the primary cause of climate change, are at high risk from climate change and have a vested interest in tackling it.

If runaway global climate change is to be avoided it will be at the city level that this will be achieved, through cities innovating and progressing measures to tackle climate change.

This chapter is a personal perspective on my experience as a practitioner implementing practical strategies for a sustainable energy society which derives its initial energy needs from low carbon energy resources whilst at the same time establishing a sustainable energy infrastructure to enable future energy needs to be derived from wholly renewable resources via a hydrogen economy. These concepts can be applied to any community in the UK or indeed in the world.

Woking leading the way

Background

Woking is a large town 23 miles (37km) south-west of central London. It lies just outside the boundary of the Greater London Authority, in the South East Region of England. The local authority, Woking Borough Council, serves a population of 90,700. Modern Woking is formed around the railway station, built over 150 years ago, but the town dates back to the 8th century and was recorded in the Domesday Book in 1086. It claims fame to a number of firsts: the first crematorium (1889), the first mosque (1889) and one of the first public electricity supplies

(1889) in the UK. However, Woking's claim to recent fame is based on its unique achievements in local energy and tackling climate change.

Energy efficiency policy

When I joined Woking Borough Council in 1989 (originally as Building Services Manager and later as Energy Services Manager and Director of Thameswey Ltd) I introduced a vision for an effective and economic low carbon energy system for Woking as a blueprint for the UK. My challenge was to catalyse the necessary political commitment at cross party level to implement this vision including energy and water efficiency, decentralized and renewable energy systems, hydrogen fuel cell technology and low carbon transport systems.

For any local authority to take on such a challenge it needs three things – political support, chief officer support and someone inside the organization able and qualified to deliver the vision. Many local authorities have one or two of these things but rarely all three as in the case of Woking. However, it was not all plain sailing at Woking, particularly in planning, but due to persistence and well publicized achievements the programme continued to even greater success. Woking's Local Plan was eventually revised to incorporate the policies necessary to support energy efficiency, combined heat and power and renewable energy. In July 1990 (two years before the Rio Earth Summit) as part of a borough-wide environmental audit, I submitted a report on global warming to the Council. It would be true to say that the Council had never heard of global warming, its causes, effects or potential solutions but by October of the same year the politicians were so enthused that they agreed an energy efficiency policy which was the catalyst for all that Woking has since achieved. Initially, the key lever was ongoing financial savings but I had taken the precaution of monitoring and reporting on energy and water consumption and emission savings right from the beginning and these over time became more important than the financial savings and provided the stimulus for the Council's climate change strategy.

On this foundation, the Council implemented a series of energy and water efficiency measures and decentralized energy projects, including the UK's first small-scale combined heat and power (CHP)/heat fired absorption chiller or trigeneration system, first local authority private wire residential CHP and renewable energy systems, the largest domestic combined photovoltaic/CHP installations, the first local decentralized community energy systems, first fuel cell CHP system and first public/private joint venture energy services company, or ESCO, in the UK.

The Council is now recognized as the most energy efficient local authority in the UK, having achieved the greatest percentage reduction in both energy consumption and CO_2 emissions in the UK. It also supplies itself with the highest proportion of onsite or local decentralized energy supplies in the UK.

Climate change strategy

With the achievement of its target to reduce energy consumption by 40 per cent within ten years, the Council changed the emphasis from reducing energy consumption to reducing CO_2 equivalent (CO_2e) emissions by adopting a climate change strategy in 2002. Key features of the strategy were the adoption of the target to reduce CO_2e emissions for the whole of the borough from 1990 levels by 60 per cent by 2050 and by 80 per cent by 2100 (as recommended by the Royal Commission on Environmental Pollution, 2002), adaptation to climate change and a new 'environmental footprint' target for development and land use which set the overall objective of reducing CO_2e emissions by 80 per cent compared to the previous land use. In other words, if the previous land use in 1990 was a factory and was to be replaced by a housing estate the emissions of the housing estate had to be 80 per cent less than the emissions of the factory. For a greenfield site the emissions reduction would effectively be 100 per cent (i.e. zero carbon) since the greenfield site was deemed to be zero emissions. This latter principle was repeated five years later by the UK government's proposals for zero carbon communities (DCLG, 2007).

By the time I moved on to set up the London Climate Change Agency in 2004, the Council had reduced its own energy consumption by 49 per cent, water consumption by 44 per cent and GHG emissions by 77.5 per cent from 1990 (the base year) as well as receiving more than 93 per cent of its electrical and thermal energy requirements from onsite low or zero carbon decentralized energy sources. In addition, the Council had improved the energy efficiency of the housing stock in its area by more than 30 per cent, achieving its Home Energy Conservation Act (HECA) target one year early. The HECA figures are for mainly private sector housing, with 90 per cent of housing in Woking owned privately. Emission savings for the borough as a whole, including Thameswey and private sector schemes, was 18 per cent in 2004. Since 2004, the Council has continued to deliver its climate change strategy programme with further reductions/improvements (Box 21.1).

London: City leadership in tackling climate change

Background

With a population of around 7.4 million citizens London's energy consumption equals that of countries such as Greece or Portugal, leading to carbon emissions above 44 millions tonnes of CO_2 per annum. Its population is expected to grow by 700,000 by 2016 (Mayor of London, 2002), requiring an increase in residential accommodation of more than 30,000 units per year. This growth represents a great opportunity to transform the way the city works and ensure it moves to a low carbon development path. As a large and wealthy world city, London has taken the commitment to lead and show by example in taking action to avert catastrophic climate change. Key drivers to achieve this ambitious objective are political leadership and effective partnerships on the ground. In this context, Mayor Ken Livingstone made addressing the causes of climate change one of the main priorities of London strategies and set ambitious targets and policies for both mitigating and adapting to climate change. Through its sustainable policies – the original Mayor's Energy Strategy, the new London Plan, the Mayor's Climate Change Action Plan and the Mayor's Climate Change Duty, London put climate change at the core of its development vision.

Greater London Authority Act 1999

The Greater London Authority Act 1999 established the Greater London Authority (GLA). In addition to the GLA, the Mayoral or GLA Group also comprised Transport for London, Metropolitan Police Authority, London Fire and Emergency Planning Authority and the London Development Agency (LDA). The first Mayor, Ken Livingstone, was elected in May 2000, and re-elected in 2004, and oversaw the establishment of the institutional arrangements to tackle climate change in London. The new Mayor, Boris Johnson, was elected in 2008.

Congestion Charge and Low Emission Zone

The road transport sector in London accounts for just over 20 per cent of CO_2 emissions, 50 per cent of NO_x emissions and 66 per cent of particulates. London also has some of the worst traffic congestion and is amongst the most polluted cities in Europe with over 1000 premature deaths per annum due to poor air quality (Mayor of London, 2006). Tackling road transport congestion and pollution therefore, was a priority policy area in the former Mayor's first term of office. The Congestion Charge was introduced in London in 2002 and is currently £8 per day in 2008. Congestion within the congestion charging zone has reduced by 22 per cent compared to pre-charging levels. The original congestion charging zone was almost doubled in size in 2006. Low and zero emission vehicles are exempt from the Congestion Charge, which has seen an increase in the number of such vehicles leading to a 16 per cent reduction in CO_2 emissions in the congestion charging zone (Mayor of London, 2007).

The Low Emission Zone was introduced in 2008, and covers the Greater London area just inside the M25. Non-compliant diesel HGVs, coaches and buses are charged £200 per day and from 2010 the scheme will be extended to the non-compliant heaviest LGV's (including minibuses) at £100 per day.

London Plan 2004

The Mayor is responsible for strategic planning in London and the GLA Act requires that the London Plan only deals with matters of strategic importance to London. The GLA Act also requires that the London Plan takes account of three crosscutting themes:

- the health of Londoners;
- equality of opportunity;
- its contribution to sustainable development in the UK.

In terms of climate change the 2004 London Plan (Mayor of London, 2004a) was an energy led plan not a climate change led plan and simply adopted UK national targets for reductions in carbon emissions. For renewable energy, the Plan required major developments to show how the development would generate a proportion of the

site's electricity or heat needs from renewables, wherever feasible. It promised that policies for climate change would be developed and addressed in its first review. At this time the London Plan was no stronger on energy and climate change than any other local plan in the UK.

Mayor's Energy Strategy

The Mayor's Energy Strategy was published at the same time as the London Plan (Mayor of London, 2004b). Although the Strategy was non-statutory it did assist planners and developers in determining compliance with the London Plan. For renewable energy, the Strategy did not require a specific target for each development but an overall London target was set to generate at least 665GWh of electricity and 280GWh of heat, from up to 40,000 renewable energy schemes by 2010 along with specific targets for particular renewable energy technologies. In practice, the 'Merton Rule' of 10 per cent of a development's energy needs being met from on-site renewable energy was used by GLA planners as a benchmark for major developments referred to the Mayor under the GLA Act 1989 (see Chapter 14). In addition, the Mayor gained further powers under the GLA Act 2007 to intervene in the planning process where the issues were of a strategic importance to London.

For CHP, the Strategy adopted a target to double London's 2000 CHP capacity by 2010. For climate change, the Strategy adopted a target to reduce CO_2 emissions by 20 per cent, relative to the 1990 level, by 2010, as the crucial first step on a long-term path to a 60 per cent reduction from the 2000 level by 2050. The Strategy also introduced the concept of energy services and ESCOs to help deliver these targets. Woking Borough was an active consultee on the draft Strategy and examples of the work of Woking and Thameswey are detailed in the Strategy.

The non-statutory Mayor's Energy Strategy will be replaced by the statutory Energy and Climate Change Strategy under the Greater London Authority Act 2007.

LDA 'Green Alchemy' report

The LDA report Green Alchemy Turning Green to Gold (LDA, 2003) concluded that the sustainable energy market was set for substantial growth. The potential market generated as a direct result of deploying technologies set out in the Mayor's Energy Strategy could be worth around £3.35 billion by 2010 and employ between 5000 and 7500 people. The LDA report highlighted Woking as one of its international case studies and concluded that the development of an ESCO had been instrumental in the development of a series of high profile projects and that there was great potential for replicating this model in London.

London Climate Change Agency

Part of Mayor Ken Livingstone's 2004 election manifesto was a commitment to establishing a climate change agency for London. There was a recognition that although the Mayor had introduced robust policies and strategies on decentralized energy and climate change, which would be further developed during his second term in office, delivery of these policies and strategies remained at risk without a body to stimulate, develop, enable and/or deliver projects on the ground.

In particular, there was no engineering or climate change technical resource to advise and work with the GLA Group, London boroughs, property developers/owners and consultants to help deliver the Mayor's targets. In addition, there was also market failure in that there were no ESCOs operating in London to design, finance, build and operate decentralized energy systems that were long-term projects, typically financed and operated over 25 to 35 years.

Following the Mayor's re-election, I was appointed in November 2004 to set up and run the London Climate Change Agency (LCCA) to transfer the experience I had gained in Woking to the task of transforming London into a leading low carbon sustainable city.

The LCCA was established as a municipal company in 2006 to develop and implement projects in the sectors that impact on climate change, especially in the energy, water, waste and transport sectors. It was wholly owned by the LDA, a regional development agency whose purposes and powers include economic development and contributing towards sustainable development.

In addition, it was financially supported by the LCCA Founding Supporters – BP, Lafarge, Legal & General, Sir Robert McAlpine, Johnson Matthey and the Corporation of the City of London. The LCCA was also later supported by LCCA Supporters – BSkyB and the Carbon Trust, and has been further supported by the Rockefeller Brothers Trust, Energy Saving Trust, KPMG, Greenpeace and the Climate Group.

The Mayor was appointed as chairperson and, as CEO, I was appointed as the executive director of LCCA Ltd. We aimed to implement a project-led strategy to deliver real carbon reductions in London rather than a carbon offset strategy that would not deliver real carbon reductions locally. In the ensuing two years the LCCA began to establish a new energy infrastructure landscape in London which tackled all sectors that impact on climate change, including energy, transport, waste and water, by taking advantage of all available mechanisms to create sustainable systems for London's businesses and communities.

The LCCA adapted the Woking strategy for a sustainable energy community to the specific characteristics of a large city such as London. This included getting as many of the boroughs, major developers and landlords on board with the tackling climate change agenda. The building blocks of the LCCA strategy were designed to enable or implement projects and to act as catalysts promoting the development of low carbon decentralized energy systems across London.

The LCCA strategy embedded future proofing into this new energy infrastructure landscape. A prime example is CHP where the fuel may initially be a low carbon fuel such as natural gas which can be replaced later by a renewable fuel or hydrogen. The important issue here is to future-proof the heat, chilled water and electricity infrastructure, serving buildings on local decentralized energy systems. Those systems have a lifetime many times longer than the CHP plant. Such future proofing enables the easy replacement or refuelling of the CHP plant in the future with alternative fuels without impacting on the energy distribution infrastructure.

The sustainable energy community approach also incorporates energy efficiency as an integral part of the strategy. Other key elements are the capturing of sustainable waste resources to provide opportunities to produce hydrogen-rich renewable gases and fuels for both buildings and transport and the collection of sustainable water (through the dewatering of waste and other currently uncaptured water resources) for non potable water resources. These systems also need to be future-proofed to allow for the emergence of renewable hydrogen, in the form of renewable gases and liquids, as the common energy carrier of the future for both buildings and transport.

Hydrogen is important as it is zero carbon in use but its overall climate impact depends on the carbon intensity of the energy source used in its generation Hence, its future significance will depend on combining hydrogen with renewable energy, allowing energy to be easily stored and utilized to overcome the intermittency of some forms of renewable energy. Only in this way are renewable sources likely to be able to offer a genuine replacement to fossil fuels.

The LCCA prioritized its work on macro decentralized energy/low carbon zones where large scale decentralized energy is designed to serve whole areas or zones, predominantly existing public sector/large private sector development. This recognized that new development represents only 1 per cent of development a year and the 25 per cent decentralized energy (about 1200MW) by 2025 target will never be achieved unless macro decentralized energy is implemented on an intervention basis, through, for example, 24 × 50MW systems or 60 × 20MW systems for both new and existing development. This approach will make it easier for new development to connect to or catalyse large-scale decentralized energy. This raises the imperative to remove the regulatory barriers to such decentralized systems. For example, the current exempt licensing regime (which Woking used for its decentralized energy systems) enables up to 100MWe of generated electricity per site to be supplied to non residential customers over private wires, whereas only 1MWe (about 1000 households) of generated electricity per site can be supplied to residential customers over private wires. This was not a problem for Woking as individual residential developments were never greater than 1000 households or could be broken down to fit within the exemption, but was a very big problem for a city the size of London where individual housing developments could be 25 times bigger.

The ability to sell decentralized electricity directly to consumers at competitive retail prices is a major part of the economics (about 70 per cent) of decentralized energy since wholesale electricity prices are less than 25 per cent of retail prices. If decentralized electricity was traded through the national grid it would attract grid transmission and distribution losses and use of system charges (which is the majority of the cost of electricity) even though it would make little use of the distribution network and no use at all of the national grid transmission network since under the laws of physics electricity will always flow to the nearest load – that is, the local community.

In order to overcome this regulatory barrier, modifications to the electricity supply licence (which enable decentralized electricity to be sold cost-effectively to local residential and non residential consumers over the local public wires distribution network, only paying a distribution use of system charge) were implemented by Ofgem (the UK Energy Regulator) in March 2009. Ofgem has the statutory power to make such modifications without further regulation or primary legislation.

As a public sector body with strong private sector support, the LCCA sought to integrate public and private sector strengths to ensure the efficient implementation of the LCCA's strategy. Among the key elements of the LCCA strategy was the establishment of the London ESCO: a public/private joint venture Energy Services Company or ESCO, with a large private sector partner with the experience and the capability to

---

Box 21.2 London Climate Change Agency demonstration, implementation and supporting actions

Creation of a centre of climate change and energy engineering excellence advice and information to public and private sectors

Advice and input to mayoral and central government policy and strategy and Stern Review

Demonstration projects:

- London Transport Museum photovoltaic system;
- City Hall photovoltaic roof and solar shading;
- Palestra photovoltaic roof and wind turbines;
- Palestra fuel cell CHP trigeneration system.

Better Building Partnership – *BBP Ltd incorporated as a company limited by guarantee with major property owners in 2008*

Green Concierge Scheme pilot – *handed over to LDA to deliver*

Renewable gases and liquid fuels from waste and biomass

LDA development projects, climate change advice and support

Thames Barrage renewable energy/flood prevention study

LED lighting – *purchase and trademark licence scheme*

City of London CHP

South Kensington carbon reductions project

London Underground cooling via trigeneration

Energy Efficiency Revolving Fund – *joint venture with the Carbon Trust*

---

implement decentralized low carbon energy systems. A similar strategy was followed in Woking with the creation of Thameswey Ltd (Woking's municipal company) and Thameswey Energy Ltd (Woking's public/private joint venture ESCO).

London ESCO

The LCCA procured the private sector partner for the London ESCO via a competitive negotiated procedure tendering process. Nine major energy and utility companies tendered for this. EDF Energy plc won the tender and the London ESCO Ltd was established in 2006.

The London ESCO was established between the LCCA Ltd (19 per cent shareholding) and EDF Energy (Projects) Ltd (81 per cent shareholding) to design, finance, build and operate local decentralized energy systems for both new and existing developments. Investment in ESCO projects is in the same shareholding proportions but because of the 20 per cent equity, 80 per cent

loan project finance formula the LCCA's equity investment is always covered by the LCCA's share of the project fee, whatever the size of the ESCO project.

It was important to procure a large private sector partner in the London ESCO as subordinated debt may be required to cover the loan finance, at least for the first tranche of ESCO projects until the project portfolio is large enough to be refinanced by loan funders.

The shareholding and project investment ratio ensures that the company is not controlled or influenced by the public sector and remains an unregulated private company under the Local Authorities (Companies) Order 1995, a key criteria for private sector participation.

Although the ESCO is run by the private sector as the majority shareholder in the company, it is important to remember that the public sector was the procuring agency and selected the private sector partner against a specification, which was embedded into the shareholders agreement. The LCCA was not a sleeping partner: it participated fully in the

London ESCO in proportion to its scale. Full participation by the LCCA was necessary for the London ESCO to take full advantage of both public and private sector experience, expertise, technical know-how, stakeholder engagement/management and attractiveness for lenders (i.e. lower loan interest rates).

The London ESCO tackles climate change by developing and implementing local decentralized energy solutions to London's electricity, heating and cooling needs. It is also able to catalyse the waste, transport energy and water low- and zero-carbon sectors leading to the establishment of Special Purpose Vehicle Companies. Utilizing a completely different technical and commercial approach to such projects, originally developed in Woking, the London ESCO is able to identify and develop sites across London where investment in sustainable and decentralized energy technology would reduce CO_2 and other GHG emissions. The London ESCO has a project portfolio of circa 50 potential short-, medium- and long-term decentralized energy projects with some 34 different parties included in its shareholders agreement. Not all of these projects are to be developed by the London ESCO but it does give an indication of the potential development of CHP/renewable energy projects in London. Projects take approximately 18 months from expression of interest to financial close. The London ESCO's initial operating plan will see investment of £200 million in projects reducing CO_2 emissions by 310,000 tonnes pa. However, the downturn in the property market may delay the implementation of these projects.

The decentralized energy and energy services markets were also catalysed by the establishment of the London ESCO which saw the ESCO market in London increasing from having no ESCO players in 2006 to having 12 ESCO players in 2007. London ESCO priority has recently turned towards developing macro decentralized energy for existing development and refurbishment schemes.

This approach to project delivery is intended to take megatonnes of CO_2 emissions out of the atmosphere as well as catalysing the ESCO market in London, the UK and even interna-

Table 21.1 Progress towards London's 25 per cent decentralized energy by 2025 target[2]

CHP/CCHP installed by 2008	205MW$_e$
CHP/CCHP under construction from 2008	63MW$_e$
CHP/CCHP new development consented to 2017	349MW$_e$
CHP/CCHP capacity required from macro decentralized energy	583MW$_e$
Total	1200MW$_e$

tionally as others seek to replicate the Woking/LCCA model. In this way global climate change is tackled through rewriting the rules of commercial engagement and capturing the true economic benefits of low and zero carbon technologies which are currently lost in trading within a vested interest fossil fuel grid and fossil fuel extraction, delivery and supply market.

Mayor's Climate Change Action Plan

The Mayor's Climate Change Action Plan was published in 2007 following a detailed analysis of London's GHG emissions and the action that would be required to avoid catastrophic climate change which growing scientific consensus had determined would occur at atmospheric CO_2 concentrations beyond 450 parts per million (ppm) rather than the Royal Commission on Environmental Pollution target of 550 ppm (Mayor of London, 2007). This implies a target of stabilizing London and UK emissions much earlier than previously envisaged at 60 per cent below 1990 levels by 2025 – *not* by 2050. However, 50 per cent of this target was dependent on additional action by central government such as action on aviation, carbon taxes and the removal of the regulatory barriers to decentralized energy.

The UK is the world's 8th largest emitter of CO_2 emissions. London accounts for 8 per cent of these emissions, producing 44 million tonnes of CO_2 pa. Unless action is taken, London's emissions are projected to increase by 15 per cent to 51 million tonnes pa by 2025.

In order to deliver this target the action plan set the actions that would be needed to tackle London's CO_2 emissions in the following sectors:

- existing homes;
- existing commercial and municipal activity;
- new build and development;
- energy supply;
- ground transport;
- aviation.

In addition, the Mayoral Group was also required to reduce its emissions of around 226,000 tonnes of CO_2 pa, 0.5 per cent of London's total emissions, through its own direct activity on a 'showing by doing' principle as the Mayor cannot expect others to reduce emissions if the Mayoral Group does not aggressively reduce its own emissions.

Of all the sectors, emissions from energy supply are by far the biggest. London's centralized energy plants cause emissions of 35 million tonnes of CO_2 pa: 75 per cent of London's emissions. This is set to increase by 15 per cent by 2025 (compared to 2006 under the business-as-usual scenario). This underlined the importance of the LCCA developing the decentralized energy and ESCO markets.

London Plan 2008

Following the work on the Mayor's Climate Change Action Plan, a review of the London Plan was undertaken. The new London Plan was published in 2008 and included a completely new section on climate change, aligning energy and water policies with the Climate Change Action Plan.

The revised London Plan (Mayor of London, 2008) is a climate change led plan, not an energy led plan, and is radically different to UK national policy/strategy and associated targets, benefiting from the detailed work of the Mayor's Climate Change Action Plan. It adopts the target of working towards the long-term reduction of CO_2 emissions by 60 per cent by 2050 but because 50 per cent of this target is dependent

on central government action the Plan sets a phased minimum 30 per cent reduction target against the 1990 baseline for development by 2025, which will be monitored and kept under review.

Being climate change led, the London Plan now requires developments to make the fullest contribution to the mitigation and adaptation to climate change and minimize emissions of CO_2. To achieve this, it sets a hierarchy that is used to assess applications:

- using less energy, in particular by adopting sustainable design and construction measures;
- supplying energy efficiently, in particular by prioritizing decentralized energy;
- using renewable energy.

There is a completely new policy on decentralized energy – combined cooling, heat and power (CCHP or trigeneration) and combined heat and power (CHP or cogeneration) – and the renewable energy target is increased from 10 per cent of energy to 20 per cent by reduction of CO_2 emissions.

In other words, under the hierarchy, CO_2 emissions are first reduced by energy efficiency beyond the Building Regulations baseline, then by decentralized energy, and finally a further 20 per cent reduction by using renewable energy. This has the effect of making a higher renewable energy target more economic since it will be more economic to reduce CO_2 emissions by energy efficiency and decentralized energy first, leaving a smaller CO_2 target to be met by renewable energy, as well as supporting all three policy objectives in a much more economic way.

This new policy also makes such developments more attractive to ESCOs which will be able to reduce most if not all of the risk of developments not being connected to decentralized energy.

Other new policies address the hydrogen economy, adaptation, non-potable water infrastructure and using waste as a renewable gas or liquid fuel to supply both decentralized energy and transport.

Climate change duty

Under the GLA Act 2007 the Mayor is now subject to a duty to address climate change and issue a climate change mitigation and energy strategy and an adaptation to climate change strategy. The new climate change duty was sought and obtained from central government by the former Mayor Ken Livingstone and made law just prior to the mayoral election in May 2008. This is the first time that any politician in the world has a duty to tackle climate change.

The new Mayor Boris Johnson committed to the 60 per cent reduction in CO_2 emissions by 2025 target shortly after his election. However, the new Mayor will not be proceeding with the proposed charge on high CO_2 emission vehicles, and is to axe the existing London congestion charge zone extension, although he is to retain London's LEZ for non compliant HGVs. However, the London ESCO and the Better Buildings Partnership should survive the disestablishment of the LCCA as they were set up to be independent bodies.

World cities tackling global climate change

C40 Climate Leadership Group

The C40 Climate Leadership Group is a coalition of international cities[3] committed to tackling climate change by reducing GHG emissions. The principle of the C40 is that individual cities acting on their own may have a significant impact on tackling climate change in their own countries but their achievements are substantially diluted by worldwide emissions. However, large world cities acting together are not inhibited by weak international targets, and can effectively tackle global climate change since cities are responsible for 80 per cent of worldwide GHG emissions.

The C40 was originally convened in London in 2005 (1st summit). It agreed to reduce GHG emissions by creating a purchasing consortium, mobilizing the best experts in the world to provide technical assistance, creating and deploying common measurement tools and practical steps to implement projects in the energy, transport, waste and water sectors, including energy efficiency, local clean generation, intelligent city electric grids, renewable fuels from waste and biomass for both buildings and transport, tackling traffic congestion and efficient water systems.

Summits are held every two years, whereas intermediate meetings are also held on specific issues such as Transport and Congestion (London, 2007), World Ports (Rotterdam, 2008) and Climate Change Adaptation (Tokyo, 2008).

Clinton Climate Initiative

In 2006, as chair of the C40, the former Mayor of London, Ken Livingstone signed an agreement with former US President, Bill Clinton, which saw the Clinton Climate Initiative become the implementing partner of the C40. The first scheme, announced at the 2nd C40 summit in New York, was a $5 billion building energy efficiency retrofit programme which brought together 5 multinational ESCOs, 5 global banks and 30 city partners to retrofit buildings and reduce CO_2 emissions around the world. This scheme alone has the potential to reduce global CO_2 emissions by as much as 10 per cent and is the first of many initiatives that the C40 will implement in partnership with the Clinton Climate Initiative.

Cities including London, Houston, Johannesburg, Melbourne and Seoul are already implementing their projects to retrofit more than 300 municipal buildings, with private sector building owners such as GE Real Estate and Merchandise Mart undertaking commercial retrofit in more than 180 million m^2 of floor space. Paris has also announced its intention to retrofit 660 schools.

Other examples of city climate change action plans:

•   Los Angeles climate change action plan, Green LA, is based on reducing CO_2

emissions by 35 per cent below 1990 baseline levels by 2030 and includes phasing out coal-fired power generation as contracts expire and implementing 35 per cent renewable energy by 2020.

- New York's climate change action plan, part of PLANYC, is based on reducing CO_2 emissions by 30 per cent below 2007 baseline levels by 2030 (with an accelerated target for municipal government to reduce CO_2 emissions by 30 per cent below 2007 baseline levels by 2017) and includes creating an energy efficiency authority for New York City and increasing the amount of decentralized energy by 800MW.
- Paris's climate change action plan is based on reducing CO_2 emissions by 75 per cent below 2004 levels by 2050.
- Sydney's climate change action plan Vision 2030 is based on reducing GHG emissions by 70 per cent below 2006 levels by 2030 and includes implementing at least 330MW of decentralized energy (trigeneration) by 2030, renewable energy, waste to energy and water efficiency.
- Toronto's climate change action plan, Climate Change, Clean Air and Sustainable Energy Action Plan, is based on reducing CO_2 emissions by 30 per cent below 1990 levels and 80 per cent by 2050 (see www.c40cities.org/ccap/).

Sustainable low carbon energy future

In the long term, we will only be able to meet our energy needs at minimal emissions through the use of hydrogen and renewable energy. Hydrogen will be the energy carrier of the future, deriving its energy from renewable fuels. If cities are able to work together on this in an innovative way, as Woking and London have done, then they will see that climate change can be tackled as well as setting the foundation for a sustainable low carbon future.

There is nothing new that needs to be invented or discovered to tackle climate change. All of the technologies and systems that we need already exist, but this is not just about technology, it is also about politics and mindsets easily swayed by vested interests. Politics can be a very influential agent and lead the way in tackling climate change as we have seen in Woking, previously in London and in other world cities, but it can also be an agent for delay and procrastination in the fight to tackle climate change.

Despite recent events I am still hopeful that the world cities will succeed in tackling climate change. The genie is out of the bottle now and the success of what was achieved in Woking, in London and elsewhere, coinciding with the electronic age and the world wide web, continues to inform and inspire other communities and cities around the world to take the necessary action to tackle global climate change. The real issue is not whether humankind can tackle climate change but whether there is enough time left in which to do so.

Notes

1   C40 Cities Leadership Group, www.c40cities.org
2   The figure for installed plant is extracted from the Ofgem database (www.ofgem.gov.uk). Ofgem has a statutory duty to maintain the database on behalf of HM Treasury under the CHP Quality Assurance programme as only assured quality CHP is exempt from the Climate Change Levy. This is probably an underestimate since not everyone applies for exemption from the Climate Change Levy or registers under the CHPQA.
3   The C40 Climate Leadership Group comprises Addis Ababa, Athens, Bangkok, Beijing, Berlin, Bogota, Buenos Aires, Cairo, Caracas, Chicago, Delhi, Dhaka, Hanoi, Hong Kong, Houston, Istanbul, Jakarta, Johannesburg, Karachi, Lagos, Lima, London, Los Angeles, Madrid, Melbourne, Mexico City, Moscow, Mumbai, New York, Paris, Philadelphia, Rio de Janeiro, Rome, Sao Paulo, Seoul, Shanghai, Sydney, Tokyo, Toronto and Warsaw.

References

DCLG (2007) 'Eco-towns prospectus', July, Department for Communities and Local Government, London

London Development Agency (2003) 'Green alchemy turning green to gold: Powering London's future – a study of the sustainable energy sector', LDA, London

Mayor of London (2002) 'Planning for London's growth', GLA, London

Mayor of London (2004a) 'The London Plan: Spatial development strategy for Greater London', GLA, London

Mayor of London (2004b) 'Green light to clean power: The Mayor's energy strategy', GLA, London

Mayor of London (2006) 'The Mayor's transport and air quality strategy revisions: London's low emission zone', GLA, London

Mayor of London (2007) 'Action today to protect tomorrow: The Mayor's climate change action plan', GLA, London

Mayor of London (2008) 'The London Plan (consolidated with changes since 2004)', GLA, London

Royal Commission on Environmental Pollution (2000) 'Twenty-Second Report: Energy – the changing climate', The Stationery Office, London

Woking Borough Council (2008) 'Service and performance (best value) plan 2008–2009', WBC, Woking

22

Planning and Governance of Climate Change

Harriet Bulkeley

Climate change poses many significant technical challenges for spatial planning, including, for example, how to plan adaptation responses with uncertain knowledge of potential impacts or how to include alternative forms of energy supply within local developments. Although significant, such technical issues are only part of a broader set of issues related to how spatial planning is framed, practised and implemented. In short, climate change has the potential not only to reshape what it is that spatial planning does, but how it is done. In this chapter, the governance challenge of addressing climate change in the spatial planning system is considered. Governance is a slippery term with almost as many definitions as advocates but in essence it relates to the institutionalized processes through which collective action is defined and determined. It is argued that through the spatial planning system multiple modes of governing climate change are taking shape, creating a fragmented governance landscape, which provides both opportunities and barriers for progress in addressing this critical issue. This challenges the notion that spatial planning can necessarily provide a way of integrating climate change into other policy domains, or offer a straightforward means of delivering climate policy goals.

The first section of this chapter reflects on the emergence of climate change as an issue for planning governance, before considering how this phenomenon might be analysed. Taking a multilevel governance perspective, the second section explores the emergence of planning governance in relation to three critical climate change policy areas – energy supply, energy demand and adaptation. In conclusion, the implications for conceiving the governance of climate change planning are considered.

Planning, climate change and multilevel governance

Over the past decade there has been a growing recognition by the UK government of the potentially significant role that planning can play in addressing climate change. As the Planning Policy Statement: Planning and Climate Change, makes clear, 'used positively' planning has a 'pivotal and significant role' in meeting the challenge of climate change (Department for Communities and Local Government (DCLG), 2007a, p9). This raises the critical question of how spatial planning is indeed being used in the governance of climate change and how such processes might be conceptualized.

Planning for climate change?

Since the early 1990s various commentators have documented the increasing salience of sustainable development in spatial planning in the UK (Healey and Shaw, 1994; Owens, 1994; Owens and Cowell, 2002). As national planning guidance began to engage with principles such as mixed use development, reducing the need to travel, and better design, so too the rhetoric of needing to address climate change began to permeate the planning system. Revisions of planning policy statements provided further opportunities to embed climate change as a central issue for planning. As Table 22.1 illustrates, most, if not all, national planning guidance

Table 22.1 Planning policy and climate change

Planning policy statement	Policy/guidance with an impact on climate change
PPS1: Delivering sustainable development	Address causes and potential impacts of climate change. Reduce energy use. Reduce emissions. Promote renewable energy use. Location and design of development.
PPS3: Housing provision	Delivery of homes that are well-designed. Making the best use of land. Making use of new building technologies to deliver sustainable development.
PPG 4: Industrial, commercial development and small firms	Reduce the need to travel. Location of development.
PPS6: Planning for town centres	Reduce the need to travel. Encourage use of public/alternative transport. Facilitate multi-purpose journeys.
PPS7: Sustainable development in rural areas	Planning applications should recognize the need to protect natural resources. Provide for sensitive exploitation of renewable energy sources.
PPS9: Biodiversity and geological conservation	Account for climate change on distribution of habitats and species, and geomorphologic processes and features.
PPS10: Planning for sustainable waste management	Encouraging more sustainable waste management, which respects the waste hierarchy (reduce, re-use, recycle, energy recovery, disposal).
PPS11: Regional spatial strategies	Addressing climate change and energy in regional spatial strategies.
PPS12: Local development frameworks	Act on a precautionary basis to reduce the emissions that cause climate change and to prepare for its impacts.
PPG13: Transport	Reduce the need for travel, especially by car, by influencing the location of development, fostering development which encourages walking, cycling or public transport etc.
PPG20: Coastal Planning	Identify areas likely to be at risk from flooding.
PPS22: Renewable Energy	Increased development of renewable energy resources through regional spatial strategies and local development documents.
PPS23: Planning and Pollution Control	Planning should reduce greenhouse gas emissions and take account of potential effects of climate change where possible.
PPS25: Development and Flood Risk	Planning policies should reflect the increased risk of coastal and river flooding as a result of climate change.

Source: DCLG (2006a, p56)

can now be read as having some bearing on the issue of climate change.

Despite the significance of placing climate change on planning agendas, the dispersed and indirect nature of the guidance involved meant that the issue languished towards the bottom of most planning priorities. Since the mid-2000s, however, there has been a more explicit and concerted effort to place climate change at the centre of spatial planning. In the strategy for sustainable development, 'Securing the Future' (Department for the Environment, Food and Rural Affairs (Defra), 2005), climate change is considered to be 'the greatest threat' and is placed squarely at the heart of the strategy. 'Securing the Future' goes on to argue that:

> *The land-use planning system provides the key framework for managing development and the use of our land in ways which take into account the sustainable use of our natural resources; for example, by promoting or encouraging the use of renewable energy in new developments and reducing the use of non-renewable resources (and emissions) by locating development where it can be accessed by means other than private car.*
>
> (Defra, 2005, pp88–89)

The explicit role for spatial planning as a mechanism for addressing climate change has been recognized in the UK Climate Change Programme (Defra, 2006) and has led to a suite of new planning policies. Planning Policy Statement 1 (PPS1), which provides the framework for spatial planning in the UK, specifically states that:

> *Regional planning bodies and local planning authorities should* ensure *that development plans contribute to global sustainability by addressing the causes and potential impacts of climate change – through policies which reduce energy use, reduce emissions (for example, by encouraging patterns of development which reduce the need to travel by private car, or reduce the impact of moving freight), promote the development of renewable energy resources, and take climate change*

> *impacts into account in the location and design of development.*
>
> (Office of the Deputy Prime Minister (ODPM), 2005a, p13 (ii), emphasis added)

While previous planning guidance suggested, for example, that the potential for regions to mitigate climate change or their vulnerability to impacts should be 'considered' (PPG11, in ODPM, 2004), or that planners should 'promote the energy efficiency of new housing where possible' (Department of the Environment, Transport and the Regions (DETR), 2000, p3), the language of PPS1 is clearer – planning bodies and authorities need to ensure that both the causes and impacts of climate change are addressed. In order to provide further guidance on what this entails, a supplementary planning policy statement, 'Planning and Climate Change' has been published (DCLG, 2007a). Here, the role of spatial planning in addressing climate change is seen to be five-fold:

> *secure enduring progress against the UK's emissions targets [...] deliver the Government's ambition of zero carbon development [...] shape sustainable communities that are resilient to and appropriate for the climate change now accepted as inevitable [...] create an attractive environment for innovation and for the private sector to bring forward investment [...] capture local enthusiasm and give local communities a real opportunity to influence, and take, action on climate change.*
>
> (DCLG, 2007a, p9)

In essence, the culmination of engagement with the principles of sustainable development and more recent explicit integration of mitigating and adapting to climate change has placed this issue as one of fundamental importance for spatial planning. As can be seen in the quote above, the policy architecture through which these principles have been established and operate is one in which planning is an essential delivery mechanism for national climate change policy. In the main, the role of planning is seen in these policy documents as one of translating national policy goals – for emissions reduction or renewable energy generation, for example – into regional and local realities.

There is, however, an alternative framing of the role of spatial planning in climate change policy. Rather than being a delivery mechanism, planning is seen as 'part of the problem' as numerous communities and local authorities challenge developments taking place in the name of addressing climate change. A recent report by the Local Government Association Climate Change Commission notes that 'there have […] been cases of councils objecting to the development of renewable energy supplies and over-riding the Environment Agency's advice on development on flood plains' (Local Government Association (LGA), 2007, p35). Numerous academic studies have documented the ways in which the objections of local communities to the development of renewable energy schemes are played out in the planning process (Bell et al, 2005; Toke, 2005; Haggett and Toke, 2006; Upham and Shackley, 2006; Eltham et al, 2008). Such conflicts start to reveal that an understanding of the governing of spatial planning for climate change, which reads this process as essentially top–down, misses the complexities of how spatial planning is becoming enmeshed in the multilevel governance of climate change responses.

Planning and multilevel governance

The shift in academic attention from the dynamics of government to governance is visible across the social sciences. While there are a multitude of definitions, in its most encompassing form governance 'relates to any form of creating or maintaining political order and providing common goods for a given political community on whatever level.' (Risse, 2004, p289). In this approach, governance is not seen as a new and distinct way of governing collective affairs. Rather, traditional modes of governing, through hierarchy and elected government are seen as but one means of achieving collective action alongside various forms of market and network.

Moreover, governance can involve diverse collections of institutional actors, not necessarily limited to state institutions (Kooiman, 2003). Such an approach avoids rather unhelpful discussions as to whether a shift from 'government' to 'governance' has taken place, while also recognizing the multiple actors and forms of authority brought to bear in the processes governing collective affairs for the public good. Within this broad approach, one particularly useful set of debates has revolved around the emergence of 'multilevel governance'. Here, the argument goes, the restructuring of the state has led to shifts in the role and functions of the nation-state: upwards, to international and transnational organizations and institutions; downwards, to cities and regions; and outwards, to non-state actors (MacLeod and Goodwin, 1999; Bulkeley et al, 2005). Hooghe and Marks (2001) have argued that through these processes two related forms of multilevel governance have emerged. Type I refers to the 'dispersion of authority to a limited number of non-overlapping jurisdictions at a limited number of levels,' which are essentially based on territorial entities (e.g. local, regional, national governments) (Betsill and Bulkeley, 2006; Smith, 2007). Type II multilevel governance 'captures both the multiple levels at which governance is taking place, and the myriad actors and institutions which act simultaneously across these levels.' (Bulkeley and Betsill, 2003, p29; see also Betsill and Bulkeley, 2006; Smith, 2007).

The extent to which the concept of (multi-level) governance provides an adequate framework for the analysis of contemporary spatial planning in the UK is moot. Cowell and Murdoch (1999, p655) have argued that 'the particular circumstances which surround land-use regulation may mitigate against a shift towards the flexible coalitions of partnerships deemed characteristic of governance.' As they go on to suggest, in relation to housing and minerals, 'the two arenas remain largely unmoved by any general shift towards governance […] characterized by strong national-to-local 'chains of command' which ensure that a 'dominant line' – associated with national demands for certain forms of development – is disseminated to a multitude of local decision-making bodies' (Cowell and Murdoch, 1999, p663). In their careful analysis of these two sectors, they suggest that through this 'dominant line' central government continues to configure spatial planning in terms that make little room for the inclusion of non-state actors in processes of deliberation and implementation. In another arena of spatial planning, waste, Davoudi and Evans (2005) reach different conclusions. They suggest that 'such a dominant strategic line imposed by government has historically been absent in the waste-policy area' (2005, p297), so that new spaces of governance are emerging with new policy architectures for managing waste at regional and local levels. However, in both cases it is the presence or absence of a 'dominant line' that shapes the governance possibilities for planning.

In an alternative analysis, Bulkeley et al (2007) suggest that the expression of a 'dominant line' in government policy for planning does not necessarily mean that 'government', in the traditional sense, actually is the dominant institutional form of governing. Rather, in relation to municipal waste policy, they identify multiple modes of governing, sets of practices and techniques 'deployed through particular institutional relations through which agents seek to act on the world/other people in order to attain distinctive objectives in line with particular kinds of governmental rationality' (Bulkeley et al, 2007, p2739). In this perspective, modes of governing are orchestrated by particular conceptions of the policy problem, institutionalized relationships, and clusters of policy interventions and techniques. Within these modes of governing, government plays a variety of roles. This approach allows for the possibility of recognizing the plurality of governing relations and arrangements that take shape around particular issues (Jessop, 1997; Cowell and Murdoch, 1999, p655). It also enables an acknowledgement of the multiple ways in which states undertake governing and may be involved in different modes of governing simultaneously (Bulkeley et al, 2005). In short, states may be involved in hierarchical forms of government, but also, for example, in processes of network management and market design (Jessop, 2002, p241). At the same time, the

distinction between the ideal types of multilevel governance is broken down, to reveal a more fragmented governance landscape in which multiple modes of governing are enacted simultaneously. In this view, spatial planning is neither the delivery mechanism for national policy goals nor an arena for conflicts over climate change, but a contested set of discourses and processes variously enrolled into, in this case, alternative modes of governing for climate protection.

Governing planning for climate change[1]

An analysis of the processes and practices of planning for climate change in relation to three critical areas – energy supply, energy demand and adaptation – suggests that in this arena the lines between 'Type I' and 'Type II' governance are blurred. Rather, a complex and fragmented landscape emerges where dominant strategic lines emanating from national government, alternative planning rationalities established from the 'bottom up', and various partnerships are entangled in responding to climate change through spatial planning.

Energy supply

At first glance, the arena of renewable energy planning bears all the hallmarks of 'strong national-to-local 'chains of command' which ensure that a 'dominant line' […] is disseminated to a multitude of local decision-making bodies' (Cowell and Murdoch, 1999, p663) which would in turn seem to emphasize the importance of government as the dominant mode of governing. The 2003 Energy White Paper reiterated the Labour government's target that 10 per cent of electricity should be generated from renewables by 2010, with the aspiration, recently cemented in European and UK policy, of achieving a rate of 20 per cent by 2020. In order to achieve this, the 2003 Energy White Paper called for the planning system to be 'streamlined and simplified' and for the (then) Office of the Deputy Prime Minister to revise PPG22 in order that the inclusion of renewables (and energy efficiency) in developments could be taken into account in the planning process (DTI, 2003, para 4.3). The revised Planning Policy Statement 22 (PPS22) on Renewable Energy takes the ambitions of the Energy White Paper to heart and states that:

> *Regional spatial strategies and local development documents should contain policies designed to promote and encourage, rather than restrict, the development of renewable energy resources*
> (ODPM, 2005b, para 1.2)

and further, that:

> *The wider environmental and economic benefits of all proposals for renewable energy projects, whatever their scale, are material considerations that should be given significant weight in determining whether proposals should be granted planning permission.*
> (ODPM, 2005b, para 1.4)

To ensure that this proactive approach to planning for renewables is undertaken, regional spatial strategies are required to include a target for the minimum amount of renewable energy generation for the region, to be monitored and increased as capacity is generated, and which 'where appropriate [...] may be disaggregated into subregional targets' (ODPM, 2005b, para 2.5). Regional and local planning authorities are then charged not only with encouraging the development of renewable energy, but also with meeting specific targets for the creation of new renewable energy capacity. As Smith (2007, p6267) argues, 'through the authority of regional strategies, there has been an attempt to institute regional renewable energy governance within a Type I multi-level governance model,' where this involves the transmission of national policy goals down to localities. In keeping with the arguments offered by Cowell and Murdoch (1999) and Davoudi and Evans (2005), here the suggestion is that the 'dominant line', or governmental rationality of renewable energy policy, configures the planning system as a delivery

mechanism for predetermined policy goals. The proposal in the Planning Bill 2008 for the formation of an Infrastructure Planning Commission to determine the outcome of planning applications in areas of national interest – including large scale renewable energy projects and new nuclear power stations – further supports this argument. It suggests that new planning mechanisms, controlled by the centre, will be imposed where the existing planning system is perceived to have failed in its delivery of adequate forms of alternative energy supply.

However, two aspects of the governing of renewable energy supply challenge the notion of a strong dominant line that it is determined from the centre. First, while changes to PPS22 and PPS Planning and Climate Change explicitly make space for the development of embedded energy generation, the growing popularity of targets for on-site renewable energy generation has emerged from the 'bottom up'. The London Borough of Merton is credited with the invention of 'The Merton Rule', whereby developments of over 1000m^2 are required to incorporate renewable energy generation of at least 10 per cent of predicted energy requirements (Friends of the Earth (FoE), 2005, p7). Following this lead, in London's 2004 Energy Strategy, the Mayor required 'applications referable to him to incorporate renewable energy technologies' and expected 'major developments to generate at least 10% of their energy needs from renewable sources' (Mayor of London, 2004). The draft Further Alterations to the London Plan specifically require new developments to have energy supplied by combined cooling, heat and power (CCHP) wherever feasible and to reduce their CO_2 emissions by a further 20 per cent through the production of on-site renewable energy generation (Mayor of London, 2006). There are now an estimated 100 councils who have implemented the Merton Rule (LGA, 2007, p34). Rather than encompassing renewable energy in general, this particular governmental rationality focuses on specific sites and technologies through which renewable energy might be generated. Here, then, is an example of an alternative 'dominant line', emanating from the 'bottom up' and adopted across local governments in the UK on a voluntary basis. This suggests that there is more to the governing of climate change planning than that which rests on central dictat, for 'there are other modes of power exercised on a regular basis in the process of governing at a distance that do not at all resemble the types of deference and recognition that accompany authority' (Allen, 2004, p27).

Second, despite the planning targets for renewable energy supply, in itself the spatial planning system has little leverage in bringing forward proposals for projects to meet these ambitions. In short, 'regions have very little direct control over the energy infrastructures in their territory. At best they can contribute to favourable contexts, but they do not take the key decisions that have long-term consequences' (Smith, 2007, p6273). Delivering renewable energy capacity, then, appears a Type II governance challenge (Smith, 2007, p6268). This is made clear in government policy: '*Planning and Climate Change* sets out a clear and challenging role for regional and local spatial strategies. They are expected to help shape the framework for energy supply in their area' (DCLG, 2006a, p4). The capacity for spatial planning to deliver changes in the energy supply system is limited without the engagement of a range of other partners – from business to community groups. This enabling role for planning is explicitly recognised in PPS Planning and Climate Change, where it is suggested that, amongst other things, the contribution that planning should make to addressing climate change includes 'bringing together and encouraging action by others [...] create an attractive environment for innovation and for the private sector to bring forward investment in renewable and low-carbon technologies and supporting infrastructure; and give local communities real opportunities to influence, and take, action on climate change' (DCLG, 2006a, pp13–14). The LGA (2007, p35) suggests that new partnerships may be required 'to exploit low-carbon opportunities, for example, with the Forestry Commission to support the use of biomass in boilers and heating systems' (LGA, 2007, p35). This emphasis on the role of spatial planning in

'enabling' action for climate protection is reflected more broadly in local climate policy (Bulkeley and Kern, 2006). Such approaches move beyond a 'dominant line' to negotiations over the distribution of costs, benefits, responsibilities and risks in the generation of new forms of energy supply. At the same time as opening up productive ground for new governance roles for planning, such debates create space for conflicts over whether new forms of energy generation are warranted and, if so, where they should be located.

Managing energy demand

The impact of planning on the form and design of urban areas and consequently on energy use has attracted sustained attention. While clearly the location, density and design of development alone cannot reduce energy use in urban areas, how developments are designed and planned will have a significant impact on future emissions of greenhouse gases (GHGs) (Bulkeley and Betsill, 2003). In recognition of this potential influence, the argument that the planning system should take into account the energy implications of the form and placement of new development was articulated throughout the 1990s in various guidance documents produced by central government (Bulkeley and Betsill, 2003). Spatial planning was seen to have two key roles to play in shaping energy demand. First, through the design of new developments. Second, through policies on location and access (Bulkeley, 2006). As climate change has risen up planning agendas it has been the former issue which has been the policy focus.

Throughout the 1990s, policies that promoted energy efficiency, passive solar gain, and the use of brownfield sites, to name but a few, were frequently couched in terms of their potential impact on climate change (see Table 22.1). However, such examples remained isolated from the mainstream of planning practice, where the national position that building regulations were being improved sufficiently to address issues of energy use in new developments continued to dominate. By the mid-2000s, this tone had changed. The introduction of the 2006 Code for Sustainable Homes (DCLG, 2006b) acknowledged that Building Regulations, however much they may have recently been improved, are still not delivering sustainable buildings. In particular, although voluntary, the Code was seen to 'signal the future direction of Building Regulations in relation to carbon emissions from, and energy use in homes, providing greater regulatory certainty for the homebuilding industry' (DCLG, 2006b, p5). The Code was one element of a suite of policies, including the PPS Planning and Climate Change and a consultation, 'Building a Greener Future' (DCLG, 2007b), on moving towards zero-carbon standards for housing in the building regulations by 2016. Through this package of policy documents, a dominant government line concerning the significance of housing in relation to climate change and the importance of reducing emissions in this sector through spatial planning and building regulations emerged:

> If we build the houses we need, then by 2050 as much as one-third of the total housing stock will have been built between now and then. So we need to build in a way that helps our strategy to cut carbon emissions – both through reducing emissions of new homes and by changing technology and markets so as to cut emissions from existing homes too.
>
> (DCLG, 2007b, p5)

However, unlike the case of energy supply discussed above, here the governmental rationality is not intended to be cascaded through layers of government. Rather, the roles of different private and public sector actors are recognized at the outset. For example, 'Building a Greener Future: Policy statement', recognizes that 'housebuilders will need to look into zero and low carbon sources of electricity supply, an area currently outside Building Regulations' and that 'zero carbon homes will also require new partnership working between housebuilders and energy companies' (DCLG, 2007b, p15). Local government, and the planning system, are also seen to be critical here, for example, in 'bringing

together interested parties and facilitating the establishment of decentralized energy systems' (DCLG, 2007b, p19).

This mode of governing domestic energy demand through partnerships has several interesting facets. First, as set out in the quote above, the tightening of regulation in respect to new development is seen as a key means of 'governing' the existing housing stock, from which the vast majority of domestic emissions of GHGs emanate. Through creating new forms of technology and new markets, the expectation is that there will be a knock-on effect across the existing stock. Here, the mode of governing is seen to work precisely because of its ability to spill out and network across public/private, current/future boundaries. Second, and related, concerns have been expressed that the focus, in the suite of current planning policies for climate protection, on new development, as the key to unlocking potential energy savings in the existing stock has caused other approaches to governing current demand to be sidelined. In particular, responses to the consultation on the draft PPS on Climate Change suggested that it 'focuses on new-build residential developments and fails to recognise the significant reductions in emissions that can be achieved through refurbishing and reusing older buildings and areas.' (DCLG, 2006a, p9). There are alternative approaches through which the planning system is being brought to bear on the existing housing stock. In a case-study highlighted in 'A Climate of Change' (LGA 2007), Uttlesford District Council use a supplementary planning document to require 'cost-effective energy efficiency measures to be carried out throughout the existing building as a condition of planning consent for a home extension' (LGA, 2007, p34). Here, spatial planning is being used more directly to govern the existing stock, but such examples are, to date, few and far between. More radical suggestions focus on the role of planning not in improving existing housing, but in demolishing it in order to make way for more efficient buildings. Boardman (2007) suggests that in order to meet the national target of a 60 per cent reduction in GHG emissions in the domestic sector, some 80,000 properties per year will need to be demolished. Although demolition is back on the policy agenda in the UK, with the establishment of nine areas of housing market renewal, the conclusion that this has been achieved 'whilst supporting and sustaining local communities' (Boardman, 2007, p371) is moot. Given that the planning system has traditionally been geared towards control or prevention, rather than proactive planning in which some forms of development can be promoted over others, it is unclear whether a new emphasis on planning as an 'enabling' mode of governance will be able to address this challenge.

Moreover, whilst the role of planning in governing energy demand appears cut loose from the tight moorings of Type I multilevel governance, which characterize at least one mode of the governing of energy supply, further examination reveals that national government requirements are never far from the surface. In particular, despite the over-riding message of current policy that climate change should be fundamentally integrated into planning and development policy, the need to secure economic growth through the provision of adequate (and affordable) housing remains paramount. For example, while local planning may seek to go beyond national standards and targets in certain circumstances, the housing market drivers prevail:

> *where there are demonstrable and locally specific opportunities for requiring particular levels of building performance through the planning system these should be set out in advance in a development plan document. In so doing, local authorities would need to have regard to a number of considerations, including whether the proposed approach is consistent with securing the expected supply and pace of housing development shown in the housing trajectory required by PPS3*
> (DCLG, 2007a, p18)

Despite significant changes in policy rhetoric, the importance of other drivers in shaping the potential policy space and realities of everyday practice for governing climate change through the planning system should not be underestimated.

Adaptation

One final means through which spatial planning is involved in climate protection is in terms of developing resilience to the predicted impacts of climate change. As stated in PPS1, the planning system is charged with ensuring that development 'contribute to global sustainability by addressing the [...] potential impacts of climate change' (ODPM, 2005a, p6). To date there is limited evidence as to how far this responsibility is being taken up in the planning system. A survey conducted in 2004 found that while threats of flooding and water supply issues were commonly mentioned by respondents from the land-use planning sector, 'surprisingly, few local authority respondents identified specific planning implications for the location of development' (SEEDA, 2004, p15). More recently, 'A Climate of Change' reports that 80 per cent of respondents felt that 'their local authority had been not very effective, or not at all effective, in adapting to climate change' (LGA, 2007, p26). Here, in a similar fashion to the issues of energy supply and demand, multiple modes of governing planning for climate change are apparent.

First, in the area of flood risk, a 'dominant line' exists. Planning Policy Guidance 25 on Flood Risk, was approved in 2001 and makes it clear that the 'the susceptibility of land to flooding is a material planning consideration' and that planning authorities should 'consider how a changing climate is expected to affect the risk of flooding over the lifetime of developments' (DETR, 2001, p4). A recent report by the Environment Agency (EA) found that 'almost all LPA development plans now include flood risk statements or policies, and the newer plans are beginning to reflect the content of PPG25' (EA, 2004, p3). Nonetheless, although the ODPM maintained that the number of applications which go ahead against the advice of the EA had halved since the introduction of PPG25 (ODPM, 2005c), 22.5 per cent of those applications to which the EA had objections on flood risk grounds (EA, 2004) continued to be built in 'at risk' areas, suggesting that at least in a good number of locations factors other than flood risk are driving the development process. Indeed, in recent consultation response to the proposed PPS Planning and Climate Change, the view was voiced that 'in terms of controlled development on floodplains, it was felt that the PPS was too restrictive, and the PPS should provide a mechanism for providing greater local flexibility for areas that have limited land available for development' (DCLG, 2006a, p14). In other words, the dominant line that development on floodplains should be avoided is being challenged in some quarters by reference to the importance of providing sufficient housing development to meet the Government's other stated policy ambitions. This suggests that even 'dominant lines' are contested and can unravel when it comes to the governing of particular places.

Second, in other areas of potential climate change vulnerability – including biodiversity, infrastructure and water supply – there is little by way of national policy guidance in relation to spatial planning. The recent advice on best practice, The Planning Response to Climate Change (ODPM, 2004), illustrates that while some planning bodies are taking account of the risks of climate change across a diverse spectrum of issues, much is still to be done. It is in this governance gap that 'bottom–up' approaches for addressing adaptation are emerging. One significant case in point is the response of the Greater London Authority. In 2001, the GLA established the London Climate Change Partnership (LCCP), a 'stakeholder group coordinated by the Mayor of London, consisting of over 30 key organisations with representation from government, climate scientists, domestic and commercial development, transport, finance, health, environment and communication sectors'.[2] The aim of the LCCP is to prepare London for the impacts of climate change through raising awareness in key sectors and by embedding adaptation measures in planning and development decisions.[3] To this end, the LCCP has commissioned various reports and guidance documents, including a 'checklist for development' which is:

> aimed at helping developers and their design teams, allowing them to incorporate the appropriate measures at the design stage of developments. It should also help planners make any necessary modifications to their local planning documents and to incorporate appropriate checks in their scrutiny of planning applications. The aim is to future-proof developments and to build-in resilience to climate change impacts now and in the future.
>
> (GLA, 2005)

Here, adapting to climate change is taking place on the fringes of the spatial planning system as a partnership of government and private actors seek to develop guidance with the ambition of governing private and public actors. Within London, authorities are also seeking to work outside the traditional boundaries of the planning system. In 'Rising to the Challenge' (CLC, 2007), the City of London Corporation provide a catalogue of potential measures to address climate change, including encouraging 'businesses to consider relocating flood-sensitive IT equipment and archives out of London to areas with negligible flood risks' (CLC, 2007, pii) and encouraging developers 'to install sustainable drainage systems and green roofs in targeted flash flood "hotspots" for new developments, redevelopments or major refurbishments' (CLC, 2007, pii). In the absence of a 'dominant line' concerning adaptation to climate change, novel approaches involving, but not exclusive to, spatial planning are emerging from the 'bottom up', creating alternative rationalities about why and how climate change should be governed.

Conclusion

This brief analysis of planning, governance and climate change suggests that the conclusions reached by Cowell and Murdoch (1999) that land-use planning remained an area of policy relatively immune to the 'new wave' of governance needs to be tempered. Certainly, in relation to the critical climate change areas of energy supply, energy demand and adaptation, we can see the emergence of 'dominant lines', or rationalities, and strong 'national to local chains of command' (Cowell and Murdoch, 1999, p663).

For some analysts, such arrangements are not antithetical to 'governance' but rather provide one form (Type I) of governance, albeit relatively centralized, through which collective action is organized and achieved (Smith, 2007). Moreover, the policy areas examined above also reveal a multiplicity of modes of governing alongside the top–down invocation of the planning system as a delivery mechanism for predetermined goals. Here, for example, we witness alternative rationalities shaping the planning of energy supply. At the same time, spatial planning is increasingly held up as a means through which to engage proactively with a range of different partners – in particular private sector and community-based organizations – in order to achieve climate change objectives. In short, Type II governance arrangements are both emerging spontaneously and being summoned into existence by central government. The result is a fragmented landscape of planning for climate change in which central dictat and grassroots innovation sometimes sit uncomfortably.

This analysis suggests that spatial planning should not be considered as a delivery mechanism for climate change policy. Rather, what it means to respond to climate change is defined, contested and made material through processes of negotiation and conflict (Owens, 2004; Bulkeley, 2006). Those modes of governing that deploy the familiar tools of central government – targets, reward and sanction – may seek to reconfigure the planning system to minimize the chances of central goals being lost in such processes of translation. In contrast, those modes that depend upon the emerging arsenal of governance tools – persuasion, inducement and 'generative power' (Coafee and Healy, 2003; Allen, 2004) – may find that the space for compromise leads only to lowest common denominator solutions. Here then lies the paradox. While Type I governance may be more effective at 'getting the job done' in terms of goal setting, without Type II governance realizing this potential may be little more than a pipe dream. Rather than seeking to give coherence to the governance landscape, the challenge for spatial planning is perhaps one of finding a means of productive accommodation in particular places.

Notes

1 This section draws on Bulkeley (2006).
2 See: www.london.gov.uk/climatechange partnership/index.jsp, accessed 30 June 2008.
3 See: www.london.gov.uk/climatechange partnership/aims.jsp, accessed 30 June 2008.

References

Allen, J. (2004) 'The whereabouts of power: Politics, government and space', *Geografiska Annaler*, vol 86, pp19–32
Bell, D., Gray, T. and Haggett, C. (2005) 'The "social gap" in wind farm siting decisions: Explanations and policy responses', *Environmental Politics*, vol 14, no 4, pp460–477
Betsill, M. M. and Bulkeley, H. (2006) 'Cities and the multilevel governance of global climate change', *Global Governance*, vol 12, no 2, pp141–159
Boardman, B. (2007) 'Examining the carbon agenda via the 40% house scenario', *Building Research and Information*, vol 35, no 94, pp363–378
Bulkeley, H. (2006) 'A changing climate for spatial planning?', *Planning Theory and Practice*, vol 7, no 2. pp203–214
Bulkeley, H. and Kern, K. (2006) 'Local government and climate change governance in the UK and Germany', *Urban Studies*, vol 43, no 12, pp2237–2259
Bulkeley, H., Watson, M. and Hudson, R. (2007) 'Modes of governing municipal waste', *Environment and Planning A*, vol 39, no 11, pp2733–2753
Bulkeley, H., Watson, M., Hudson, R. and Weaver, P. (2005) 'Governing municipal waste: Towards a new analytical framework', *Journal of Environmental Policy and Planning*, vol 7, no 1, pp3–25
Bulkeley, H. and Betsill, M. (2003) *Cities and Climate Change: Urban Sustainability and Global Environmental Governance*, Routledge, London
CLC (2007) 'Rising to the challenge: The City of London Corporation's Climate Adaptation Strategy', www.cityoflondon.gov.uk/ Corporation/living_environment/sustainability/ climate_change/, accessed 30 June 2008
Coafee, J. and Healy, P. (2003) '"My voice, my place": Tracking transformations in urban governance', *Urban Studies*, vol 40, no 10, pp1979–1999
Cowell, R. and Murdoch, J. (1999) 'Land use and the limits to (regional) governance: Some lessons from planning for housing and minerals in England',

International Journal of Urban and Regional Research, vol 23, pp654–669

Davoudi, S. and Evans, N. (2005) 'The challenge of governance in regional waste planning', *Environment and Planning C: Government and Policy*, vol 23, pp493–517

DCLG (2006a) 'Consultation – Planning Policy Statement: Planning and Climate Change, Supplement to Planning Policy Statement 1', HMSO, London

DCLG (2006b) 'Code for Sustainable Homes: A step-change in sustainable home building practice', www.planningportal.gov.uk/uploads/code_for_sust_homes.pdf, accessed 30 June 2008

DCLG (2007a) 'Planning Policy Statement: Planning and Climate Change, Supplement to PPS1', HMSO, London

DCLG (2007b) 'Building a Greener Future: Policy Statement', www.communities.gov.uk/publications/planningandbuilding/building-a-greener, accessed 30 June 2008

DEFRA (2005) 'Securing the Future: Delivering UK sustainable development strategy', www.sustainable-development.gov.uk/publications/uk-strategy/index.htm, accessed 30 June 2008

DEFRA (2006) 'Climate Change: The UK Programme 2006', Department for the Environment, Food and Rural Affairs (DEFRA), HMSO, London

DETR (2000) 'Planning Policy Guidance 3: Housing', HMSO, London

DETR (2001) 'Planning Policy Guidance 25: Development and flood risk', HMSO, London

DTI (2003) 'Our Energy Future: Creating a low carbon economy', Energy White Paper, HMSO, London

Eltham, D., Harrison, G. and Allen, S. (2008) 'Change in public attitudes towards a Cornish wind farm: Implications for planning', *Energy Policy*, vol 36, no 1, pp23–33 Environment Agency (2004) 'High Level Target 12: Development and flood risk 2003/04', Joint Report to DEFRA and ODPM by the Environment Agency and Local Government Association, www.environment-agency.gov.uk/commondata/103599/final_hlt12_952545.doc, accessed 30 June 2008

Friends of the Earth (FoE) (2005) 'Tackling climate change at the local level: The role of local development frameworks in reducing the emissions of new developments', www.foe.co.uk/resource/briefings/ldf_climate_briefing.pdf, accessed 30 June 2008

GLA (2005) 'Adapting to climate change: A checklist for development. Guidance on designing developments in a changing climate', Greater London Authority, www.london.gov.uk/climatechangepartnership/docs/adapting_to_climate_change.pdf, accessed 30 June 2008

Haggett, C. and Toke, D. (2006) 'Crossing the great divide: Using multi-method analysis to understand opposition to windfarms', *Public Administration*, vol 84, no 1, pp103–120

Healey, P. and Shaw, T. (1994) 'Changing meanings of "environment" in the British planning system', *Transactions of the Institute of British Geographers*, vol 19, pp425–438

Hooghe, L. and Marks, G. (2001) 'Types of multi-level governance', *European Integration Online Papers*, vol 5, no 11, eiop.or.at/eiop/texte/2001-011.htm, accessed 30 June 2008

Jessop, B. (1997) 'The governance of complexity and the complexity of governance: Preliminary remarks on some problems and limits of economic guidance', in A. Amin and J. Huaner (eds) *Beyond Market and Hierarchy: Interactive Governance and Social Complexity*, Edward Elgar, Cheltenham, pp95–128

Jessop, B. (2002) *The Future of the Capitalist State*, Polity, London

Kooiman, J. (2003) *Governing as Governance*, Sage, London

LGA (2007) 'A Climate of Change: Final report of the LGA Climate Change Commission', LGA, London

MacLeod, G. and Goodwin, M. (1999) Space, scale and state strategy: Rethinking urban and regional governance, *Progress in Human Geography*, vol 23, no 4, pp503–527

Mayor of London (2004) 'Green light to clean power: The Mayor's energy strategy', www.london.gov.uk/mayor/strategies/energy/docs/energy_strategy04.pdf, accessed 30 June 2008

Mayor of London (2006) 'Draft further alterations to the London Plan' (Spatial Development Strategy for Greater London), www.london.gov.uk/mayor/strategies/sds/further-alts/docs.jsp, accessed 30 June 2008

ODPM (2004) 'The planning response to climate change: Advice on better practice', ODPM, London

ODPM (2005a) 'Planning Policy Statement 1: Delivering sustainable development', www.communities.gov.uk/planningandbuilding/planning/planningpolicyguidance/planningpolicystatements/planningpolicystatements/pps1/, accessed 30 June 2008

ODPM (2005b) 'Planning Policy Statement 22: Renewable energy', www.communities.gov.uk/planningandbuilding/planning/planning policyguidance/planningpolicystatements/planningpolicystatements/pps22/, accessed 30 June 2008

ODPM (2005c) 'Planning for flood risk: The facts', ODPM, London

Owens, S. (1994) 'Land, limits and sustainability: A conceptual framework and some dilemmas for the planning system', *Transactions of the Institute of British Geographers*, vol 19, pp439–456

Owens, S. (2004) 'Siting, sustainable development and social priorities', *Journal of Risk Research*, vol 7, no 2, pp101–114

Owens, S. and and Cowell, R. (2002) *Land and Limits: Interpreting Sustainability in the Planning Process*, Routledge, London

Risse, T. (2004) 'Global governance and communicative action', *Government and Opposition*, vol 39, no 2, pp288–313

SEEDA (2004) 'Meeting the challenge of climate change: Summary of the south east climate threats and opportunities research study (SECTORS)', Atkins plc and Oxford Brookes University on behalf of the South East Climate Change Partnership and the South East England Development Agency

Smith, A. (2007) 'Emerging in between: The multi-level governance of renewable energy in the English regions', *Energy Policy*, vol 35, no 12, pp6266–6280

Toke, D. (2005) 'Explaining wind power planning outcomes, some findings from a study in England and Wales', *Energy Policy*, vol 33, no 12 pp1527–1539

Upham, P. and Shackley, S. (2006) 'Stakeholder opinion on a proposed 21.5 Mwe biomass Gasifier in Winkleigh, Devon: Implications for bioenergy planning and policy', *Journal of Environmental Policy and Planning*, vol 8, no 1

23

Public Engagement in Planning for Renewable Energy

Claire Haggett

Introduction

This chapter is about public consultation, participation and engagement in spatial planning decisions. While the discussion focuses on renewable energy, these issues have resonance far beyond this topic. Consideration of the possibilities and procedures for involving people in the spatial planning agenda are important for many aspects of a response to climate change, from debates over governance to topics such as flood management. This chapter addresses the rationale, practicalities and difficulties of engaging people in planning, highlighting the directly transferable implications for other sector objectives, through an examination of renewable energy.

Indeed, it is impossible to think about the implementation of renewable energy without addressing the involvement and impact of the public in these processes. While fiscal regulations and subsidies, technical efficiency and political deliberations all affect the deployment of renewables, the stark fact remains that all of this matters little if there is no public support for a development. This is demonstrated by the success rate for wind farm applications in England and Wales, a mere 40 per cent through the normal procedures of the planning system, and a low rate compared with other forms of development (see Toke, 2003, 2005). As Toke (2005) and Haggett and Toke (2006) have explored, the key reason for wind farms being rejected is opposition from local people. People are protesting against renewable energy and they are doing so very effectively.

This might at first seem both odd and easily explained. Clean, green energy from a limitless supply could be the ideal 'win–win' option – for addressing carbon emissions from fossil fuel burning, securing energy sources that are not dependent on international political agreements or from areas of unrest and war, and for ensuring a continued and limitless supply. Local opposition seems to contradict the apparently widespread support for renewable energy, and opinion polls which consistently report three-quarters of people being in favour of more energy from a variety of renewable sources (for example, Tyndall Centre and MORI, 2006, www.ipsos-mori.com/polls/2005/uea.shtml). Any objections against such a solution are simply categorized as 'Nimbyism' writ large – people who are more concerned about their 'backyard' than the greater national and international good.

However, a growing body of research from around Europe has indicated that the reasons for protest might not be so straightforward and crucially, in terms of the purposes of this chapter, they depend on where, when and how people are able to engage effectively in the planning processes for renewable energy. Of course, what is meant by 'the public' and 'engagement' is complex as these are not homogenous concepts and, as will be discussed throughout this chapter, these definitions are at the very heart of the disputes over renewable energy. This chapter will consider the reasons for engaging people, and then different forms of engagement and the impact they have. It will conclude by discussing some of the difficulties of conducting appropriate and meaningful engagement, and the role that spatial planning can play.

The rationale for engaging the 'public' in renewable energy planning

The encouragement of public involvement in planning procedures and decision making about renewable energy has been well documented. There is both a long tradition of involving people in planning processes and indeed, a requirement in UK government policy to discuss decisions with local communities (Rydin and Pennington, 2000). There is also evidence to suggest that there is public support for efforts to involve local people in decisions, as Devine-Wright (2005) shows in his case study of wind farm development in South Wales.

There are a number of reasons for involving the public in decision making. Yearley et al (2003), for example, identify three key objectives behind encouraging greater participation which are useful to consider here. The first of these is a pragmatic approach, where public involvement is seen as a way of increasing the likelihood of a successful siting. At the very least there is perhaps the hope that involvement of the public may lead to 'better' or more competent decisions.

The second reason is because people have a right to participate in decisions that affect them, and involvement of the public may be an end in itself, rather than being intended to deliver better decisions. Different methods of participation may overtly or implicitly adopt this rationale (Fischer, 2000) and, in her study of renewable energy implementation, Gross (2007, p2734) unpacks this concept further to explore the associated issues of trust and fairness in participation. She makes a distinction between perceptions of fairness of *outcomes* and fairness of *process*, and in her interesting discussion argues that both of these are vital for encouraging engagement and acceptance. For some, a fair process is most important because it 'will allow discussion of the merits and impacts of the proposal, thereby helping determine what a good outcome is'. Gross concludes that people should therefore be allowed to participate so that they 'have the opportunity to speak and be heard'. Trust in decision-making processes about renewables is crucial (Jobert et al, 2007). Indeed, establishing 'fair' processes may be one way to restore diminished trust in authorities and institutions, as Healey (1996, p213) describes the potential for the 'building of social capital of trust, and the intellectual capacity of understanding, even across deep divides and tensions' through public consultation. The recent report by the Institute of Public Policy Research (IPPR) on attitudes and behaviour in response to climate change states that public engagement with energy issues is beneficial because 'empowering people to exert control and resolve problems is a good thing in its own right: improving governance, deepening democracy and rebuilding trust' (IPPR, 2007, p4).

The third reason identified by Yearley et al for encouraging the participation of the public is because their rich and full understanding of their local environment may differ from an outside 'expert' view. The validity of this approach has been demonstrated by Irwin (1995) and Wynne (1989). Cass (2006) notes that decisions based purely on rational, technical, 'objective' and 'scientific' assessments have been called into question and may not be generally accepted. Instead of trying to produce value-free judgements, the importance of incorporating values and beliefs is recognized (see Owen et al, 2004).

All three of these rationales resonate with the broader ethos of 'collaborative' forms of planning, influenced by Habermasian ideas, which opposes 'rationalist' models of land-use planning. Pennington (2002, p187) describes the latter as 'based on a technocratic conception of decision making, whereby public managers in possession of objective knowledge make decisions on the basis of maximising social welfare'. Collaborative planning, on the other hand, regards knowledge as being socially situated, not objective or solely the preserve of the scientific or technical domain. Such a focus values rather than ignores tacit understandings and everyday knowledge. Rationalist planning methods that do not place any emphasis on this form of knowledge not only may lead to poorly developed policies but also to disempowerment from and distrust of decision makers. This reflects what Habermas identified as a 'legitimation crisis' for those in power, who are distant from their electorate and make decisions without involving them on the basis of knowledge that has little relevance for people's lives.

Habermas proposed the concept of 'ideal speech communities', where participation for all is possible, and undistorted communication can take place (Habermas, 1976, p484). Rather than a focus on the achievement of rational, instrumental ends, communication can be based on mutual trust and comprehension, and attempts can therefore be made to harmonize different objectives through negotiation. Some of the key concepts in this communicative approach include recognizing and including all stakeholders, spreading ownership, building strategy and recognizing diverse interests (Harris, 2002). Habermas' theorizing has certainly been influential in planning theory and practice, with concepts of 'collaborative action', 'communicative planning' and 'inclusionary discourse' all attempting to apply Habermas' ideas to the arena of planning.

Engagement in practice

So, there is a well established theoretical rationale for engaging people in decision-making processes. But is this carried through into practice? Indeed, how often are efforts to engage people in planning for renewable energy actually made? And what impact do such attempts have?

Cass's (2006) comprehensive review of public engagement methods distinguishes between different means of 'engaging' people. Engagement, as he points out, can refer to formal, statutory processes of consulting and communicating with the public, or much more involved and inclusive efforts. Behind these are different rationales about the involvement and influence that the public can have, the sorts of responses that are intended to be elicited, and the action that is planned on the basis of them. In their useful mapping of methods for engaging the public and stakeholders in broadly defined energy issues, Chilvers et al (2005) distinguish three types of engagement: providing information and educating the public, consultation and deliberation. Each of these will now be considered.

Engagement as 'information provision'

One method of engaging the public is to provide information, and even attempt to educate. This may be about a particular renewable energy development or the need for renewables in general. The objectives of this form of 'engagement' seem most focused on pragmatic attempts to win support for an application, and to avoid the 'problems' of opposition (Cowell, 2007). It is in keeping with the 'decide–announce–defend' tradition, informing people of plans that have been made, and involves such methods as distributing leaflets, advertising and providing exhibitions and displays.

There are two points to make about this. Firstly, the public may well not need 'education' or even 'information' about a proposal. There is not necessarily a direct correlation between information and attitudes, nor is it sufficient or accurate to say that the people who oppose a development are uneducated or misinformed. More often, protesters are very familiar with the details of a particular proposal (Wolsink, 1994).

The UK government's planning policy on renewable energy, PPS22, exhorts local authorities to 'promote knowledge of and greater acceptance by the public of prospective renewable energy developments' (Office of the Deputy Prime Minister, 2004, p8, section 1: viii). It may well be that some concerns, such as the effect on local bird populations of a wind turbine or the fire risks from a hydrogen filling station, can be addressed by independent research from sources that people will trust. However, the statement in PPS22 implies that opposition can be addressed by closing an information gap through 'education'. This is the 'public deficit' model, based on the idea that 'if only people knew better' they would support renewable energy. But this form of engagement does not value people as local experts and may not even listen to them. An example of this is the public 'engagement' that took place about the offshore wind farm planned off the coast of North Wales (Haggett, 2008). While the developer did hold a series of meetings and open days at sites along the coast, and their representatives were available to answer questions from those who attended, the flow of information was one-way only. Points raised by people attending were not responded to, acted upon or even recorded. These were sessions designed to give information only, not to engage in dialogue, and local people criticized the process, feeling that the decisions about the wind farm had already been made and that there was a lack of 'real' consultation.

Secondly, 'engagement as information provision' may actually encourage protest. The 'decide–announce–defend' procedure through which most planning applications are made and decided means that people are not involved, or even informed about, a proposed development until many of its major features (location, size, dimensions) have already been determined. They are then forced into a position of protesting against it, as a way of having their opinions and concerns raised. This approach, with its minimal public involvement, has been repeatedly shown to generate public mistrust, concern and conflict (Walker, 1995; Wolsink, 2000; Haggett and Vigar, 2004; Breukers and Wolsink, 2007; Agterbosch et al, 2009). Public opposition may therefore mean that people are protesting about their lack of meaningful involvement in the planning process,

as much as the actual application. The approach also has the effect of problematizing opposition to proposals and fails to communicate with a representative balance of opinion. It also does nothing to encourage supporters to express their views. If people support a development, they rarely bother to write to their council and tell them so, and Pasqualetti (2001, p69) has discussed the difficulties of interpreting this 'public silence'.

So, engagement as 'information provision' is unlikely to be effective in terms of encouraging public support and trust, both for the particular proposals, and for the planning process as a whole. What is important to note however is that this form of consultation is the most frequently used, by both government and industry, when attempting to engage the public about renewable energy. Chilvers et al (2005, p28) describe how this 'bottom-line' approach to engagement, 'the minimum level allowed by law' ignores the success of and importance accorded to more thorough forms of engagement. These will now be considered.

Engagement as 'consultation'

A different form of engagement, rather than just providing information to a passive public, is to actively elicit their responses. This may help to address the reasons for what Bell et al (2005) have described as 'qualified support' for renewables: if people support renewables in general but oppose particular schemes, their support may depend on certain conditions being met. Engagement as 'consultation' provides the opportunity to discuss with people what their reasons for 'qualified support' are. These reasons may be grouped under four broad categories:

1   Support or opposition may depend on the ascribed aesthetic value of the particular landscape or seascape where a development is planned. This may form the basis of concern, rather than because it happens to be local, and as Toke et al (2007) point out, the attachment to landscape in the UK should never be underestimated (see also Wolsink, 2007a,

2007b). The specifics of an application will also be significant in shaping responses – size, layout and design of the development, and any mitigation of its impact (Wustenhagen, et al, 2007). This is a consideration for any renewable energy development, from a wind farm (Wolsink, 1996) to a hydrogen filling station (Ricci et al, 2007). The point is that people can be involved with discussions about both location and design specifics – as will be illustrated shortly.

2   Protests may also have roots in the perceived use of any location and its social, political and historical context (Haggett and Vigar, 2004). This will depend on which 'local people' are involved. They could, for example, be long-standing residents or 'incomers' to an area. This will affect their particular conception of the locality: for example, whether it is valued most for leisure or economic development, as a rural idyll or a place of livelihood. For example, protests about the offshore wind farm off the coast of Teesside have caused surprise, when the area is seen as 'already' scarred by the petrochemical industries – but these industries form the economic heartland of the area, and (unlike the wind farm) provide jobs and income for local people (Phillimore and Moffatt, 2004). The social context of the area means that factories might be acceptable to some, while the wind farm is not.

3   Renewable energy conflicts epitomize the disjuncture between the local and the global. While issues of global warming may be far removed from everyday life, fears of house prices falling or the impact on local businesses are not. If the concerns of local people are not taken seriously they are likely to become the focus of opposition. Further, while there may be national and international benefits from a reduction in the use of fossil fuels, the proportional reduction in carbon dioxide emissions for each person who lives near a renewable energy development may be a small and intangible compensation for the local impacts of the development. As Agterbosch et al (2009) state: 'local residents are inclined to oppose a project when they feel the decision-making serves the external economic interests or the global environmental interests by ignoring local aspects such as hindrance and risks for citizens'. Engaging with people, and finding out what these local aspects are, how they feel they may be disadvantaged by a development, and any possible remediation strategies can begin to address some of these issues. In South Wales, a community renewable energy initiative has aimed to provide solutions to issues that local people felt were important – be those free energy-saving light bulbs, photovoltaic panels on village halls or a solar powered system for recharging one resident's electric wheelchair (www.awelamantawe.org.uk). The point is about engaging people to find appropriate answers for their locality, and the challenging but not impossible task of doing this.

4   Related to this are issues over the ownership of a development. Indeed, people may position themselves as much against the developers as the development itself (Wolsink, 1996). Jobert et al (2007, p2758) discuss the importance of the 'local integration of the developer in terms of proximity, knowledge of the context, contacts with authorities and the media, and the ability to create a network of local actors around the project'. They highlight a case where a renewable energy developer was perceived as a profit-motivated outsider, with no interest in promoting or protecting the region, and the local opposition that occurred.

These broad issues can be explored through an example of renewable energy planning with seemingly meaningful public consultation: the Middelgrunden offshore wind farm project, in Copenhagen harbour (Jessien and Larsen, 1999; Soerensen et al, 2000, 2001). The prominent location of the development was used in the design to create a feature in the seascape, and to reflect the historical defences of the city. There was a high level of information output from the developers, with leaflets, public meetings, news articles and television coverage, and the open planning process invited a broad spectrum of people to participate. From this, an 'understanding' (Soerensen et al, 2001, p329) was achieved between the developers and local people, and the process generated widespread appreciation and social acceptance. Crucially, the developers responded to public concerns and action was taken. For example, after public criticisms, the number of turbines in the plan was reduced from 27 to 20 and the layout changed from 3 rows to a sweeping curve, although the size and capacity of each turbine was increased slightly so that the same total amount of electricity could be generated. The conclusion that Soerensen et al (2000, 2001) draw is that while public involvement is challenging, it yields confidence, acceptance and support.

Engagement as 'deliberation'

A third way of envisaging engagement is as 'deliberation', where the public are not just permitted to discuss any plans, but are more thoroughly involved in developing them, along with wider policy, in the first place. The approach is based on public participation rather than public consultation, and necessitates a shift in emphasis from competitive interest bargaining to collaborative consensus building, recognizing and including all interests (Harris, 2002).

This may overcome what Bell et al (2005) have described as a 'democratic deficit' in renewable energy decisions, where the minority (the 25 per cent who do not support renewable energy in opinion polls) is able to impose its will on decision making processes, leading to the low success rate for applications. As Bell et al discuss, it is important to assess what the majority thinks, and more deliberative processes about renewable energy may achieve this.

Examples of this kind of approach include citizens' juries, interactive panels, workshops and conferences, where issues are broadly considered and recommendations for decision makers discussed. However, this approach is rare. One example is the work carried out by Landscape Design Associates (2000), where efforts were made to build 'consensus' about the cumulative

effects of wind turbines. The process included a workshop, with a wide range of wind energy stakeholders to 'identify and tackle key areas of conflict' (2000, p20), and public focus groups, to understand people's concerns and develop best practices for improved communication between developers and communities. The outputs were intended to inform policy and practice. A second example, and the only instance that Chilvers et al (2005) could find in their review of deliberation processes, relates to broader level energy issues. The UK government conducted a wide-ranging engagement process about the Energy White Paper in 2003, involving 'all levels of engagement strategy, from simple information provision to complex deliberative processes' (Chilvers et al, 2005, p24). The process was aimed at both stakeholders and the public, and intended to be as open and inclusive as possible. It focused on a variety of issues relating to energy in the UK, and sought to 'understand public perceptions of energy and their energy concerns' (Chilvers et al, 2005, p25). The process included widespread dissemination of material, road shows, focus groups, deliberative workshops and a final integrating conference. What Chilvers et al note from this process is that the commitment shown by the government to public engagement was welcomed, and that the key concerns that were raised by the public were largely incorporated and addressed in the White Paper.

Issues for engagement practice

The success of the above example suggests that deliberate forms of engagement should be used as widely and frequently as possible. The planning process would be less about deciding, announcing and defending, and more about local people and decision makers working together. Views would be sought, and listened to, and outcomes that were satisfactory to all would be negotiated. There are, however, a number of issues, both procedural and practical, to take into account.

For example, it has been pointed out that communicative planning ideals do not fully recognize conflicts and power relations. Any deliberatively developed policies may be disrupted by manipulation, control, confusion and exclusions (Richardson, 1996): a 'sobering' realization (Flyvbjerg, 1996, p389). It is debatable whether local people and decision makers can ever be in a situation where power relations do not shape both process and outcome. Even if collaborative planning methods are being attempted, people will have differential access and influence, through factors such as language, education, social position, ethnicity and gender, despite any compensatory measures (Tewdwr-Jones and Thomas, 1998). Powerful interests may, for instance, go 'over the heads' of those in the collaborative process to directly influence the decision maker. There are also questions to be asked about who should or can be involved, and what influence they should have. It has been mentioned that 'the public' is hardly a homogenous group. How is it possible to stop some sectors, groups or members having more power than others? Or should they be allowed to? In the example from Middelgrunden, Soerensen et al (2001, p329) discuss the large numbers of people who actively supported the wind farm; and only as an aside mention the 'relatively small group of yachtsman, fishermen, individuals and politicians [who] remained in opposition'. While they are a 'relatively small' group, as frequent sea users who will be more affected by the development than those in support, should their views carry more weight?

One way to decide who should be involved would be to allow 'local' people priority over local decisions. In their study of biomass plants. Upham and Shackley (2006) argue for negotiated agreements between regional renewable energy agencies, local authorities and local people 'on the nature and limits of renewable energy within a locality' (2006, p60). But who determines what a 'locality' is and who those 'local people' affected are? And how might this apply to other forms of renewable energy, such as tidal power and offshore wind farms, where there is an increased spatial separation between 'local' people and any development? Gross (2007) has shown that although decisions that involve communities are laudable, 'local people'

are not a unified group, and decisions perceived to benefit some sections of a community over others will cause protests and disputes (a point, reiterated by Walker and Devine-Wright, 2008). If energy is also a national and international issue, can it be argued that there is a strong role for representation beyond the immediate area?

Not only are there different publics to involve, but they may prefer or need different forms of involvement. In their study of the engagement between offshore wind power developers and fishing communities, Gray et al (2005) discuss how both of these groups had a very different view of the processes. The developers had held a series of public meetings and felt they had made every feasible effort to consult with the fragmented fishing industry. However, meetings are not an appropriate form of communication for the informal, non-hierarchical culture of the fishing communities, and were subsequently not attended by the fishers. These issues compounded the distance between the two groups, and led to scepticism, distrust and a seemingly entrenched divide, and demonstrate the difficulties of achieving mutually appropriate forms of engagement.

As well as criticisms that have been made at the theoretical level, questions have also been asked of the practicability of collaborative planning. Indeed, while attempts to genuinely involve the public in planning are generally acknowledged to be valuable for the reasons discussed above, they are rare, often tentative, and subject to constraints. They are certainly time-consuming and costly, although Toke (2003) suggests that this is 'money well spent'.

Another consideration is the way in which the conclusions of any engagement are incorporated into decision making. If collaborative planning around a particular set of issues is taking place, there is the danger that it may be circumvented in an overall framework that places greater emphasis on a different set of issues. The comments and suggestions made by local people may be about things that are beyond the remit of local planners but if people feel that their views, once elicited, are not being listened to, they may become disillusioned with the process as a whole (Tewdwr-Jones and Thomas, 1998).

There is also the need to integrate flexibility with a necessary procedural and strategic framework. So while Wustenhagen et al (2007, p2688) discuss the need for 'the right balance between territorial planning and room for open participation' for renewable energy – a sentiment echoed by Nadai (2007) – this leaves the dilemma of devising practical strategies to enable this to happen (Healey, 1997, p276).

The Planning Bill

So far this chapter has focused on engagement in development proposals, and the possibilities for doing so. These considerations have been thrown into sharp relief by the UK government's Planning Bill. This was introduced as a White Paper in November 2007 and presented to the House of Commons in June 2008[1] and contains radical proposals to overhaul the planning system. These include a new system for the approval of major infrastructure projects, encompassing water, waste, transport and energy – which include nuclear power stations and large wind farms. National Policy Statements will be introduced, setting out a framework for infrastructure development and establishing the need for such projects. Decisions on individual proposals would then be taken by an independent body, the Infrastructure Planning Commission (IPC), within the context of these policy statements.

The intention is to streamline decisions and avoid long public inquiries. The existence of a policy framework would mean that applicants would not be required to establish the broader need for a proposed development, and the permitting process will only be concerned with the specifics of an application. There are doubts about a number of issues to do with the Bill, questioning its presumption in favour of development, whether it will safeguard the principles of sustainable development, and the power that unelected officials on the IPC will have. For example, a coalition of many of the major UK environment and conservation groups has formed to campaign against the Bill on these grounds, uniting under the banner 'Planning

Disaster'. In terms of this chapter, the public consultation and engagement that will be allowed under this new system is particularly of interest. There will be public consultation on the development of the National Policy Statements; and when a development is proposed within a particular location, there will be a 'duty to consult the local community' (Planning Bill, 2008, section 46:1). The responsibility rests with the applicant to ensure that they have publicized their proposals and consulted the relevant stakeholders and local people, and the IPC will have the power to refuse the application if they deem that the applicant has not complied with these procedural requirements.

However, concerns are being expressed that local people will have no say about developments that will affect them. At inquiries into specific developments, the public will not be about to call into question the principles of the National Policy Statements; and one of the groups campaigning against these proposals, Friends of the Earth, believes that this is intrinsically problematic: 'Attempts to limit people from questioning national statements at public inquiries will be controversial and bring the process into disrepute, further distancing people from Government decisions and increasing the risk of conflict' (FoE, 2007, p3). Indeed, the only chance that local people will get to speak at an inquiry into a proposal will be at the end, in what is being termed an 'open floor' session. This will take place after the members of the IPC have examined the evidence, leading to suspicions that decisions will already have been made by then. The Planning Bill states that after publicizing their proposals and consulting with the local community, applicants must 'have regard to any relevant responses' (Department of Communities and Local Government, 2008, section 48:2); but it is not clear which responses this might be; or how responses will or must be incorporated into plans.

The intention is to incorporate public responses earlier on in the process – when the National Policy Statements are written. It remains to be seen whether the public involvement in the development of the statements is based on information provision, engagement or deliberation; whether the public will be involved because they are seen as local experts or to ease the path of the statements; and whether this involvement will be meaningful if there is an existing presumption in favour of development. However, in light of the issues discussed in this chapter, it is questionable whether this will be a valid and meaningful way to engage people. Wolsink (1994), Bell et al (2005) and Haggett and Vigar (2004) have shown that people's attitudes change when thinking about an idea in the abstract to being faced with an imminent development in reality. It seems likely therefore that the new proposals will only distance people from the planning process, because by the time they are effectively engaged with it – and have a proposal intended for their locality – the decisions about its necessity will already have been made.

Conclusion

This chapter has addressed the apparent low public support for renewable energy applications by focusing on the amount and form of engagement that the public are able to have in the planning for such applications. Indeed, while the reasons for protest against renewables are not straightforward, what underlies them, and is at the heart of any 'democratic deficit', are the opportunities people have for meaningful engagement in the decision-making process. This relates not just to renewable energy, but development and project management much more broadly. Crucially, Gross (2007) has shown that if people feel that there have been fair and just processes that lead to an outcome, they are more likely to support that outcome, whatever it may be.

The different means of engagement will affect the trust people have in both those outcomes and process. Information provision alone may not address any concerns or needs, and may even incite protest if these are ignored. Consulting people can be workable and effective and can lead to mutually beneficial outcomes. 'Ideal' deliberative processes, where people are involved with the generation of policy, rather than commenting on plans after they have been made, can mean that issues and values are incorporated into strategic and detailed decisions. However, such examples are rare and are certainly not simple or straightforward to deliver.

What does this mean for planning? If renewable energy is a core part of tackling carbon emissions and climate change, issues of public support and opposition have to be addressed. Seeing the public as a 'barrier' or as ignorant about renewables (or any other issues) is unlikely to be helpful. Instead, a more thorough approach to understand the concerns that people have, and how these might be mediated, is required. This will be possible only through a more thorough approach to engagement. How successfully this might be achieved depends on a number of theoretical and practical difficulties. These range from the conceptual framework, within which engagement processes are adopted, to the practical difficulties of gathering diverse interests together, encouraging them to express their views and genuinely incorporating their concerns and interests into policy. Such a task may be made even more difficult by what Harris (2002) describes as the need for a fundamental change in planning systems to allow the development of collaborative forms of planning. Adapting its processes to more effectively incorporate local issues and concerns will be a complex task. Finally, while the UK government's Planning Bill has emphasized the need and importance of involving the public in decisions, quite how this might be achieved remains to be seen.

Note

1 The Bill became a Planning Act in November 2008.

References

Agterbosch, S., Meertens, R. M. and Vermeulen, W. J. A. (2009) 'The relative importance of social and institutional conditions in the planning of wind power projects', *Renewable and Sustainable Energy Reviews*, vol 13, no 2, pp393–405

Bell, D., Gray, T. and Haggett, C. (2005) 'Policy, participation and the social gap in wind farm siting decisions', *Environmental Politics*, vol 14, no 4, pp460–477

Breukers, S. and Wolsink, M. (2007) 'Wind power implementation in changing institutional landscapes: An international comparison', *Energy Policy*, vol 35, pp2737–2750

Cass, N. (2006) 'Participatory-deliberative engagement: A literature review', School of Environment and Development, Manchester University

Department of Communities and Local Government, Department for the Environment, Food and Rural Affairs, Department of Trade and Industry and Department for Transport (2007) *Planning for a Sustainable Future: White Paper*, The Stationery Office, London

Chilvers, J., Damery, S., Evans, J., van der Horst, D. and Petts, J. (2005) 'Public engagement in energy: Mapping exercise', report for the Energy Research Public Dialogue Project, Research Councils UK, www.epsrc.ac.uk/CMSWeb/Downloads/Other/EnergyMappingExerciseBirmingham.pdf

Cowell, R. (2007) 'Wind power and "the planning problem": The experience of Wales', *European Environment*, vol 17, pp291–306

Department of Communities and Local Government (2007) Planning Bill 2007/08, section 48:2, http://services.parliament.uk/bills/2007-08/planning.html

Devine-Wright, P. (2005) 'Local aspects of renewable energy development in the UK: Public beliefs and policy implications', *Local Environment*, vol 10, no 1, pp57–69

Fischer, F. (2000) *Citizens, Experts, and the Environment: The Politics of Local Knowledge*, Duke University Press, Durham

Flyvjberg, B. (1996) 'The dark side of planning: Rationality and *Realrationalitat*', in S. Mandelbaum, L. Mazza and R. Burchell (eds) *Explorations in Planning Theory*, Centre for Urban Planning Research, New Jersey

Friends of the Earth (2007) 'A Better Plan: An alternative view of the land use planning system', Friends of the Earth, London

Gray, T., Haggett, C. and Bell, D. (2005) 'Wind farm siting: The case of offshore wind farms', *Ethics, Place and Environment*, vol 8, no 2, pp127–140

Gross, C. (2007) 'Community perspectives of wind energy in Australia: The application of a justice and fairness framework to increase social acceptance', *Energy Policy*, vol 35, pp2727–2736

Habermas, J. (1976) *Legitimation Crisis*, translated by Thomas McCarthy, Beacon Press, Boston

Haggett, C. (2008) 'Over the sea and far away? A consideration of the planning, politics and public perception of offshore wind farms', *Journal of Environmental Policy and Planning*, vol 10, no 3, pp289–306

Haggett, C. and Toke, D. (2006) 'Crossing the great divide: Using multi-method analysis to understand opposition to wind farms', *Public Administration*, vol 84, no 1, pp103–120

Haggett, C. and Vigar, G. (2004) 'Tilting at windmills? Understanding opposition to wind farm applications', *Town and Country Planning*, vol 73, no 10, pp288–291

Harris, N. (2002) 'Collaborative planning: From theoretical foundations to practice forms', in P. Allemendinger and M. Tewdr-Jones (eds) *Planning Futures: New Directions for Planning Theory*, Routledge, London

Healey, P. (1996) 'Consensus-building across difficult divisions: New approaches to collaborative strategy making', *Planning Practice and Research*, vol 11, pp207–216

Healey, P. (1997) *Collaborative Planning: Shaping Places in Fragmented Societies*, MacMillan, Basingstoke

IPPR (2007) *Positive Energy: Harnessing People Power to Prevent Climate Change*, Institute for Public Policy Research, London

Irwin, A. (1995) *Citizen Science: A Study of People, Expertise, and Sustainable Development*, Routledge, London

Jessien, S. and Larsen, J. H. (1999) 'Offshore wind farm at the bank Middelgrunden near Copenhagen harbour', paper presented at European Wind Energy Conference, Nice, France, 1–5 March

Jobert, A., Laborgne, P. and Mimler, S. (2007) Local acceptance of wind energy: Factors of success identified in French and German case studies, *Energy Policy*, vol 35, pp2751–2760

Landscape Design Associates (2000) 'Cumulative effects of wind turbines: Report on the preparation of a planning tool by means of consensus building', ETSU (Environmental Technology Support Unit) and the Department of Trade and Industry, London

Nadai, A. (2007) 'Planning', 'siting' and the local acceptance of wind power: Some lessons from the French case, *Energy Policy*, vol 35, pp2715–2726

Pasqualetti, M. J. (2001) 'Wind energy landscapes: Society and technology in the California desert', *Society and Natural Resources*, vol 14, pp689–699

Pennington, M. (2002) 'A Hayekian liberal critique of collaborative planning', in P. Allemendinger and M. Tewdr-Jones (eds) *Planning Futures: New Directions for Planning Theory*, Routledge, London

Phillimore, P. and Moffatt, S. (2004) 'If we have wrong perceptions of our area, we cannot be surprised if others do as well: Representing risk in Teesside's environmental politics', *Journal of Risk Research*, vol 7, no 2, pp171–184

Office of the Deputy Prime Minister (2004) *Planning Policy Statement 22: Renewable Energy*, The Stationary Office, London

Owen, S., Rayner, T. and Bina, O. (2004) 'New agendas for appraisal: Reflections on theory, practice and research', *Environment and Planning A*, vol 36, pp1943–1959

Ricci, M., Newsholme, G., Bellaby, P. and Flynn, R. (2007) 'The transition to hydrogen-based energy: Combining technology & risk assessments and lay perspectives', *International Journal of Energy Sector Management*, vol 1, no 1, pp34–50

Richardson, T. (1996) 'Foucauldian discourse: Power and truth in urban and regional policy making', *European Planning Studies*, vol 4, no 3, pp279–292

Rydin, Y. and Pennington, M. (2000) 'Public participation and the local environmental planning: The collective action problem and the potential of social capital', *Local Environment*, vol 5, no 2, pp153–169

Soerensen, H. C., Larsen, J. H., Olsen, F. A. and Svenson, J. (2000) 'Middelgrunden 40 MW offshore wind farm: A prestudy for the Danish offshore 750MW wind program', *Proceedings of the 10th International Offshore and Polar Engineering Conference*, Seattle, 28 May to 2 June

Soerensen, H. C., Hansen, L. K., Hammarlund, K. and Larsen, J. H. (2001) 'Experience with strategies for public involvement in offshore wind projects' *International Journal of Environmental and Sustainable Development*, vol 1, no 4, pp327–336

Tewdwr-Jones, M. and Thomas, H. (1998) 'Collaborative action in local plan-making: Planners' perceptions of "planning through debate"', *Environment and Planning B: Planning and Design*, vol 25, pp127–144

Toke, D. (2003) 'Wind power in the UK: How planning conditions and financial arrangements affect outcomes', *International Journal of Sustainable Energy*, vol 23, no 4, pp207–216

Toke, D. (2005) 'Explaining wind power planning outcomes: Some findings from a study in England and Wales', *Energy Policy*, vol 33, no 12, pp1527–1539

Toke, D., Breukers, S. and Wolsink, M. (2007) 'Wind power deployment outcomes: How can we account for the differences?', *Renewable and Sustainable Energy Reviews*, vol 35, no 3, pp2737–2750

Upham, P. and Shackley, S. (2006) 'Stakeholder opinion of a proposed 21.5 MWe Biomass gasifier in Winkleigh, Devon: Implications for Bioenergy Planning and Policy', *Journal of Environmental Policy and Planning*, vol 8, no 1, pp45–66

Walker, G. (1995) 'Renewable energy and the public', *Land Use Policy*, vol 12, no 1, pp49–59

Walker, G. and Devine-Wright, P. (2008) 'Community renewable energy: What should it mean?', *Energy Policy*, vol 36, pp497–500

Wolsink, M. (1994) 'Entanglement of interests and motives: Assumptions behind the NIMBY-theory on facility siting', *Urban Studies*, vol 31, no 6, pp851–866

Wolsink, M. (1996) 'Dutch wind power policy: Stagnating implementation of renewables', *Energy Policy*, vol 24, no 12, pp1079–1088

Wolsink, M. (2000) 'Wind power and the NIMBY myth. Institutional capacity and the limited significance of public support', *Renewable Energy*, vol 2, no 1, pp49–64

Wolsink, M. (2007a) 'Planning of renewable schemes: Deliberate and fair decision-making on landscape issues instead of reproachful accusations of non-cooperation', *Energy Policy*, vol 35, pp2692–2704

Wolsink, M. (2007b) 'Wind power implementation: The nature of public attitudes: Equity and fairness instead of "backyard motives"', *Renewable and Sustainable Energy Reviews*, vol 11, pp1188–1207

Wustenhagen, R., Wolsink, M. and Burer, M. J. (2007) 'Social acceptance of renewable energy innovation: An introduction to the concept', *Energy Policy*, vol 35, pp2683–2691

Wynne, B. (1989) 'Sheep farming after Chernobyl: A case study in community scientific information', *Environment*, vol 31, pp10–15

Yearley, S., Cinderby, S., Forrester, J., Bailey, P. and Rosen, P. (2003) 'Participatory modelling and the local governance of the politics of UK air pollution: A three-city case study', *Environmental Values*, vol 12, pp247–262

Index

(n after page number indicates footnote.)